The Art of Scientific Innovation

CASES OF CLASSIC CREATIVITY

Syed V. Ahamed and Victor B. Lawrence

PEARSON
Prentice Hall

Upper Saddle River, New Jersey 07458

Library of Congress Cataloging-in-Publication Data

Ahamed, Syed V., 1938-
The art of scientific innovation : cases of classic creativity / Syed V. Ahamed and Victor B. Lawrence
p. cm.
Includes bibliographical references and index.
ISBN 0-13-147342-5
1. Technological innovations. I. Lawrence, Victor B. II. Title
T173.8.A32 2004
608–dc22

2004044276

Vice President and Editorial Director, ECS: *Marcia J. Horton*
Editorial Assistant: *Carole Snyder*
Vice President and Director of Production and Manufacturing, ESM: *David W. Riccardi*
Executive Managing Editor: *Vince O'Brien*
Managing Editor: *David A. George*
Production Editor: *Daniel Sandin*
Director of Creative Services: *Paul Belfanti*
Art Director: *Jayne Conte*
Cover Designer: *Bruce Kenselaar*
Art Editor: *Greg Dulles*
Manufacturing Manager: *Trudy Pisciotti*
Manufacturing Buyer: *Lisa McDowell*
Senior Marketing Manager: *Holly Stark*

PEARSON Prentice Hall

Pearson Prentice Hall
Pearson Education, Inc.
Upper Saddle River, NJ 07458

Printed in the United States of America

10 9 8 7 6 5 4 3 2 1

ISBN 0-13-147342-5

Pearson Education Ltd., *London*
Pearson Education Australia Pty. Ltd. *Sydney*
Pearson Education Singapore, Pte. Ltd.
Pearson Education North Asia Ltd., *Hong Kong*
Pearson Education Canada, Inc., *Toronto*
Pearson Educación de Mexico, S.A. de C.V.
Pearson Education—Japan, *Tokyo*
Pearson Education Malaysia, Pte. Ltd.
Pearson Education, Inc., *Upper Saddle River, New Jersey*

Dedicated to the International Community of Our Doctoral Students

Contents

Foreword

We are at present witnessing a major revolution in telecommunications. When it is over, practically everything about how we telecommunicate will have changed. Advances in computers and communications during the last century no doubt have played a critical role in spurring this revolution. Nevertheless, we might examine the inventions and innovations that were fundamental to this revolution and the type of research environment that led to these major developments.

During the last century, most industries provided strong support for their research laboratories; in particular, there was a rapid expansion of research facilities after World War II. While a number of important innovations occurred in industrial laboratories, in recent years the support for industrial research has been shrinking. This book, *The Art of Scientific Innovation: Cases of Classic Creativity*, thus has come out at an appropriate time.

In their book, the authors have addressed important and underlying issues of human creativity. The authors attempt to capture the individual motivations that led researchers towards their major contributions and to examine the social, academic, and corporate opportunities that foster and encourage these gifted individuals. The book includes a range of inventions, such as the invention of the telephone, the wireless, the transistor, and the laser. The authors correctly point out that no single invention triggered the Internet revolution. Such vast insight and trailblazing of the still undocumented path to inventions can be undertaken by only a few people in this world. I commend the authors for taking a bold step in writing a book that is likely to become useful to a large community of readers from both corporate and academic circles.

The book emphasizes the role of corporate and academic research in accelerating the strides of technology and provides researchers and students with insights into the cooperative role of the machines and networks that invite the human spirit to the realm of innovation and creativity.

I am especially delighted that Professor Syed Ahamed and Dr. Victor B. Lawrence have taken the time and initiative to write this book. Syed has many years of experience in working with research students at the University of Hawaii, Polytechnic University, and the City University of New York. Victor has gained vast corporate research and development experience at AT&T Bell Laboratories and now Bell-Labs Lucent. Both Syed and Victor are prolific writers and inventors themselves. Syed has made contributions ranging from techniques for accelerating the convergence of nonlinear vector potential problems to the development of the architectures of wisdom machines. Victor has made contributions related to data communications, signal processing, speech codecs, and sensing networks for artificial intelligence.

Both authors are eminently suited to addressing the paradigm of change that is at work in making research and development a delicate and sensitive issue. These changes in society make computer systems, networks, and the working research environment essential in paving the way for new inventions and innovations. The ideas presented in this book will prove valuable to students, teachers, and leaders in industry in managing research to keep it productive.

Bishnu S. Atal
Laureate (2003), Franklin Institute Award
Retired Director, AT&T Labs Research
and Affiliate Professor
Department of Electrical Engineering
University of Washington
Seattle, WA

Preface

Creativity can be a controversial topic. It has drawn considerable attention during the latter half of the 20th century. Stemming from imagination, creativity extends into the realm of finding new principles, concepts, relationships, and ways to solve problems. There is no precise scientific definition of the underlying concepts dealing with creativity, nor is there a standard operating procedure for the practice of creativity. As a result, we have used utmost discretion in keeping subjective opinions out of the material presented here. However, we were constrained by our own perceptions and judgments in selecting the breakthrough patents included in this book.

There is good reason for stressing the breakthrough nature of the earlier patents and the nature of innovations commonly documented in the later patents. Whereas we live in an information-rich society influenced by the existence of high-speed networks, the Internet, and instant access to information, earlier scientists lived in an information-poor, but imagination-rich, society. This leads to basic differences in the nature and thought embedded in the creative process. This shift in paradigm is a major focal point of the book. It is possible to find more diversity in the thought process than what meets the eye on an initial reading of the patents presented in the book. The comparison of fundamental concepts on which these patents are based provides substantial insight into the inventors' thought processes and intentions.

It is our objective to make this material available to those students and those individuals starting new careers in research and development who intend to make a genuine contribution to knowledge. Such contributions naturally vary from area to area, specialty to specialty, and discipline to discipline. However, the creative process as a broad concept unifies all scientific activity. We attempt to capture one such portrait of creative activity in the area of technology and present it as a possible guide and strategy in the evolving knowledge society. In addition, in the context of mentoring others who are working towards their advanced degrees, we feel we can substantiate the statements in the patents presented in this book and discuss the circumstances that led to these inventions.

We feel certain that many other authors writing on the art of innovation would present the material in a different manner. Nevertheless, it is our hope that the book will serve to help students and those dedicated to enhancing the contour and structure of relational knowledge by exploring the creative frontiers of great scientific minds. This is a highly variable and personalized activity and there is no intellectual highway. Many times, it takes many decades of effort and sometimes it requires a quantum leap. The mental process in this area is too complex to be described with mathematical precision or to be documented in complete detail. However, a certain amount of logical and inductive reasoning applies, and sometimes extrapolation and predictive reasoning become the only way out. Imagination, the use of creative graphics, portrayals of scenarios, and the consolidation of mental objects encountered in analyzing and solving mathematical equations, expressions, and identities all can help. It is the structure and the interaction that become the solution to the riddle of scientific invention.

Even so, scientific invention is more than a riddle. Numerous factors combine to provide the circumstances and the environment that surround the time and location of an invention. The personalities and genetic predispositions of the scientists, their sensitivities, their acquired training and traits, and their opportunities and interactions all play a decisive role in the chain of events that culminates in an invention. The most sophisticated software provides no match for the mental activity that leads to invention. Otherwise, inventions would become programmable.

However, today the sophistication of software and of the processing power of computers does allow investigating patterns and recycling ideas. The underlying similarities in syntactic pattern recognition are within striking distance of the door to creativity. In a sense, the definition of creativity defies programmability. However, the element of chance that used to be buried in the subconscious of scientists now becomes an exhaustive search in a very close region of investigation. Then again, the close man–machine interplay may just be the initial barrier, and the machine may end up exploring the most promising venues. Two such examples that have substantially influenced the research community are Atal's algorithms for the predictive coding of human speech and Karmakar's algorithm for the efficient solution of large linear programs. In both instances, the convergence of human ingenuity and the (almost) boundless computational power of the computer are evident. The outcome of both of the contributions is remarkable.

In the first case, the predictive coding techniques for speech retrieve the symbolic characteristics of the vocal tract that produced the initial segment of speech. The temporary storage of these characteristic parameters and their subsequent use in digitally encoding the speech greatly reduce the bit-rate requirements for encoding the rest of the speech. Predictive coding is successfully deployed to carry telephone conversations over wireless media, where the bandwidth is somewhat limited.

In the second example, the search for optimized solutions is based on Karmarkar's algorithm. The process uses the interior-point method to solve very large linear-programming problems. This algorithm provides solutions much more quickly than earlier ellipsoid methods. Very large optimization problems are quickly solved. In a sense, part of the solution lies in the correct formulation of the problem and in the correct choice of variables for which the optimization problem in polynomial time is going to be solved. When the initial knowledge bases/databases are empty, the ingenuity of the human is the last and only recourse for innovation or invention. Sometimes the genius or the circumstance lies here. In a very realistic sense, Atal's predictive technique and Karmarkar's computational approach would assume a highly different flavor of human thought from Louis Pasteur's observation of the interdependence between cowpox and smallpox.

Computer-aided creativity, though not an impossibility, is a contradiction in terms. On one hand, creativity is the highest state of intellectual activity. On the other hand, modern machine intelligence is presently competing with the intricacy of the human thought process. Random coincidences, correlation, and events are more systematically and expeditiously analyzed by machines than by humans. The machine searches are far more extensive and global than a chance phenomenon in the human mind. Knowledge banks derived from machine searches and pattern recognition are much more systematic and programmable than the scrap papers on which human beings write. There is good reason to believe that the nature of research will undergo an evolution of the type that computer-aided design (CAD)/computer-assisted manufacturing (CAM) has brought to industry.

In academia, the scenario is as follows: Whereas the research student will still perform the intellectual activity, the machines will provide the tools and process knowledge and information based on documented and available literature. The reward is the increased number of inventions and level of innovation, at the expense of not challenging the documentation and literature itself or the manner in which inventions are developed and deployed. A short spurt of pseudo-innovations may become eminent. The surest dead end of long-term creativity is an overreliance on computer-aided creativity and machine-assisted pattern recognition without an awareness of the genesis of the patterns being investigated and matched. Computer-generated music, for example, can drown the senses rather than incite the genius that lies dormant within the mind.

The adaptive processing of knowledge, tempered by the judicious use of wisdom, is likely to be a human activity for a long time. Whereas computer-based innovation is likely to become boring and mundane, the human mind can "invent" new dimensions in creativity that intensify the process. We dare these students pursuing an advanced degree to overcome the boredom of computer-aided creativity. We believe that reinventing the innovative process can be at least twice as challenging and can lead to a unique mode of thinking that still needs investigation. The patterns of thought from one inventor can become manifest as the program in an integrated circuit (IC) chip in another inventor's device. In some instances, social and technological changes force past inventions (e.g., the abacus) to return as reincarnated superinventions (e.g., handheld computers).

SYED. V. AHAMED
Graduate Center
of the City University of New York

VICTOR B. LAWRENCE
Vice President
Advanced Communications Technology
Bell Laboratories
Lucent Technologies

Synopsis of the Book

This book deals with cases of classic creativity and emphasizes the proven contributions of scientists from the late 19th and 20th centuries. These inventions have left an imprint on the direction of science and technology of later years. Our focus is on the fields of computers and communications stemming from their precursors: physics and electrical engineering. The topics are vast, and many of the major inventions in other sciences are not covered here intentionally.

This book also deals with some of the more recent patents and stresses the emerging inventions that may affect the direction of change for the 21st century. The influence of the information age on creativity is analyzed in the context of the instant access to thousands of very well organized databases and information bases around the world. The inventions we present in this volume have not weathered the test of time. However, it is an educated guess on our part, and we hope that these inventions will be educational to students on their journey into the realm of creativity. We have limited some of these inventions to those of our colleagues because we know a great deal about these inventions and about the circumstances that led from the realization of a concept to the development of a device intended to serve a commercial or a social need.

As educators, we have attempted to make this book meaningful to doctoral students in four distinct ways and by consolidating the most important topics within five chapters. It is our intent that students become familiar with the methodology and drive behind the major inventions of the last century and become effective contributors to the sciences and disciplines of the 21st century.

First, the key attributes of our best students are presented in the context of the academic and intellectual environment of most modern universities. While the personal attributes of each advanced student are unique, general shared attributes can be isolated and identified. The object is to identify attributes that students can make a conscious effort to capitalize on and enhance. From our experience, attempting to encourage unwilling or unmotivated students to attain an advanced degree can be futile. On the other hand, willful direction from a mentor and students' self-imposed strengthening of these individual attributes can improve students' creative talent and intellectual capacities. This could be a lifelong contribution that a mentor can make for the student. This material is presented in Chapter 2.

Second, we have included patents in this book either in their entirety or in edited versions. It is our hope that readers will analyze the inventors' words with a view toward understanding the personal goals and motivating factors that underlie these patents. Study of these patents can lead to an understanding of the native reasoning behind the inventions, the inventions' connections to society, and, in some cases, the mathematical foundation of the inventions, stated in these patents in the words of the inventors. It is also our intent that students gain insight into what makes inventors invent by reading these original patent applications. The 16 patents presented in the book provide a representative sampling of the different formats of creativity of the individual scientists.

Third, we differentiate between unstructured creativity and structured creativity for our students. Whereas a flash of inspiration (in the mind of the gifted inventors) may bring the full-fledged patent and its final embodiment to some special people, most scientists struggle to translate an idea into a tangible product or a concept that can make a worthwhile contribution to society. If the former flash of inspiration is unstructured creativity, the later process of conceiving something novel and toiling ceaselessly towards perfecting it is the structured creativity. All the scientific discipline and mathematical precision would be needed in making a brilliant invention. Generally, there are no significant patents at the two extremes of creativity. The relative proportion of the two types of creativity, unstructured and structured, differs from one inventor to the next and from one patent to the next. A portrait of this blending process is captured in the words the inventors chose to include in their patents.

Fourth, we present the major commonality in the patents: the creative force within the inventors' personalities and the social pressures of their own times, which together have had a profound effect on the scientific community and have brought innumerable rewards to society. However, more than that, we try (as we understand it) to tie generic and scientifically proven axioms of science or mathematics to the inventions themselves. Making these connections is quite challenging, and we suggest that readers not take our word for it, but instead reinvent the inventions as a painter would repaint or as a poet would rerhyme.

In many cases, the creativity of scientists becomes apparent in numerous dimensions. For the gifted, the desire to create something is a strong drive that cannot be abated. In a sense, such a drive can become an obsession. The undisciplined and creative mind is a gift, and scientists who understand how best to use this gift can provide great benefits to their contemporaries and to future generations. The blend between the (instinctual) drive and the formality of its presentation to a scientific community is a personality attribute that can result from rigorous training. The role of the mentor also becomes crucial in making the dissertation a scientific contribution, in addition to making the product a financial success. It is our hope and intent that this book will provide a sense of direction to students who hope to unearth their own gifts and then develop an ethical ability to deploy these gifts as many scientists of previous generations have successfully done. A sense of excitement accompanies the invention or the discovery with an equally strong drive to perfect the representation (or the symbolism) of truth manifest as invention. In some notable cases, the truth lives on long beyond any product of the invention.

CHAPTER 1

CONCEPTS AND PERSPECTIVES

Creativity has received considerable interest [1.1] from both psychologists [1.2, 1.3, 1.4] and sociologists [1.5, 1.6, 1.7]. In an attempt to understand the process and thus alter it, the experimental and deductive approach is sometimes used. Most of these studies generalize the factors and the processes that lead to creativity. However, scientists and inventors break the laws of logic and coherence to create the novelty germane to invention. By and large, the attempts to systematize creativity are somewhat futile. For this reason, we present the actual inventions to portray the creative thought of the scientist rather than to dwell on the logic behind the invention. We do claim the logic plays an important role in invention, but we make this claim in hindsight. We stress the role of logic [1.8, 1.9], pattern analysis [1.10, 1.11], and sophisticated algorithmic analysis [1.12, 1.13] in derived inventions (or innovations).

In the early days, the two latter methodologies were generally unavailable in the development of the fundamental concept that broke the scientific domain the first time. In modern days, these logical, analytical, and algorithmic steps are thorough and precise in the computing environment [1.13], but the intuitive leaps are still the privilege [1.14] of human beings. One harmonious blend of the two equally powerful approaches in the scientific domain is evident in two instances. In Atal's contributions to speech coding and synthesis work [1.15], the Linear Predictive Coding (LPC) techniques are refined and optimized [1.16] as Code-Excited Linear Predictive (CELP) to encode human speech for bandwidth-limited applications. In the optimization techniques presented by Karmarkar [1.17] to solve very large linear-programming problems in polynomial time, the approach is based on the interior-point method, thus offering a far less computationally intensive solution than the traditional ellipsoid method.

Early inventions were far fewer than recent inventions. An "invention" would be device, contrivance, or process rather than an innovation or an improvisation. The genius of invention was truly practiced at the concept level rather than at the implementation stage. In a sense, early inventions came out of a vacuum. The lack of knowledge surrounding a concept was a good basis for an invention. For example, the "eureka" of Archimedes [1.18] was a genuine cry denoting the realization of a concept. The notion of specific gravity was thus invented. The force resulting from the influence of current on a magnetic needle [Oersted, 1822] was not lost like a needle in a haystack, but instead it became the foundation for understanding electromagnetic fields.

Many significant concepts and inventions thus resulted from this chance discovery or a coincidental observation. Luigi Galvani (a physiologist) discovered the effect of current electricity as he noticed the muscular twitch of a frog leg [1.19] due to an electric shock, Oersted [1.20] discovered the effects of electric current on a magnetic needle in 1822, Rontgen [1.21] discovered the penetration of X rays accidentally, and so on. In the modern world and knowledgeable society, discoveries are more in the conceptual domain than in the physical domain. Yet, it would be very noteworthy to invent/discover a principle/concept as generic as electromagnetism (Faraday [1.22]) or as universal as Einstein's theory of relativity [1.23] in the 21st century. The parallels of such major concepts still need to be traced in the knowledge domain if they exist (see Section 5.1.3).

Today's environment has evolved, and the precursors to almost any discipline are already encoded in the Library of Congress [1.24]. In a sense, the boundaries of knowledge overlap. The laws of electromagnetism [1.25] and thermodynamics [1.26] are universal in all of the engineering applications. In some cases, today's innovations may surpass the inventions of a century ago. For example, the basic patents for transmitting light to enhance the purity and clarity of crystalline glass (see Section 3.9.3) demand far fewer inventive skills than the innovative synthesis of fused glass fiber (see Section 3.9) [1.27, 1.28] to carry numerous wavelengths of light for the dense wavelength division multiplexing (DWDM) applications [1.29] in lightwave systems. In the same vein, Einstein's headache from deriving his general theory of relativity [1.23] would be greater than the pain that Newton [1.30] would have felt when the apple dropped on his head!

In summary, the chance and coincidental effects that have played significant roles in the past are fewer today. Instead, the rigor, the determination, and the scientific methodology of modern research (see Sections 2.3 through 2.10) have yielded strings of related inventions [e.g., from semiconductors to ultra large-scale integration (ULSI) [1.31] and from local area networks (LANs) [1.32] to the Internet (see Section 3.11)]. Such a continuum of inventions has helped the sciences of computers and communications immensely. The more recent patents bear a strong aura of the information age and of Internet access to global knowledge banks. The breadth of expertise and coverage of the more recent patents are indications that the technological revolution of the information age has altered the dimensions of human thought for good.

1.1 CREATIVITY AND INVENTION

Concepts that influence creativity are discussed in this chapter. We present the influence of colleagues, coworkers, supervisors and others on the "motivation and selfhood" of the individual. We also suggest some of the means to sharpening the learning process, which leads to the development of external influences and circumstances that provoke a creative mode of thinking.

In the modern climate of research, six conscious modes of mental activity are categorized and discussed in Section 2.5 to force the mind towards optimality in the search of new solutions. The numerous directions of mental activity leading to creativity, while keeping track of the incremental gain or concession at each step lie hidden in the patents. In studying the evidence from history, it becomes increasingly evident that there are no mathematical and logical ties to elaborate this phase of the individual's effort. In fact, the effort and results can differ significantly from one individual to the next.

1.1.1 Evidence from History

Ten major inventions (from Telephony to Intelligent Networks) and 16 patents have been selected to represent the early inventions. The patents for these inventions are presented in Chapter 3 to indicate how and why society has changed because of them. The patents issued by the United States Patent and Trademark Office [1.33] document the inventions. They also offer a sense of proprietary ownership of the idea to the individual. The individual, then, enters into a business/legal framework to bring the benefit of the invention to society.

Most of Chapter 3 contains the original patents, thus showing how those patents relate to the inventors in a personal sense. Chapter 3 also provides insight into the thinking that resulted in the patent (see Sections 3.1, 3.2, 3.5, and 3.7). Using a conceptual framework, we suggest that readers attempt to perceive the material in Chapter 1 in relation to how it became important to the inventions documented in Chapter 3. This process is quite individual to the earlier inventors. In modern days, an attempt to link the personality attributes of early scientists with their inventions generally sharpened the perception of the inventions in a conceptual rather than an Internet environment.

1.1.2 Generic Innovations from Recent Times

The inventions from the last 3 decades bring with them an evolved methodology of doing research and of scanning the vast amounts of scientific data that are available on the modern information highways. These inventions assert the nature and mode of inventing in the 21st century. Some of the differences between modern inventions will be readily evident to readers. For example, the depth and detail of coverage in recent patents are significantly greater, the area of invention is far more specific, the interdependence and the differentiation between patents is taken quite seriously by inventors and examiners, the claims and details are so precise as to be almost mathematical, and the mathematical reasoning and content are significantly greater. We also suggest that readers systematically dissect the contents of these patents and compare the systems approach in constructing and reengineering the "graph" or genesis of modern inventions with the "graph" of the older inventions. It is our hope that readers will "reinvent" the inventions in Chapter 3 and then reconstitute them in current technologies and in view of newer social needs. Changes in current technology may make some of the older patents virtually worthless in the next few decades, but the methodologies the scientists used live on.

The most recent inventions indicate how and why major changes in society demand the aforementioned type of modern patents. The relatively short life of patents and the impact of refined computational and simulation techniques make it possible to bring in a tide of insignificant patents rather than the more generic and longer lasting patents of the last century. These more recent patents issued by the United States Patent and Trademark Office document the invention in the context of the technology of the current times. They also offer a sense of global ownership of an idea to the individual and enter into a worldwide business/legal framework to bring the benefit of the invention to the entire scientific community.

In comparing the earlier patents with the more recent patents, two dominant instances of the synergy of patents start to surface. First, in the broadband revolution [1.34] arising from fiber-optic technology, the laser technology of the 1960s and the success in material compositions for glass in the 1970s brought about the first phase of revolution. Second, as the fiber-optic fabricators have tried to extend the 1550 nanometer wavelength region to encompass more digital bands to carry additional information, the laser designers have designed a more stable source of coherent light,* resulting in a second phase of the revolution. The synergistic effects are thus compounded by the feedback.

1.1.3 Innovations, Adaptations, Improvisations, and Reengineering

We allude to the methodology to decompose inventions from the past and then recompose them in different configurations to suit new technology, the new set of social needs and social tastes, new materials, etc., as the modern era unfolds before us. In a sense, this conscious "work" is as close to the practice of creativity as possible without containing a spark of genius. However, then a spark of genius could be no more than the recognition of patterns or the execution of artificial intelligence in the mind of a human.

As the mode of mental activity falls into these last phases of work, the computer becomes quicker and more exhaustive. For this reason, we suggest that the knowledge worker supported by any fine computing platform is likely to outperform the inventor who is limited to working at the lower levels of improvisation, reengineering, or CAD [1.34]. However, there are no constraints for truly creative individuals. In fact, the personal traits of this type of individual (see Sections 1.4 and 1.5) will drive away such artificial constraints.

1.1.4 Deathtrap for the Creative?

In dealing with inventions, we detect one possible scenario where reduced social demand and the innate laziness of human beings can together bring about the stagnation and even the death of creativity. However, this possibility of such an occurrence is as likely as the realization of the predictions of Karl Marx, who loudly proclaimed the death of capitalistic society. In reality, the creative genius of a society is unlikely to be set back by High Definition Television (HDTV) [1.35], baseball games or the legalization of marijuana. Such a situation is possible only if the guardians of society assume a melodrama of passivity and national isolation (e.g., India during the birth of Buddhism or China from 1950 through 1970).

Given that such a possibility can exist, we suggest to our students and fellow researchers that the practice of creativity is voluntary at low levels. It is not unlike the satisfaction of our more basic needs, such as safety and physiological needs. Only a structured and dedicated effort towards their satisfaction leads to effort necessary for the higher level needs, such as the ego, social and realization needs. In a sense, the needs of truly creative people are at the top spectrum of the realization needs. In fact, an enormous amount of discipline may be necessary to transform an invention into a breakthrough or even to transform an innovation into an invention. Motivated and self-driven individuals drive society rather than being driven by it. We urge students to seek out the creative talent within themselves during their early research years at the university and then help them to refine these basic skills very carefully and precisely, as Maxwell [1.36] would have done while building the mathematical platform of his generalized electromagnetic theory or as Einstein [1.23] would have done while generalizing his theory of relativity.

1.2 CONCEPTS IN CLASSIC INVENTIONS

Most inventions have stemmed from the contributions of scientists seeking generic and

*For a detailed discussion of the DWDM technology, see [1.27].

global solutions. The existence of a real application or even a problem was sometimes beside the point. For example, binary arithmetic [1.37] was written long before computers came into existence, antibiotics [1.38] were developed before they began to be used in medicine, roentgen rays [1.21] were identified before they were deployed as X-rays in medicine, noise contamination was identified long before it could be evaluated as a source of errors in communication theory [1.39], etc. In the past, concepts often were investigated in order to assess their validity rather than for their use in a commercial venture. In some cases, concepts were analyzed in a pursuit of knowledge and wisdom. Many times, the validity of concepts was established first, and then the application of those concepts was sought in order to make them useful to society by entrepreneurs seeking to profit from the inventions. Notable exceptions include Bell (see Section 3.1), Edison (see Section 3.2), and Marconi (see Section 3.3.1).

The generic concept that force is equal to mass times acceleration (in classical mechanics) was conceived before astronomers actually used it to accurately predict the orbits of planets. The principles of thermodynamics [1.40] were formulated before the steam engine [1.41]. The rules of electromagnetism [1.38] were written before the invention of the transformer [1.42], the induction motor [1.43], or turbine generators [1.44] or both, and so on. It appears that scientists frequently pursued fundamental knowledge and truth for its own sake rather than for the purpose of deriving a financial profit from the knowledge. In modern days, it stands as the anticlimax that paid scientists who fill the technological gap between one line of products and the next.

True creativity is generally well crystallized in the mindset of the creator. It cannot be hastily extinguished in creative people. It appears as if truly creative people seek the truth and strive to attain perfection in the practice of their work. Such attitudes have played a fundamental role in the development of many remarkable inventions. New concepts emerge from validated concepts (e.g., gravity from force, energy from mass, electricity from magnetism, and lightning from clouds), and thus the cycle continues.

In essence, nature has forced humankind to seek answers to many basic questions. Earlier generations invented wheels to move objects, dams to hold water, and bridges to cross rivers. More recently, new inventions and technologies have also emerged. For example, air-conditioning and refrigeration have benefited from the principles of thermodynamics, the power industry has benefited from electromagnetism, the aviation industry has benefitted from aerodynamics, etc.

In the more recent past, the pursuit of knowledge has not necessarily been the catalyst for invention and innovation. In many cases, a social need exists, and an invention (innovation) follows. From the writings of Einstein [1.45], far more problem solvers are now labeled as inventors rather than as fundamental thinkers. The imagination that finds the answers and invents as well can also create the need for the invention (e.g., the electric car or the AIDS vaccine). It is a paradox that, in the modern world often times society and science, after new progress by solving the problems that they themselves created!

1.3 CONCEPTS IN MODERN ADAPTATIONS

From a distant perspective, modern routine inventions seem less impressive than many of the breakthroughs of the past. The importance of these breakthroughs is proven by survival of various inventions and of the technology that has resulted from these inventions. Typical examples are technologies related to the telephone, electrical, semiconductor, and computer industries. More recent patents need to weather the test of time to prove their importance. More recently, the patents are profuse and less far reaching than the very early small number of generic patents. Recently, the technology has been more intricate and complex, thus leading to a new breed of novelty that could spur an innovation or an adaptation. Today, there is more opportunity to develop a greater number of highly specific inventions than in the past, where inventions frequently served a more fundamental purpose and lasted longer. In this book we present a set of typical patents that are restricted and specialized. In a great majority of the cases, the sciences cover the areas of electrical engineering, computers, and communication. Naturally, it would be possible to write similar books for any other given area of expertise.

REFERENCES (C-1)

[1.1] Industrial Research Institute, "Bibliography on Creativity." Prepared by the Subcommittee of the Research Personnel Committee, 1955.

[1.2] Guilford, J. P., "Creativity." *American Psychologist*, September 1950, p. 444.

[1.3] Andrews, Michael F., (Editor)., *Creativity and Psychological Health.* Syracuse University Press, Syracuse, NY, 1961.

[1.4] Bergler, Edmund, "Unconscious Mechanisms in Writer's Block." *Psychoanalytical Review*, April 1955, pp. 160–167.

[1.5] Foundation for Research on Human Behavior, "Creativity and Conformity." Foundation for Research on Human Behavior, Ann Arbor, MI, 1958.

[1.6] Hart, Hornell, and Others, "Preliminary Conclusions from a Study of Inventors." *Proceedings of the American Sociological Society*, 1927, pp. 21, 191–194.

[1.7] Veatch, Jeannette, "The Structure of Creativity." *Journal of Educational Sociology*, November 1953, pp. 102–107.

[1.8] Johnson, Donald M., "A Modern Account of Problem-Solving." *Psychological Bulletin*, 1944, pp. 201–229.

[1.9] Conant, J. B., *On Understanding Science. An Historical Approach.* Oxford University Press, London, 1947. Also see Vallery-Radot, R., *Life of Pasteur.* Constable & Co. Ltd., London, 1948.

[1.10] Reid, John W., "An Experimental Study of 'Analysis of the Goal' in Problem Solving." *Journal of General Psychology*, 1952, p. 44.

[1.11] Duda, Richard O., Hart, Peter E., and Stork, David G., *Pattern Classification* 2d ed. Wiley-Interscience, New York, 2000. Also see Cilfillan, S. C., "The Prediction of Invention." *Journal of the Patent Office Society*, September 1937, pp. 623–645.

[1.12] Ballard, Dana Harry and Brown, Christopher M. (Contributor), *Computer Vision.* Prentice-Hall, Englewood Cliffs, NJ, 1982.

[1.13] Knuth, Donald E., *Art of Computer Programming.* Volume 1: Fundamental Algorithms, 3d ed. Addison-Wesley, Reading, MA, 1997. Volume 2: Seminumerical Algorithms, 3d ed. Addison-Wesley, Reading, MA, 1997. Volume 3: Sorting and Searching, 2d ed. Addison-Wesley, Reading, MA, 1998.

[1.14] Conan, James B., "Scientists, Inventors, and Executives." *Chemical and Engineering News*, June 4, 1951, pp. 2262–2264.

[1.15] Atal, Bishnu S., and Hanauer, S. L., "Speech Analysis and Synthesis by Linear Prediction of the Speech Wave." *Journal of the Acoustical Society of America*, Vol. 50, 1971, pp. 637–655.

[1.16] Atal, Bishnu S., "Franklin Institute Awards Week." Electrical Engineering Symposium. Villanova University, Connelly Center, Villanova, PA, April 21–22, 2003.

[1.17] Karmarkar, N., "A New Polynomial-Time Algorithm for Linear Programming." *Combinatorica*, 4, pp. 373–395.

[1.18] Stein, Sherman, *Archimedes: What Did He Do Beside Cry Eureka?* Mathematical Association of America, Washington, DC, 1999.

[1.19] Beveridge, W. I. B., *The Art of Scientific Investigation.* 2d ed. William Heinemann Ltd., Melbourne, Australia, 1953.

[1.20] Frederick, Gregory, "Oersted and the Discovery of Electromagnetism." An excellent portrayal of Hans Christian Oersted in Episodes in Romantic Science, 1998.

[1.21] Garcia K., *Wilhelm Roentgen and the Discovery of X-Rays (Unlocking the Secrets of Science)*. Mitchell Lane Publishers, Inc., Bear, DE, 2002.

[1.22] Tyndall, J., *Faraday as a Discoverer*. Longmans, Green & Co., London, 1868.

[1.23] Einstein, Albert, *The Origin of the General Theory of Relativity*. Jackson, Wylie & Co., Glasgow, 1933

[1.24] Ganendran, Jacki, *Learn Library of Congress Subject Access*. Scarecrow Press, Lanham, MD. Also see Mortimer, Mary, *Learn Dewey Decimal Classification*. Ed. 21, Scarecrow Press, Lanham, MD, 1999.

[1.25] Kraus, John Daniel, and Fleisch, Daniel A., *Electromagnetics*. 5th ed., McGraw-Hill Science, New York, 1999.

[1.26] Incropera, Frank P., and Dewitt, David P., *Fundamentals of Heat and Mass Transfer*, 5th ed., John Wiley & Sons, New York, 2001.

[1.27] Born, Max, and Wolf, Emil, *Principles of Optics: Electromagnetic Theory of Propagation, Interference and Diffraction of Light*, 7th ed., Cambridge University Press, New York October 1999.

[1.28] Becker, P. C., Olsson, N. A., and Simpson, J. R., *Erbium-Doped Fiber Amplifiers*, 1st ed., Academic Press, San Diego, CA May 15, 1999.

[1.29] Bates, Regis J., *Wave-Division Multiplexing and Dense-Wave Division Multiplexing*. Telecom Series Adobe e-Book, McGraw-Hill, New York, 2002.

[1.30] Cohen, I. Bernard, and Smith, George E. (Editors), *The Cambridge Companion to Newton*. Cambridge University Press, New York, 2002

[1.31] Sze, S. M. 2d ed., *VLSI Technology* New York, McGraw-Hill Science, 1988. Also see Chang, C.Y. and Size, S. M. (Editors), *ULSI Devices*. Wiley-Interscience, New York, 2000.

[1.32] Lowe, Doug, *Networking For Dummies*, 6th ed., John Wiley & Sons, New York, 2002.

[1.33] *OG—Official Gazette of the United States Patent and Trademark Office: Patents*. Government Printing Office, Washington, DC, subscription edition, June 1986.

[1.34] Ahamed, Syed V., and Lawrence, Victor B., *Design and Engineering of Intelligent Communication Systems*. Kluwer Academic Publishers, Boston, 1997.

[1.35] Netravali, Arun N., and Haskell, Barry G., *Digital Pictures: Representation, Compression, and Standards (Applications of Communications Theory)* 2d ed. Plenum Publications Corp., New York, February 1995.

[1.36] Maxwell, James Clerk, *A Treatise on Electricity and Magnetism*, 3d ed. Dover Publications, New York, January 1991.

[1.37] Mendelson, Elliott, *Schaum's Outline of Boolean Algebra and Switching Circuits*. McGraw-Hill Trade, New York, 1970. Also see Solomon, Alan D., *The Essentials of Boolean Algebra*. Research and Education Association, Piscataway, NJ, 1990.

[1.38] Bankston, John, *Alexander Fleming and the Story of Penicillin*. Mitchell Lane Publishers, Inc., Bear, DE, 2001.

[1.39] Ziemer, Rodger E., and Tranter, William H., *Principles of Communication: Systems, Modulation and Noise* 5th ed. John Wiley & Sons, New York, 2001.

[1.40] Hatsopoulos, George N., *Principles of General Thermodynamics*. Krieger Publishing Company. Huntington, NY, 1981.

[1.41] Hills, Richard L., *Power from Steam: A History of the Stationary Steam Engine* reprint ed. Cambridge University Press, New York, 1989.

[1.42] Kaiser, Joe, and Kaiser, Joseph, *Electrical Power: Motors, Controls, Generators, Transformers*. Goodheart-Willcox, Tinley Park, IL, 1991.

[1.43] Hughes, Austin, *Electric Motors and Drives: Fundamentals, Types and Applications*, 2d ed. Butterworth-Heinemann, Woburn, MA, December 1993.

[1.44] Sill, Udo, and Zorner, Walter, *Steam Turbine Generators: Process Control and Diagnostics: Modern Instrumentation for the Greatest Economy of Power Plants*. John Wiley & Sons, New York, 1996.

[1.45] Einstein, Albert, "Preface." In *Where Is Science Going?* by Max Planck. Translation by James Murphy, George Mien & Unwin Ltd., London, 1933.

CHAPTER 2

ACADEMIC SETTING

In this chapter, we propose that the intellectual activity is a continuous and contiguous realm between mindless manipulations to scientific breakthroughs. In addition, we propose that the many factors that facilitate the mind's exploration of the domain of inventions be characterized in accordance with the strongest attributes of the individual scientist. In this vein, internal motivations and the external climate for creativity are identified to help students and scientists capitalize on the social/corporate setting of the workplace. While there is no standard operating procedure (SOP) [2.1] for the process of invention, the individual student or scientist can enhance the probability that an invention will result. Historically, we find that many scientists' novelty of thought combined with their thoroughness of effort can lead to an invention or contribution. Thomas Alva Edison (see Section 3.3.1), Albert Einstein, and Alexander Graham Bell (see Section 3.1) are noteworthy examples.

There is order and rigor in the methodology used to transition from intellectual drifting to inventing. In an ideal world, the drifting of the mind can be directed (see Section 5.1.1) in much the same way that the movement of photons in optical waveguides and the flow of electrons in plasma are directed.

In the past, when logical relationships between pieces of fragmented knowledge were not completely understood, many breakthroughs occurred by chance or as a result of the work of a genius. However, more recently, correlation between disparate ideas can be tracked more accurately than ever before, and every possible venue toward novelty can be explored. The basis for such activity results from the recognition of patterns that occur naturally or that are artificially created. Pattern recognition (PR) [2.2] is an expanding and evolving discipline in computer science. In fact, such activity is routinely carried out in the synthesis of new drugs, and such an extension of logical patterns has been detected and documented in the past. New antibiotics, antiviral drugs, and some new device methodologies (e.g., the fabrication of multiple quantum-well devices [2.3] and new DWDM fibers [2.4]) are technological innovations that greatly benefit society. To a large extent, the extrapolation of an invention [e.g., the transistor (see Sections 3.6.1 and 3.6.2)] by more and refined technologies (e.g., Silicon [2.5], Gallium Arsenide GaAs [2.6], and Pentium [2.7]) has kept the semiconductor industry alive for five decades. Of course, there are many other examples of industries riding the coattails of a breakthrough.

Unfortunately, creativity is not an industrial goal; instead, profits and products are the primary motivation of industry. For this reason, creativity can be abandoned when economic goals take precedence. Only committed individuals continue to strive to satisfy their own intellectual curiosity. For example, photonic waveguides [2.8] and very large scale integration (VLSI) chips [2.9] need to be reengineered and reconsidered in the realm of innovative thinking to meet the current needs of society. Much like genetic engineering which creates new creeds and biological structures, the process of fine tuning can also lead to creativity rather than to mere rationality. Gigantic leaps rather than microscopic steps become feasible. It is not that gigantic leaps are not feasible in quantum mechanics, but quantified and structured creativity can lead to immense social and economic strides.

Money, power, and prestige are some of the intermediate by-products for those dedicated to pursuing the research/creative path. Many times, these by-products are distractions that can sway the scientist's intent. Here, the personality of the individual becomes the guiding force, and an obsession for truth becomes the drive. For such individuals, the pursuit of perfection is an end unto itself. The beauty lies in striving to reach the ultimate truth and to treat inventions as by-products. In many of the long careers of renowned scientists, it is the person (e.g., Bell, Edison, or Einstein) who has created the inventions, rather than the inventions that have created the personality.

Such drives are also evident in the negative direction. They may well be the instinct of the psychotic, the compulsive drive of a gambler, or the cravings of an alcoholic. The acceptance of social values and the utmost rigor of discipline are absolutely essential to launching a career in research. Money and prestige can become milestones in a career, but they are not necessarily the ultimate purpose of the career itself. Perhaps, the career lies within Maslow's need hierarchy [2.10], which lists as its highest goal the seeking of the ultimate truth that becomes the basis of invention. The harnessing of the truth is perhaps the documentation of the patent that results from the invention.

For most scientists, the joy of research lies in the pursuit of ever-illusive truth. Many times, universal truth lies hidden even after the limited truth within the context of a particular search has been found. It can benefit students to realize that, whereas an advanced degree may be career goal, such a degree is only the beginning of a research career. Sometimes the advanced research enhances and brightens the personality. Both student and mentor may rejoice from the intellectual rewards of discovering an underlying concept, a commonality in the principles of nature, or a freak coincidence, the kind of which that can make innovations rich and rewarding. In social and classroom settings, the same concepts and co-relations make lives challenging and interesting. For a considerable number of seasoned researchers, such contributions to science as theses or dissertations are only a stepping-stone. In fact, a degree is only a social and a collegial recognition of a student's capacity to make individual contributions. It is a beginning and by no means an end.

2.1 BASIS AND MOTIVATION FOR THE BOOK

There are two basic ideas illustrated in this book: the ethics of research and the responsibilities and norms for interaction between students and mentors.

The text will help readers to become familiar with the etiquette involved in research. Cultivation of a student's research attitude is usually nurtured during the graduate years. The commitment necessary to persevere and discover new frontiers of knowledge is a trait of lifelong scientists and contributors. These cornerstones in becoming an investigator dedicated to the truth or a principle are generally based on the personality traits of the student and the mentoring skills of the faculty. These traits are discussed in Sections 2.4, 2.5, and 2.6.

The book also will help readers to acquire a more precise picture of what is expected from them as they enter a graduate program. It is equally important that the mentor should play both an educational and a supportive role in enhancing the student's ability to perform independent research. Students need to understand that the urge to receive a degree as soon as possible by any means can result in degradation of the thoroughness of research.

It is our observation that some modern students think of a graduate degree as a passport to financial well-being. In the rest of this section, we present alarming trends in graduate education that have warranted a book of this nature. If educators do not set and uphold the most rigorous of standards, then a growing number of graduate students will think it is permissible to produce sloppy research. In all dissertations, the documentation of findings should be accurate and complete. In addition, such documentation should provide enough information to allow others to reproduce the results and to confidently reach the same conclusions. Finally, it is important to note that the advanced degree is a mere acknowledgment of contributions made.

Two vital goals in a viable research program are that students must be loyal to the pursuit of scientific truth and that they must be held accountable for the methods used to attain published results documented in the literature. In the past, the pride and joy of a nation often resulted in the pursuit of graduate studies. The discovery of an unethical practice on the part of either the student or the mentor can eventually become a source of embarrassment. Thus, the responsibility for ethical behavior lies with both the students themselves and their mentors.

2.1.1 The Student Perspective

Some disturbing trends regarding graduate studies have developed over the last decade or so. First, an advanced degree has become an intermediate step toward seeking monetary gain. The degree in some cases is viewed as simply a means to an end. Students often are motivated by economic incentives, prestige, and/or the possibility of corporate promotions. In short, a passion for learning in some cases has now become secondary to the convenience and mechanics of getting a degree. Second, the concept of going to a graduate school to become an expert in a field is now often displaced by an interest in developing new social relationships. In a sense, the compromise starts in the motivation of a multiple set of goals at the same time. The allocation of resources, time, and energy is a highly directed activity that is generally practiced to (near) perfection by individuals and corporations. Third, the sense of urgency in getting an advanced degree has led to an increased level of cheating on assignments and an examinations.

Sometimes students seek ways to get rich quickly by using any means at their disposal rather than to acquire wealth ethically. This trend seems to be escalating rather than declining. Students must choose what they consider more important: being a scholar or being rich. Unethical practices are flourishing in today's academic environments. If the attitude of students is a leading indicator of the actions of professionals [2.11, 2.12], then the forecast is a gloomy one indeed.

2.1.2 The University Perspective

Sometimes the tendency of institutions is to "give away" degrees routinely after students spend several years on campus. Such actions undermine the academic commitment to scholarship, cheapen the value of the degree, and demean the university. An ideal academic collaborative effort involves the scholarship of the student, the mentorship of the faculty, and the reputation of the university. Anything less is an injustice to the student, the profession, and the university.

With regard to graduate education, intellectual honesty is a prerequisite to a career that deals with investigating and documenting scientific truth. Most scientists would gauge the honesty of scholars by their exploration and the exposition of the deepest truths. In fact, there is beauty in the pursuit of truth just as there is intellectual honesty in the pursuit of the basic methodology used to discover the truth. This type of conviction is evident from the original writings of early scientists such as Michael Faraday (the father of electricity) [2.13] and from the personal letters and writings of Albert Einstein [2.14], the founder of theories of relativity.

In the 1970s, when graduate degrees were being handed out more casually [2.15], many universities in the United States addressed the problem by tightening the standards students must meet to attain an advanced degree. When students fall short of established standards, then both the mentor and the university are held responsible. Today, the complex nature of technology leads to the interdependence of many disciplines in making inventions function optimally. The present nature of the new technology demands more from subject matter experts than was demanded of them in the past.

2.1.3 Students and Mentors

Classroom cheating is also linked in part to some students desire to finish their degree as soon as possible and in some cases, can even become a group activity. The desire to outsmart the examiner is taken as a challenge by some students.

In some instances, students can take such unethical behavior from classroom to the boardroom. In a corporate environment, opportunistic claims to someone else's genius or hard work oftentimes prevail and can be unwittingly rewarded by bosses too busy to investigate. More disconcerting is the gradual increase of such unethical behavior in the workplace.

The solution perhaps lies in firm adherence to the ethical norms of research. It appears to us that there is no excuse in the university environment and in the professional world to move down the spiral of honesty, justice and the practice of fairness. To this extent, the fundamental difference between good and bad is as universal in the research environment, as it is in the job environment. The student, the mentor, and the institution should subscribe to the movement of beauty of thought, into the virtue of wisdom, during the college years. This is perhaps a long-term effort for all the three entities.

2.2 ATTITUDE TOWARD RESEARCH

The orientation of students toward research is generally learned during the graduate years. In all likelihood, such research can become the foundation of a lifelong career. In most instances, the mentor is responsible for modeling those ethics and values deemed as important while students progress throughout their research career. To build a strong foundation and an ethical code for the conduct of research, we suggest that students and mentors work to establish a long-lasting, healthy relationship, as described in Section 2.4.

2.2.1 Motivation and Drive

At the outset, we suggest that an advanced degree should be awarded to those who do or create something new rather than merely rehash well-known literature. The criterion is fairly broad, of course, and many people have earned advanced degrees without demonstrating much creativity. Generally, one's motivation for attaining an advanced degree is complex. It is for the student to weigh and consider what we say in light with their own personalities rather than to accept or to refute what we say. However, research should begin with the personal challenge of taking on advanced work. For the initiation of the research career this book may possibly retune the thinking from being a student to being a lifelong contributor to society.

2.2.2 Doubt and Dilemma

In Section 2.4, we discuss some students' sense of doubt regarding whether or not they should be a researcher. We suggest that students work to discover what makes them genuinely happy. Students who hope to attain personal wealth would probably find business school training more consistent with their goals. Those with true commitment to the pursuit of knowledge, however, find their own rewards and try to become dedicated scientists.

Most younger students do not have clearly established long-term goals. Some of the more capable students pick up an advanced degree along the way and become valuable members of society. In some cases, the research career is initiated during the graduate years at a university.

2.3 THE FLAVOR OF CONTRIBUTION

In this section, we present seven types of contribution: breakthroughs, inventions, innovations, adaptations, improvisations, reengineering, and CAD or artificial intelligence (AI) [2.16] methods in problem solving. Every scientist sees the disparity between genuinely creative (i.e., socially disruptive) inventions and unoriginal computer-aided object manipulations [2.17]. Generally, it is for students to tackle significant problems early in life and for mentors to introduce such problems.

As outlined in Section 2.4, the flavor of contribution is greatly affected by mentors and their role. In almost all academic relationships, the student–mentor relation can be genuinely influential. The rewards can last a lifetime and should not be taken lightly by the student or the mentor. At times, the student is capable of making contributions greater than those made by the mentor, in which case the mentor's responsibility becomes even greater. The role of the members of the student's research group also becomes critical, and the mentor has the responsibility of selecting group members who possess a complimentary level of expertise in order to further enhance the student's potential.

2.3.1 The Major Breakthroughs

Most of the breakthrough patents shown in Chapter 3 have outlived the inventors and have been selected from notable scientists who have reached the Inventors Hall of Fame. We have no means of assessing the personal motivations or ambitions that might have bestowed to these scientists' ability to make such radical contributions to science and society. However, we have collected these patents as learning materials for those who hope to achieve breakthroughs of similar magnitude. We suggest that students read this material objectively so as to uncover the truth that lies within the patents. It then may become appropriate for students to rework the underlying truths they discover into new patents, innovations, or enhancements. Even though the discipline might be different, the practice of invention is similar in most areas of expertise. We hope that the students read more than the legal jargon that describes each contribution.

Many times, the figures and sketches convey important concepts that speak louder than the words.

2.3.2 Early Breakthrough Patents

In Chapter 4, we summarize the major lessons learned from the breakthrough patents documented in Chapter 3 and characterize these contributions within the limited context of computers and communications. The early contributors had the luxury of exploring and exploiting new disciplines for their inventions. Today, a massive amount of literature documents and offers analysis of every invention. For many employees of scientific institutions, specialization opens up opportunities; however, for the truly gifted, specialization can form a barrier. Today's contributors consider their contributions in light of diverse social needs and demands. In a sense, modern inventors are turning to society for guidance rather than attempting to guide society.

2.3.3 Recent Inventions

In examining some of the more recent inventions of the past two decades, we would argue that the inventions of some scientists and educators who collaborate with their research students can profoundly influence those students' attitudes toward research while they are at the university. In most cases, these scientists and inventors influence the way the students perceive the inventions. In addition, our experience working with notable contributors like B. S. Atal, J. L. Flanagan, N. S. Jayant, A. Netravali, L. Rabiner, R. Shafer, H. Seidel, and others at Bell Laboratories has shown that scientists often discuss with their students the circumstances that have led to the inventions and the associated patents. All these contributors are from the field of communications. It is also our observation that inventors-educators compare the relative advantages of the newer technologies, newer methodologies, and newer social requirements that may have led them to more recent discoveries. It is not our intent to ignore other noteworthy scholars in the fields of communications and computers. However, since such contributions already are documented in textbooks, it becomes logical to relate these patents to relevant classroom topics.

This approach can influence those students who are working to develop similar ideas or enhance students' core ideas as they progress toward attaining their advanced degrees. In this regard, teaching students how to recognize similar trends and concepts in these patents can reinforce the depth of the subject matter. It is our hope that researchers benefit from the numerous interpretations of concepts presented in Chapter 3.

2.3.4 Breakthroughs and Inventions

In this book, we present both breakthroughs and inventions. At the patenting stage, however, such a distinction between the two is not made. In fact, the only clear demarcation between the two is that those patents that endure the test of time and gain long-term acceptance from society tend to prevail as breakthrough patents. However, we make a case for the difference and present the ideological differences between the breakthroughs and an invention, simply because we have the luxury of hindsight.

In our opinion, the genesis of invention lies in the complete novelty of the entire process (e.g., the use of the properties of electrons/holes [2.18] in vacuum tubes or in solid-state devices or the use of photons [2.19] in optical media, etc., to carry information-bearing signals). Typical examples also exist in other disciplines. For instance, in device physics [2.20], the use of magnetic domains (or bubbles [2.21]) to carry and store bits of data is very similar to the use of charge-coupled devices (CCD) [2.22] to store, carry, and/or manipulate analog/digital information. Yet, CCD technology has matured [2.23] quite independently from the magnetic domain or bubble technology. On the other hand, the use of new methodologies in moving from a solid-state transistor to a planner transistor could be considered an innovation because the basis of the three-element transistor (base, emitter, and collector) was already invented. Even so, the innovation of the planar transistor has created the origin of the IC [2.9] chip revolution.

Many dominant forces collaborate in calling for recognition of the innovation or the invention. Publicity, connections, name, institution, and political mileage all influence the attention that the event receives. Media attention can also influence the perceptions of the scientific community. As a result, an invention may become a breakthrough, and an innovation may become an invention. The difference thus becomes fuzzier. The initial recognition a scientific event (e.g., an invention or an innovation) receives is generally unimportant. In a large number of cases, the difference between the two is immaterial, and the importance shifts to the social benefits that the event brings to humanity, to the economy, to the political prestige of a country, etc.

From a historical perspective, both extremes can be seen. The inventions of Tesla and his induction motor were not initially recognized nor well received, even though in modern days the polyphase induction motor [2.24] is one of the most common types of electric motors. At the other extreme, the role of magnetic-domain technology [2.21] was over emphasized and exaggerated during the late 1970s.

2.3.5 Adaptations and Innovations

Typical examples of adaptations are also evident. Adaptations are routinely implemented in industry without much ado. All the same, the social and economic impact could be enormous. The use and adaptation of the internal-combustion engine to the Gnome engine for the propeller of an aircraft was necessary before the first viable aircraft could be flown. The concept of having multiple IC engines situated around a circle to generate the power to rotate the huge propellers of the aircraft could be considered as either an innovation or an adaptation. Again, the demarcation becomes vague and often depends on individual points of view.

The use and implementation of prior art and technology might be defined as an adaptation rather than as an innovation or an invention. On the other hand, the generation of the enormous propulsion by a jet engine that uses the alternate compression ignition concept [2.25] differs radically from the use of internal combustion [2.26] to rotate the propellers of an aircraft. There are other examples, too many to be cited here.

2.3.6 Improvisations and Adaptations

In a sense, the straightforward use of newer materials and systems engineering results in an improved device, method, procedure, or even assembly of the many steps in the production process. Some of the classic cases of improvisation include the design of automobiles over the last five decades. Improvisation is a continuous and lasting process. In fact, improvisation does not always relate to an engineering goal; instead, it may relate to the desire for a product to have a more appealing or aesthetic appearance. All manufactured goods undergo a review regularly in order to ensure corporate sales goals are satisfied. Even though such improvisations do not have much or any scientific appeal, they can ultimately account for the success or failure of a corporation.

The improvisation of algorithms, routines, software, and mathematical procedures has a definite scientific basis. Here, the concept becomes more important than the end result. As we discussed earlier, the creativity of the student and the scientist is to move the task of customizing, improvising, or reengineering to an invention or even a breakthrough. In addition, we would suggest that students challenge the traditional, explore the unexplored, and use advanced research to aim at the highest possible target of intellectual endeavor—genuine creativity.

2.3.7 Reengineering and Improvisations

Using the process of reengineering to meet specific goals is considered routinely in industry. Once a concept is conceived and its commercial viability is proven, then industry, government, or any private agency may become interested in finding new applications for the concept entering new markets, etc. The proponents of the project are convinced of the success of the proposed changes in the design and in the building of the prototype. In addition, the technology may indeed have enough flexibility to allow construction of the prototype of the reengineered device, machine, or its application. In such circumstances, a systematic variation of the design rather than a completely original creative approach would yield the reengineered product.

In this context, a team of scientists/engineers might discover a more elegant way of solving a problem, a more efficient combination of resources, or a new approach to the problem that could become an invention in its own right. However, before the positive feedback effect could accrue, a certain critical mass and a certain threshold amount of venture capital is needed before such invention is adapted. Sometimes, the conviction of the original inventor provides sufficient justification for the investment of time needed to modify the original invention. In addition, sometimes scientists are not totally satisfied by the optimality of the original concept, making a refinement, enhancement, or improvement necessary.

Typically, refinement of a breakthrough concept naturally occurs in many cases. In fact, such refinement was expected after Albert Einstein published his "specialized" theory of relativity. By a similar token, after Charles Darwin published his theory of evolution, he was able to apply his work in numerous other situations. To follow these two powerful concepts, Einstein developed his "general" theory of relativity many years after he developed his special theory. The insight was presented in the special theory of relativity, and then the rigor, dedication and commitment unfolded in the general theory of relativity. Both theories stand on their own merits. The contributions of Einstein cannot be classified as extensions of Newton's work.

General concepts of scientific evolution also follow Darwin's theory of evolution. For example, other scientists document social and cultural Darwinism. In fact, some scientists can and do extrapolate the results of earlier scientists and expect the same scientist or the research establishment to come up with a subsequent string of inventions. Typical of such follow-up innovations are the development of ULSI technology [2.6] after lower-density VLSI chips [2.5] were built, the DWDM techniques [2.4] that emerged after the simpler wave division multiplexing (WDM) techniques [2.27], and the multiple quantum well devices [2.3] that came after the single quantum well devices [2.28].

The application of computer graphics to medical imaging is an improvisation. The convergence of X-ray [2.29], MRI (see Section 2.8.1), CAT-SCAN [2.30], and bone-density imaging techniques [2.31] with computer graphics [2.32] into medical-imaging algorithms has resulted in significant social and commercial benefits. Other examples of such reengineering efforts are plentiful. Computer-aided engineering (CAE) is another example, because vector graphics via direct-view storage tubes and raster-based TV tubes [2.32] preceded the computer-assisted drawing and drafting (CADD) revolution of the 1970s. The electric car [2.33], jet-powered racing vehicle [2.34], and the battery-operated wheelchair [2.35] are further examples of reengineering accomplishments.

2.4 STUDENT–MENTOR RELATIONSHIP

It is sometimes possible that an advanced degree and an association with a mentor will bring about a profound personal change in the character of a student. This is a case of interpersonal dynamics, where the role of the mentor is more than that of a teacher and the role of the researcher is more than that of a student. Such a relationship can endure a lifespan.

2.4.1 The Mentor's Roles and Responsibilities

The role of a mentor is greater than that of a teacher or a professor. The practice of research, though not standardized, is rigorous. In some cases, the unification of many methodologies into an adaptive, logical, and inferential approach involves more than mere rationality. A touch of genius, a spark of creativity, and intuition beyond the ordinary are the basis of the novelty, innovation, or invention on which the dissertation is based. The mentor should indeed teach such compositions in creativity and constituents of originality. The art of innovation, then, becomes a centerpiece of the student–mentor relationship. In dealing with a fine duet of thought, interpersonal dynamics and attractions should not clutter the rhyme, rhythm or melody for the evolving symphony of ideas.

In rare circumstances, mentors may have an economic motivation in building relationships with their students. Mentors could identify the brightest students and teach them the finest scientific methodologies for inventing, designing and then perhaps manufacturing the end product of the dissertation. In such cases, the economic stake might be sustained by the venture capital supplied by the mentor at the startup of a company. Such a scenario could result in a triple-win situation, where the student, the mentor, and society all stand to gain from the student's accomplishments. However, this scenario, though not illegal, borders on indiscretion, if the mentor does not disclose long-term plans to the student. As a result, we recommend that such plans be discussed and implemented openly. Any action on the part of the mentor to withhold the progress of the student or the graduation to coerce the student in a direction other than the best interest of the student could easily be considered unethical.

On one hand, if two people always agree, then one of them is redundant for any creativity. On the other hand, if the conflict is destructive, then the participation would become unproductive. In many cases, the association between two mature scientists can lead to a rewarding learning experience for both and to a complementary exchange of knowledge that can have a valuable outcome. In a student-mentor relationship, a mentor who plays too dominant a role could have the effect of limiting the creativity of the student. On the other hand, an overly detached mentor could waste the student's time and not reinforce the idea that a certain amount of progress is expected every week, every month, or every quarter. Mentoring is not tutoring, and a dissertation is not a report. For these reasons, flexibility in mentoring is desirable to accommodate the intellectual composition of the student. However, it is still the part of the mentor to gauge the capacity of the student and to adapt the mentoring techniques accordingly.

2.4.2 The Student's Role and Responsibilities

It is also desirable for a student to choose the most creative mentors at the university. The practice of creativity is perhaps the greatest asset that students can acquire during their tenure as a research student, yet there is no handbook or established methodology for teaching it. For this reason, we suggest that students ask one very basic question: What is the track record of the perspective mentor? We also suggest that students talk to others who have worked with the particular mentor.

There is no perfect way for students to gauge a mentor's creative abilities. However, a few guidelines may be used with some success. First, consider the creative nature of the patents that the mentor has received from the U.S. government or from the European or Japanese Patent Offices. In most cases, patents are based on the conception, creation, and the documentation of novelty. Second, consider the number of papers written by the mentor that have won prizes and honors from academic societies. To some extent, the contributions made by the mentor should be recognized and accepted beyond a doubt by professional societies and peers. Third, consider the mentor's membership in prestigious professional groups that base membership on an individual's creative contributions and accomplishments. Many such societies and honor rolls do exist for students to search on the Internet. The professional awards and the prizes, especially dealing with the patent contributions, would be a very good indication of the mentors' own creativity.

If there is a process in learning creativity, then this process is closer to osmosis of thought in communications rather than the words of rationality in conversations. A wink of the eye, a change of posture, a metaphor in logic or a twitch of the finger sometimes catalyzes the flow of ideas toward an underlying truth. Consciously seeking the truth is an art form it its own right. Moreover, this process generally lasts throughout the lives of researcher and the mentor.

This type of student–mentor relationship tends to grow stronger as the graduate students near the end of their research work. With no ulterior motive, the mentor has shared with the student a wealth of creativity that (perhaps) no one else could have offered. While the mentor thus earns the student's respect, the student also offers enough opportunity for the mentor to express the art of being creative in a very human and sensitive way.

2.4.3 The Abstraction of Creativity

We suggest that students challenge the basic skills involved in dealing with the creative process. In challenging a concept, there are two possibilities: either the concept is wrong or the truth, basis, or generalization of the concept is being misunderstood. No new idea should be blindly accepted. If there is uneasiness about the concept, then reason alone can restore the validity of the thought process. However, we also suggest that a student launch a thorough review outside the classroom before disputing any new thesis or proposition that is being discussed. This delicate balance is learned more and more judiciously as students gain additional experience as scientists after they graduate.

We suggest to our students and fellow researchers that they should sense an infinite amount of wisdom and understanding that lies unseen ahead of them. Research is a mere start of the journey toward greater knowledge to this pool of wisdom. The pool is a reservoir of the mental activity that is replenished throughout the life of the scholar and the scientist. Loosely classified, there are three basic attributes of wisdom or knowledge. The first is *(a)* the truth and its partial (if not its absolute) consistency with time, situation, and environment; the second is *(b)* the virtue that knowledge brings to society; and the third is *(c)* the beauty and the mathematical precision by which this knowledge is represented. There is a sense of perfection embedded in each of these three attributes. Attaining greater knowledge is a never-ending process. Nevertheless, the capacity of the mind is infinitely large, making this process a lifelong journey with innumerable rewards. In a sense, the student starts out balancing two sets of infinities. One is the knowledge base to which the student is trying to contribute, and the other is the bio-neurological basis in which the knowledge base resides.

Over a period of time, the imagery of the body of knowledge becomes increasingly clear in the mind of the student. The three aforementioned attributes *(a)*, *(b)* and *(c)* can be represented by the three primary colors, red *(R)*, blue *(B)*, and yellow *(Y)*. In fact, the other colors and hues that exist are derived from these primary colors. The process of linking in the mental domain becomes as easy as establishing pointers in the computational domain. In a sense, the role of the mentor becomes that of helping students use their own mental processes to investigate a problem and then refine these processes in an attempt to reach the truth *(R)*, the virtue *(B)*, and the beauty *(Y)* of greater knowledge in its true colors.

Creativity, innovation and inventions become manifestations of this pool of knowledge through small crevices of the reality. A drop (hue) of knowledge that contains a major composition of truth *(R)* may well be an equation in physics or a universal constant that can consistently and coherently withstand the test of time and space. A flash of insight with a major composition in virtue *(B)* may become a verse of scripture that will guide men and nations out of sin and cruelty. And a vision of an image or text with a major composition of beauty *(Y)* can become Picasso's art or Omar Khayyam's [2.36] poetry. There is a great amount of uncertainty and chance in the process, yet from time to time new equations are derived, physical constants are computed, classical operas are written, and masterpieces are painted.

2.5 THE CLIMATE FOR CREATIVITY

The climate of creativity is almost as abstract a concept as creativity itself. However, we would like to present a hypothesis to help describe the process of innovation. For example, the ancient Persian astronomers who charted the movement of the planets had to deal with the concept of zero in a mathematical sense. The concepts of infinities and asymptotes did not exist at that time. Yet, the notion of total absence of anything was conceived even though such absence could not measured. The bridge between concept and reality was crossed early and has since been crossed often. The creative process is the inverse in this context. The mentor and the student venture to discover something new when they have no basis for such a discovery. At the very least, they seek to find an extension of an existing concept, device, or principle that already exists and is proven and perhaps is serving society in a meaningful way. Without crossing the vague line of Heisenberg's uncertainty principle, something that is the end product of the dissertation should be sensed (if not measured), be proven, and convinced as an extension of the concept. The burden of proof falls on the student and perhaps on the mentor. The mentor's role sometimes becomes that of a skeptic who demands proof beyond a reasonable doubt that a genuine contribution has been made. In sum, the personal convictions of the student should be challenged as well as given a chance to win over the scientific community.

The climate for creativity is never perfect. It is presented here as a possible environment in which the student, the mentor, and scientists may enter by being mentally aware and alert. Such an environment paves a path toward creativity in any one of the six stages of activity: reengineering, improvisation, adaptation, innovation, invention, and breakthrough.

The boundaries between these stages are vague, and any forward leap from one stage to the next is possible and should be encouraged. In fact, the student should be willing to explore even those insights that appear illogical or irrelevant and discuss them with the mentor. The genius of the student/researcher can be a catalyst for the mentor's resolution of the problem. Conversely, the mentor can ignite the imagination of the student. After all, the first notion of the special theory of relativity [2.37] did not have any early warning.

Sometimes the support of the state, provincial, or national government can play a crucial role in the accomplishments of scientists and researchers. Such government entities can provide unique opportunities for students and mentors to make genuine contributions. However, sometimes the instinct toward the self-preservation of giant corporations can become a roadblock to scientific progress. Relevant examples abound. For example, AT&T and IBM became strange bedfellows when they worked together to block the Aloha network (see Section 3.11.1). Further, the almost universal lack of enthusiasm on the part of automobile giants in the development of the electric car or of cleaner natural-gas vehicles has provided the basis for action from government agencies. In addition, two major venues have been explored on the part of the Federal agencies: government funding of competing and viable technologies [e.g., National Science Foundation (NSF) and Aloha networks] and antitrust procedures against corporations (e.g., AT&T, Microsoft, and Polaroid). These opportunities thus created for student–mentor teams at universities, can be very conducive for inventions and breakthroughs (e.g., the seminal ideas for packet networks at the University of Hawaii in the late 1960s and ideas related to intelligent medical networks at the City University of New York in the 1990s).

2.5.1 Inherent Personality Attributes of the Student

In this section, we present some of the attitudes and attributes of the student that are favorable for scientific investigations. This list is not exhaustive, and other attitudes and attributes may also prove useful. As we pointed out earlier, human values are, in our opinion, quite important. Scientific values are generally built atop human values. It is our opinion that a perfect scientist would also come quite close to a perfect human being. However, the converse is not always true. To defend our point of view, we would need to comment on what we envision to be the qualities of a perfect human being. We refrain from undertaking this philosophic and lengthy discussion.

Two types of student traits can be identified: inherent or natural traits of the student and acquired traits. We present these to the extent that there is no clear-cut evidence that a certain amount of effort and toil will not bring about these traits in every student. The traits we describe would be desirable in every human being; however, they are particularly important to research students as they work toward attaining the advanced degree. In the rest of this section, we discuss the importance of these traits.

Freedom from Biases, Resentment, Vendettas, Negativism, Unsupported Generalities, and Hasty Conclusions. An open mind and an unbiased approach are essential to any scientific investigation. The truth that underlies the scientific process is not dependent on notions or predispositions. The noblest way to discover truth is to take the path of reason with well-placed intuition as a guide. However, here, prior knowledge and understanding is a good part of intuition. True genius is far too rare for us to generalize. Inspiration is still a small, but critical, part of innovation that can be drowned in bias.

Any resentment toward an idea or a concept usually limits viable options in an investigation. Students and scientists have to exercise their intellectual freedom in order to explore the unknown. In academic circles, the student who wears a mental straight jacket tends to be a poor researcher and a poor learner. The other extreme is equally problematic. A student who suspects every notion, concept, and theory only wastes a lot of time reinventing the wheel. A judicious choice of the topic of investigation and the ensuing expenditure of energy can facilitate the progress toward a meaningful solution to the problem. Sometimes the much sought after step forward comes from the most unexpected situation. An open mind to evaluate ideas in the context of investigation is at all times necessary.

Harboring a vendetta toward other researchers, their achievements, and their approaches can also lead to a destructive mental block. As previously mentioned, it is important to keep an open mind to new ideas and suggestions as long as they point toward the truth or at least toward the relative truth. This attitude exists in some of the researchers who would abandon any approach "not invented here" (NIH). The NIH syndrome is a serious problem even in some of the more established laboratories around the world.

Negativism toward an idea or suggestion can be equally detrimental. Sometimes negativism sneaks into research because of the attitude of jaded or cynical researchers, and it needs to be consciously eradicated by implementation of an open-minded approach toward finding the most optimal and logical solution to the problem. This cleansing process is a small price to pay in making major contributions and also in winning the cooperation of fellow researchers.

Furthermore, unsupported generalities tend to result from a lack of the rigorous work that is necessary after any new finding or unexpected result is discovered through research. Isolation of the true cause for any given effect can be a very demanding and time-consuming task. Many times, making the correct inference requires extensive experimentation or even the withdrawal of an earlier conclusion. Hasty conclusions have ramifications as severe as those of unsupported generalities. In our opinion, both need to be avoided in research.

Reasonable Freedom from Personal Conflicts. Such freedom is extremely desirable, if not imperative. Strong emotional conflicts of a personal nature can sometimes drain the creative energies of the student. When the mind seeks to resolve the personal conflict, it oftentimes cannot manage the serious concentration that is necessary to resolve the scientific problem. It is usually impossible to have total freedom from all conflicts, and the effects upon the student often depend on the degree and duration of these conflicts. In many cases, the student also requires the ability to find a balance between the professional life of being a true scientist and his or her personal life, with all of its attendant responsibilities. Sometimes both sets of responsibilities can become intense and equally demanding. However, the emphasis and priorities should permit the student to finish the advanced degree. To some extent, these priorities are acquired early in life and reflect the parental attitude toward work. The near-perfect scenario is when the student experiences reasonably a low level of personal conflict and acquires a useful research skill set from the mentor and fellow researchers.

A Balanced and Deliberate Passion for Truth. A passion for truth, an open mind, and an absense of bias can see the student through days of toil and frustration. Typically, an eager and active mind provides both the energy and the focus needed for a student to keep on track. The will to pursue the answers to questions, the conviction to offer sound reasons to the scientific community, and an almost insatiable curiosity to complete an investigation lead to major findings in the sciences.

Many scientists believe that results that disprove a hypothesis or a previously published generality are as valuable as results that extend the scope of previous work. A fairly detailed and through investigation is necessary in such cases because of the possibility that the current investigation may itself produce false results and mislead other researchers. Collaboration with fellow scientists in addition to the mentor can also help in case some unforeseen environmental effect is leading to an erroneous conclusion.

A Passion for (Not an Obsession with) Perfection. Many early efforts are prematurely aborted because of a lack of dedication to the truth and nothing but the truth. Time and energy expended in the pursuit of truth can be expensive, results also can become marginal at the very highest levels of truth, and the measure of the cause/effect relationship can become effort intensive. Under these conditions, the motivation for unbiased investigation is a passion for perfection. Theoretically, a scientist should always seek perfection that will not overlook anything. However, in view of the fact that scientific projects have a purpose and an end (e.g., the development of a device, a chip, an algorithm, tangible product, or even a dissertation), absolute perfection may not be a viable goal. Thus a passion for truth rather than a obsession with perfection can only improve the quality of a scientific project.

A Sense of Commitment. Commitment, like perfection, is best in moderation. Total commitment can become unhealthy and unproductive whereas controlled commitment can bring out the highest level of creativity in the student. If the student can clearly define the subject matter of the dissertation and the methodologies to be used in reaching the conclusions, then the commitment level is satisfactory to defend the dissertation. Typically, students reach such commitment during the latter phases of their research. At the initial stages, students' commitment should be focused on completing the research and obtaining significant results.

A Passion for (Not an Obsession with) Detail. The pursuit of detail can become time intensive. Much like the pursuit of perfection, it can become overwhelming to the extent that it can distract the primary goal of the dissertation. At the other extreme, a lack of interest in details can lead further away from the overall truth. In a sense, the student has the responsibility to look at both the forest and the trees in their own right. From our experience, the trees should be clearly visible from every perspective in the laboratory as well as in the workspace where the actual research is done, and the forest should be clearly visible to the mind in the solitude of introspection. Sometimes the mentor can help the student keep both types of vision in perspective. The depth of understanding becomes crucial in keeping both views in check and balance for the student. The achievement of both *(a)* the goal of finishing the dissertation in a reasonable time and yet *(b)* making a significant contribution to the body of knowledge in the subject matter is indeed the research and documentation for the degree.

2.5.2 Acquired Personality Attributes of the Student

A Love of the Subject Matter. We use the word "love" with some caution. In a sense, the subject matter needs to be understood and accumulated in the context of the knowledge and information gathered throughout the life of the student. In a perfect setting, new knowledge, skills, techniques, and procedures become integrated into the knowledge that the student already possesses. If the mental process of organizing all the logical steps is complete after the learning, then the new topic learned should appear much like a textbook on a shelf in a large library or like a subroutine in a large program. Both the logical (mental) pointer and the physical (instances of its use and application) location of the new book or subroutine play a part in its retrieval and its optimal use. In a sense, the student has to practice discipline (much like algebra in mathematics) to keep the mental sanctity. We suggest that the research experience be taken seriously and that students strive to perfect this particular discipline. After many years of practice, a sense of elegance and beauty in the organization of contextual linkages starts to appear. In some instances, the flow of the various steps in the process of organizing the subject matter becomes a beautiful and personal experience, much like bringing in a Dewey Decimal Classification system into a library. It is not impossible that the subject matter truly becomes the love of a lifetime for the scholar.

When the art of accumulating and positioning new topics is practiced over decades of professional life, a certain amount of beauty and order starts to appear. Students increasingly grow to respect the work of others, their own contributions, and the ever-changing landscape of the subject matter. Further, a certain amount of respect for the subject matter also continues to grow. A certain amount of appreciation develops in the mind of the scientists who hold the knowledge, for this knowledge has evolved in a systematic and methodical, and perhaps even a passionate, way.

Capacity to Concentrate. The capacity to concentrate without becoming distracted or bored allows students to direct all of their mental energy (even for a short while) to a scientific problem area. Concentration allows the mind to get a chance to absorb the problem and find an immediate solution that can solve a great many of the minor problems that may often seem like roadblocks to the research/dissertation at hand.

Furthermore, concentration allows the mind to absorb the problem, thus giving the student an opportunity to tackle the problem more effectively over time. There is a danger, however, in concentrating too much on a problem. Excessive concentration on small problems or details can lead to an unfortunate loss of perspective regarding the bigger picture.

Greatest amount of caution is necessary in this concentration approach. One strategy that helps ensure that the time spent concentrating is productive work is to allocate a period of time for the solution of the problem. If the problem still lingers after the designated time frame has passed, then students might try the other strategies described in Section 2.5.2 or revisit the problem at a later time. This is the same strategy that the network specialists use if they encounter a network problem, and that is "to try the call or connection later." In a great many cases the strategy works when the (neural) network conditions are different.

Capacity to Capture an Intuition. Intuition is a quick mode of communication for the subconscious mind. Be it a flash of genius, the structure of a verse, the architecture of a new computer system, or even an idle or unsubstantiated thought, the conscious mind cannot instantly evaluate its worth. When the subconscious mind gets increasingly refined, intuition frequently offers almost effortless solutions to both routine and novel problems. Many times, intuition is lost if the thought is not processed quickly by the conscious mind and stored into a more accessible and retrievable part of the brain. In fact, many early scientists (e.g., von Helmholtz, Darwin, Poincare, Gauss, and Einstein), inventors (e.g., Edison, Tesla, Fleming, and Townes), painters (e.g., Picasso, Leonardo da

Vinci and Rembrandt), and poets (e.g., Goethe, Tagore, Omar Khayyam, Fitzgerald, and Iqbal) have been able to capture and then capitalize on the intuitive part of their thought process. For the mature scientist, intense concentration and sleepless nights are the routine price they pay for their major contributions to the scientific endeavor. Even though it may be reasonable for those who seek to climb the intellectual ladder alone into the sky, we do not recommend it for our research students. The role of the mentor is to frequently counsel and confer with research students. Together, mentors and students can share and assess each other's intuitions. In the early stage of the research career, a wireless link to a messaging system or an electronic notepad can serve as reservoirs for intuition.

Capacity to Systematize, Analogize, Categorize, Conceptualize, Rationalize, and Generalize. Certain skill sets are necessary in order for students to adopt a methodical and ultimately successful approach to research. In some instances, these skills are essential not only for students, but also for mature scientists and others who seek knowledge and truth.

The research student also has to be able to draw upon these skills until the contribution is complete. While these personality traits often result from one's early environment and training, they may also be acquired. In any case, it is necessary to focus the mind and to follow systematic, analytical, conceptual, and rational steps in order to investigate the topic under study. As with many character traits, they can be taken to excess. The analytical student also needs to let the mind wander periodically in order to achieve moments of true creativity.

Agility of Concepts. This unique capacity involves articulation and manipulation of modules of thought that carry basic processes germane to building new inventions from old inventions. To be specific, if an invention can be decomposed into fragments of ideas that constitute the core of novelty within the invention, then re-composition of that invention is possible. In a real-life context, this practice or ability can be restated in terms of compiler concepts used by compiler designers to process higher level languages. In this context, lexical analysis of an existing invention could lead to the core concepts/objects of the invention. Syntactic analysis, in turn, can lead to an understanding of the physical and fundamental laws of nature that bind such concepts/objects and make the invention work. Semantic analysis leads to the framework of circumstances in which the invention is feasible and the constraints that will enhance the success of the invention.

Fluency in the Language of Ideas and Its Vocabulary, Grammar, and Structure. In the real world, most human beings analyze and understand their own spoken and written language extremely well and very quickly. This perception of words, their meaning, and their context occurs at a subconscious level, provided that the intellectual levels of the speaker and listener or the author and the reader are about the same. When there is a notable gap between the intellectual levels of the two, then an iterative approach often is used until the concept eventually is conveyed. In other words, the rate of communication slows down until a common mode of delivery and a common vocabulary, grammar, and structure comprehensible by both parties is developed and effectively understood.

In a research or dissertation situation, language is directed toward processing concepts/inventions rather than toward processing routine transactions wherein the words, their grammar, and their semantic context are already well established. For this reason, skills related to the language of ideas should consciously (and perhaps iteratively) be practiced by students and directed toward comprehending the unusual or novel circumstances/rules within which the invention was made. This linguistic ability permits the readers of patents and papers to recognize the novelty of inventions and to perform a conceptual scan (i.e., an analysis that is more than the lexical analysis, the syntactic analysis, and the semantic analysis of concepts germane to the invention). Mastery of the language of ideas leads to the rarest of qualities in the individual: the ability to invent on demand.

If the lexical analysis of inventions leads to fragments of concepts, then fragments of concepts become the vocabulary for the language of future inventions. It is one thing to look back at existing inventions, and it is yet another thing to foresee new inventions. In a sense, we suggest that students study inventions to build a rich vocabulary (e.g., the symbols in classic and quantum physics) and then build upon that vocabulary within a scientific/innovative context.

In the research or dissertation situation, study of the classic inventions of the past builds an intellectual knowledge base of inventions and their compositional fragments. Analysis and understanding of inventions leads to a corresponding understanding of the fragments of inventions. Reengineering these fragments with the new technologies and new materials that have become available in the last few decades can lead to the development of innovations rather than of inventions. Such innovations are discussed in Reference 2.37. To some extent, the basic laws of physics as we know them today have been explored to make major inventions of the past and are discussed in this book. When there is no technological revolution, however, inventions tend to become intellectually less provocative and innovations tend to become routine. Recombinations of fragments and subfragments sometimes can seem more like exercises than contributions. Nonetheless, such recombinations still can have vast social implications. For example, the use of synthetic oil (as a new fragment/object resulting from recent petroleum technology) instead of shale oil (a replacement) in automobiles can significantly prolong a car's engine life, increase its efficiency, and reduce pollution. As another example, electric locomotives can also lead to a cleaner environment, and so on. Today, the more challenging scenarios appear to lie in quantum computing [2.38] and in optical/photonic data processing [2.39].

Understanding of How to Create a Flow Graph of an Invention. At an expert level that deals with concepts and their recombination, a graphical representation of the whole invention starts to evolve. Such abstraction is generally not needed when dealing with routine inventions, but if an invention needs to be optimized to its locality and its immediate environment, then a flow graph of the invention should be attempted much along the lines of systems engineering and component optimization. These techniques are well documented in the field of CAD [2.27].

The mathematics and techniques of such optimization do not ever need to be formalized; instead, it is students' mental processes that need to be developed with the help of the mentor. Students should have the capacity and the inclination to deal with inventing first and then gain the capacity to optimize the invention with respect to the technology, the environment, and the material properties that will be used to implement the invention. Many classic inventors (e.g., Bell, Edison, Farnsworth, Tesla, Bardeen, and Townes) have the methodology embedded in their contributions to the society.

Understanding of the Symbolic Manipulation of Concepts. Concepts are tightly symbolized as equations. They represent a natural or a physical phenomenon and are valid within the range of parameters over which the concepts are expected to be valid. Many such generic equations (e.g., Maxwell's equations [2.40], partial differential equations [2.41], and laws of thermodynamics [2.42]) are documented and used consistently as the underlying themes of inventions. Traditionally, symbols and equations can be manipulated according to the rules of algebra that are appropriate to the particular representation.

In the looser and less exact representation of concepts, many computer programs (at the software level) and many logic diagrams embedded in the chip-set (at the hardware level) also represent the concepts that work for more restrictive computational environments. In purely research and dissertation situations, concepts are discussed at an initial level and then are processed by the mind as part of the student–mentor relationship. Then, it becomes necessary to fine tune the concepts and construct the backbone of the dissertation during the polishing and enhancement stage. During these stages, the role of the mentor is crucial in helping to make the student aware of a two-part process: the process of weaving concepts into the backbone of the dissertation and the more abstract process of compiling such concepts based on equations, observations, and the fragments of other inventions.

If the student has some capacity to process the ideas within the proposed dissertation, then a symbolic manipulation of the concepts may yield a string of inventions rather than just one invention. For example, if the student is working on the topic of knowledge processor units for knowledge-based systems, then the rules for symbolic manipulation of binary data will apply to the symbolic manipulation of objects and their attributes. Specifically, if the multiple instruction, single data (MISD) central processing unit (CPU) concept and the manipulation of the intermediate binary data are used (as is done in pipeline processors) to adjust the flow of objects and their attributes, then we would have a novel multiple instruction, single object (MISO) KPU. Other concepts such as a single instruction, multiple data (SIMD) methodology would lead to a single instruction multiple object (SIMO) type of object array processor, and so on.

Understanding of How to Link Concepts (Hierarchy and Recursion). From the perspective of a computer scientist, if fragments of inventions are considered as subroutines or utilities within a computer environment that are used to process higher-level languages, then all of the established methodologies of software design and engineering can be deployed. For instance, subroutines and tasks are invoked within each other and within one another. The forwarding of variables and the return address is routinely handled by all linkers. In the context of inventions, the hierarchies of concepts are to be arranged in much the same way as the hierarchy of subroutines or tasks. In the former case, we would have a new invention, and in the latter case we would have a software system that accomplishes a computer function. The linking of concepts is perfected by practice at a conscious level. The linking of concepts also could occur as a stroke of genius at the subconscious level. Nevertheless, the effect would be the same.

Another distinct computer tool has been implemented as recursion. At the scientific level, the concept of recursion would be similar to a scientific process that invokes itself, but with a different set of variables. For example, in wireless communication, the cellular approach divides a geographical area into "cells" and uses transmission technology repeatedly in each of the cells. At the microcell level, the same methodology and technology are used. The degree of recursion is used three times as we deploy the cell, the microcell, the picocell, and now the nanocell. At an invention level, if an invention deploys a subinvention or a fragment of another invention, then the deployment may be repeated many times (e.g., as in signal processing), or it may be deployed within itself (e.g., as a reprocessed signal is used to reprocess the object or image that contained the signal). In nature, the methodology occurs widely to process, and organize microorganisms and cells.

If recognition of a microorganism calls for boundary mapping (e.g., those deployed in image processing) of the organism and then use of the segments of contour to identify the organism, then the same tools (e.g., boundary mapping and contour algebra for identification) can also be used for identification of the supercells. In the software environment, the procedure is simply handled by the organization of the subroutines or in the design of the compilers that handle recursion (used in the recent scientific compilers for C+ and C++). In hardware environments, the duplication of fragments of the LSI chip within the large VLSI chip is a good example. In the research domain, the practice to handle recursion needs far greater care. If a process includes a novel feature, then the generality of the novelty needs to be investigated, and it needs to be determined if the novelty can be deployed outside of the original invention. One such case is presented in U.S. Patent #4,316,061, "Minimal Delay Rate Change Circuits," and then re-deployed in Patent #5,978,831, "Synchronous Multi-Processor Using Tasks Directly Proportional in Size to the Individual Processor Rates." These two different patents are essentially two innovations derived from the same basic concept that gate functions can be repeated at clock rates that have a binary (1,2,4,8,16, . . . times) relationship.

Nature uses recursion extensively. Basic building blocks and genes are only a few examples of the use of recursion, but the evolution and manifestations of species are enormous. The genome library is limited but the linkages, and recursions are abundant. In fact, the concepts of genetic engineering are founded in the discrete changes to the linkages and in the hierarchical structure of the microcells that constitute the hybrid organism.

At the scientific level, the symbols, equations, and the natural constants are few, but it is for the student–mentor team to rewrite new sets of relations and equations. Each equation becomes an invention in its own right, provided it satisfies the laws of nature within the constraints of the symbolic representation of the fragments of the prior inventions.

In most theses and dissertations, the creation of novelty does not follow the extreme rigor involved in formulating basic equations and pursuing their generalizations. However, in the creation of an entirely new philosophy of thought (e.g., the theory of relativity from Albert Einstein or telegraphic codes from Morse), symbols, equations, and their generalizations become essential.

Unusual Sensitivity to the Nature and Interplay of Basic and Derived Objects Involved in Research. Events and circumstances are intertwined when certain observations are made. Such circumstances need to be understood in much the same way as the results or the observations themselves. Many times, poor attention to detail or a disregard for the nature (or both), interaction, and interplay between the causes and circumstances involved can lead to erroneous conclusions about the nature of the process itself. When interaction between numerous elements and causes is involved, inductive reasoning and the isolation of the true cause can become a tedious step-by-step process of association and elimination. Sometimes, it is possible to throw out the baby with the bath water by not paying enough attention to detail! In the end, only very keen analysis and sensitivity can save time and embarrassment.

Capacity to Permute and Commute the Derived Objects of Research. Derived research objects are physical and mental snapshots (including concepts) of the events and circumstances that have culminated in prior inventions. At the next level of detail, the objects of research have many attributes. Whereas the process of permuting these objects and their attributes is a conscious activity, the process of regrouping them can be attempted at both levels of consciousness. While the process of regrouping these objects and their attributes can be either a conscious or machine-driven process, the globally optimal rearrangement of these research objects and their attributes to yield the desired invention defies a mathematical treatment of a new invention. The (per)mutation of compound entities can become as complex as the mutation of genes for survival.

To some extent, systems engineers deploy the modularization of functions and subfunctions whenever a new piece of software is developed. Further, code is recycled for new data or new (AI type of) objects to fabricate a new routine/library. The practice of constructing new and more powerful code is a skill that programmers develop and refine over a period of time. In the context of inventing, such a skill can become crucial in using the tools and techniques of conceptualization and subconceptualization when dealing with the data and objects of research. The data and objects of early inventors are documented in their patents and devices. The mind, being the superprocessor of ideas and concepts, can always reprocess any concept (including this concept) developed by prior inventors, thus leading to a genuine breakthrough.

Courage to Face the Challenges of Scientific Investigation. Pursuing a thorough scientific investigation of the topic and following documented steps in reaching conclusions are essential in defending a dissertation. Frequently, the courage to present and convince, a panel of examiners is based on the personal conviction that there are no errors or deception in the dissertation. The only way to achieve this level of confidence is to follow a path of total intellectual integrity throughout the tenure of the investigation. Adherence to the utmost level of total honesty is a prerequisite.

Another challenge the student faces is based on the student's ability to deal with the overall motivation that is prompting the investigation. By and large no problem should become overwhelming, and no observation should be incomprehensible. In a sense, the student should deal with the facts and truth by being sure that the overall picture is always within the realm of understanding.

Willingness to Take on a Personal Challenge without Offending Others. In many cases, the mentor is not the student's only source of new knowledge and concepts. When a suggestion is forthcoming from another authority, freedom from bias becomes essential. To be able to acknowledge the truth even if comes from the humblest of sources is a milestone in the research career. Sometimes, the examining committee belittles or humiliates the researcher, rather than truly listening to what the researcher has to say or acting in a cooperative manner in the investigation. Openness of thought starts with the freedom from bias. Appreciation for the student's efforts and accomplishment rather than undue criticism of the student can provide the student with sufficient inspiration to move forward with the dissertation.

The role of tact and diplomacy becomes crucial when the student really believes that a breakthrough has been achieved. Offensive statements often have no remedy. Many times the damage such statements cause is permanent. As is widely known, some mature and established scientists (e.g., Bell and Edison) have engaged in lighthearted criticism of each other's work. However, in such case the personal friendship between the parties often allows each to understand that the other's criticism is not meant to be destructive.

Ability to Challenge Established Schools of Thought in a Nonconflictive Way. In many cases, a breakthrough is a matter of receiving the right inspirational concept at the right time at the right university. In many cases, such breakthroughs could occur at other schools or using other approaches. For example, as Charles Darwin was contemplating evolution, other scientists were not far from documenting the role that the environment played in shaping the adaptation of species.

The researcher's personality plays an active part in his seeking the truth. Rejection of pettiness and of false hasty conclusions is the first step for all research students. The mentor has the responsibility of ensuring that dishonesty plays no role in the practice of research, and the student has the burden of proving that the dissertation does not compromise truth.

A Personal Sense of Responsibility to Complete Any Task Undertaken without Delay. It is natural for the student to feel some slight anxiety or uneasiness about finishing the process or subprocess once it has been initiated. Completion of the process or the sub-process generally leads to a milestone toward the dissertation. Nevertheless, it should also be a definite step leading toward a goal of the investigation.

When all of the work that is done for the dissertation is reevaluated, it often becomes clear that some steps are more constructive and concrete than others. It thus becomes necessary for the student to take personal responsibility for exploring the most promising venues that can lead to a breakthrough. By the same token, this sense of responsibility should also prevent the unnecessary expenditure of effort when there is no promise to be realized in pursuing the work any further. Most major decisions in research require conviction and courage. The student–mentor team should proactively take all steps necessary to make every dissertation letter perfect if it is within the realm of scientific possibility to do so.

2.5.3 External and Contributing Circumstances

External factors involving finances, family, and health can affect the continuity and the intensity of a student's research. The integrity of the research, however, should not be influenced by such external factors. When the mentor agrees to work with a student, the mentor also takes on the responsibility of supporting the student's efforts to attain the degree. In many cases, the mentor can help to shield the student against those external events that may hinder the completion of the research.

Concentration and Persistence. Committed students make it a point to remain steady and steadfast throughout the course of their work on a dissertation. Nonetheless, external and personal factors can undermine commitment and concentration. The mentor should provide a stable research environment from the university and/or corporate perspective. The student, in turn, should work to maintain the stable environment needed to complete the dissertation and to successfully reach the goal of attaining an advanced degree.

Not all dissertations are the same, and not all results are equally significant. The mentor-student team aims for the most noteworthy results in light of extraneous constraints if any (e.g., funding, laboratory facilities, etc.). The goal is to reach deeper in the search for truth or novelty by cascading the work of the students. In most corporate settings, this type of cascading is a requirement for success. For example, the development of fiber at Lucent Technologies has increased the transmission capability of a single strand of fiber from a few megabits per second in the 1960s to 400 gigabits per second in 1999.

Persistence is another asset for the student. As with any other quality, persistence should be used in moderation. For example, persistence might result from the conviction that other researchers may have overlooked a significant factor. To be wrong and persistent is foolhardy, yet to be correct and persistent is praiseworthy. In a sense, the laws of microeconomics dictate the extent of persistence. In other words, the expected reward of finding something novel should justify the time and effort required to be persistent in the investigation.

To some extent, the conclusion of research lies at the end of an investigation of various clues. At times, such clues could be obscure or they may conceal something quite novel and significant. To be able to find such clues and pursue them to a Nobel Prize or even to the development of an invention or two would be the ultimate goal. However, on the other hand, to warn other researchers about possible sources of clutter in the clues is also a valuable finding and a result in its own right.

Awareness of the Limits of Knowledge and the State of Technology. Investigations need human skills complemented by technological tools. The two are equally important. For example, the investigation of the origin of the cosmos by Arno Penzias and Robert Wilson of Bell Laboratories occurred only after the most powerful noise-free (for that time and age) feed-forward amplifiers were invented (also at Bell Labs). As another example, the signal that Harold Seidel was trying to measure was extremely small, yet it clearly exceeded the noise level in radio telescopes (including the periodic noise introduced by pigeon droppings on the parabolic surfaces of the antennas), and it exceeded the noise in the amplifiers that were used to pick up distant signals and then amplify them to convince other scientists of the source and cooling of the universe after the "big bang."

On the other hand, the apple that fell on Newton's head led to the discovery of gravity, Sir Alexander Fleming [2.43] worked with antibiotics before he found a use for them in medicine, Michael Faraday [2.44] and Karl Friedrich Gauss [2.45] did the basic legwork that allowed James Clerk Maxwell [2.46] to publish his treatise on electricity and magnetism in 1873, and so on. The path toward a significant contribution is for those who seek and strive, and the path toward an advanced degree is for those who strive and finish. Many times, the thesis or a dissertation is only the start of a research career. The life of a scientist is as dynamic as the science that is pursued, and the ultimate truth is perhaps as inaccessible as the beauty behind the truth or the truth behind the beauty. The pursuit becomes even more challenging and noble if the scientist strived to make the discovery or invention useful to the society.

Awareness of the Basic Working Tools Used in the Exploration of Knowledge: A Systematized Methodology for Discovery. Scientific tools come in many kits. The many tools used for Internet research are vastly different from those used in biomedical research. The capacity of the student to filter through the entire body of knowledge acquired from primary school to postgraduate training can facilitate the selection of tools for the subject and for the environment. The most flexible tools and approaches are perhaps stored in the basic mindset of the student. Such tools may be as rudimentary as a surgical knife or as complex as the resynchronization circuits in the HDTV environment.

The awareness and selection of tools sometimes involves common sense, improvisation and customization, or even an invention of the tools themselves that may lead to many useful patents and applications. In many instances, the commonly used tools of research are inadequate for the cause-and-effect relationship that a student is seeking to explain or for a thesis that needs confirmation. One such example that has gained international repute comes from the research at Bell Laboratories for which Arno Penzias and Robert Wilson won the Noble Prize. The tools of inference and measurements based on radio astronomy were simply inadequate for use in confirming the big-bang theory. In order to affirm or deny the big-bang hypothesis, specialized amplification devices were developed and built to infer the temperature of the universe above 0° Absolute (−273°C). Later on, such refinements were used in the recovery of the faintest signals from microwave-transmission towers and of the faintest noise from other galaxies.

An Extraordinary Need to Solve the Problem. The intellectual motivation of the student also plays a vital part of the student's success as a researcher. In the early years, the student's need to impress others can be quite strong and the student's level of aspiration generally can be fairly low. As the student gains experience, the personal goals the student sets tend to become more challenging and more targeted. The student's motivation changes from more selfish needs to the needs of the community and then of the society. Perhaps the final personal breakthrough is when ambitions turn into commitment and the search for truth becomes a life goal. In most of the significant discoveries, motivation to seek and find the truth is a complex collection of subsets of the goal. The goal is for the student to make a conscious choice to find the best solution.

Such an evolving change in attitude is documented in the biographies of Marie and Pierre Curie, Albert Schweitzer, and Charles Darwin, all of whom dedicated their lives to the pursuit of knowledge and to the betterment of humankind. Mostly, when an urge of this magnitude becomes apparent in students, it is really a time for introspection rather than pure investigation.

A Well-Defined Problem That Confronts Us All. Wars, national priorities, threats, and other major events mobilize the general population. For example, the need to land on the moon, to be the first to probe Venus, the need to find a cure for cancer, etc., has prompted ordinary individuals to become heroes and has prompted scientists to become inventors. The daily lives of students, however, tend to lack such urgency and drama. However, such a drive can be seen in students and theorists whose dissertations become Nobel Prize-winning seminal papers [e.g., Philip Anderson's dissertation on the basic notions of condensed matter physics (Cambridge University)].

Likewise, family crises and personal longings can motivate the genius that sometimes lies dormant in the student. These short-lived spurts of energy, even of energy born of frustration and anger, can be channeled in creative ways. Students should tap into their anxieties rather than become overwhelmed by them. A combination of personal and social benefits of lasting value can indeed be derived as a result.

2.6 MILESTONES ALONG THE PATH

The ideal invention would put an end to all human misery. It would also put an end to all human struggle and human life. Finding the passage to Nirvana would be as easy as taking a jet ride. In this section, we will present milestones along the path to progress that range from some very disruptive inventions to very ordinary changes that have facilitated the advance of society.

Inventions that have negative social consequences have existed throughout history. The guillotine, Thomas Edison's experimental electric chair, and the increasingly potent drugs that distort the mind fall into this group of inventions. In a sense, it might be more desirable to be an ordinary individual with dignity and integrity, rather than being a cruel or sadistic scientist. For this reason, we will restrict our discussion to those activities that facilitate beauty of life rather than destroy it.

2.6.1 The Systematic and Rigorous Rules of Engineering Sciences

Reengineering is perhaps the simplest form of making a change. Change by itself is not necessarily progress, nor is it entirely desirable. The aim in implementing a change is to achieve positive goals. The successful design and engineering of the Boeing 747 from earlier versions (727s and 737s) is an example of a much needed transition in the 1980s airline industry. However, negative results of a change could prove disastrous as the Titanic. Unless change brings about some discernible improvement, it should not be considered as dissertation material. Simple and generic CAD/CAM and CADD packages can make dimensional or environmental changes in the physical product (or both). Unless interwoven with the art of innovation, reengineering is the application of the rigorous rules of the engineering sciences. Modern machines are fully capable of performing this mundane and noncreative task.

In most corporate environments, reengineering is an ongoing activity. Corporate goals for product diversification, enhanced profit, market share, globalization, repackaging, etc., change constantly, and development teams have the responsibility to reengineer products and services to meet these corporate goals. Such reengineering activity sometimes spills over into an academic setting. As a result of oversight or through negligence, it may even become a substitute for a real innovation or invention.

Strict reengineering as a discipline occurs rarely. Strict adherence to only redimensioning could produce an inferior product or process or could even result in failure. At best, it would be a dimensional scaling exercise. Some poorly reengineered products are gapping failures. The Zeppelin, the Edsel, and the Titanic are real-life examples.

2.6.2 Slight Variations on a Basic Concept

Improvisation is a notch above reengineering. Improvisations are created in numerous ways and in many combinations. If the change is of significant social importance or has a monumental impact on humanity, the improvisation may meet the minimum standards for a dissertation. Improvisation may win prestige for the scientist, but such improvisation may fall short of academic standards. For example, the improvisation of a new network virus has novelty, but it also has no real value and it may even have a negative value; the improvisation of an industrial solvent may have large sales implications, but little or no academic value; and so on. In a sense, the examining committee, the standards of the university, and the mentor all play a role in distinguishing an innovation and an improvisation.

At times, improvisation and other simultaneous improvements are interdependent. It may be manipulated, reconstituted, and recombined in ingenious ways in order to make a profound impact on the process under investigation. At least two levels of novelty (change and its contextual or syntactic interdependence) become evident, leading to an innovation. Multiple other levels also exist (e.g., linear and nonlinear cascading of incremental changes; controlled, structured, and/or recursive changes; and compounded or convoluted improvements of the nature, constitution, and composition of objects that give rise to the end product, intermediate process, or final assembly). However, the mentor has the responsibility to see through the underlying relation and to encourage students to explore the topic in order to envision a class of such improvements, devices, or processes. Many times, students stop just short of a breakthrough in their haste to finish the degree. However, the whole contribution in the dissertation has to be examined in perspective to decide whether the all the degree requirements would be met.

2.6.3 The Relaxation of Constraints

Typically, all existing gadgets, devices, systems, and even concepts have some constraints on their use. Even ideas, notions, and thoughts have some constraints. Total freedom from all constraints, though ideal, is inconceivable in actuality. Therefore, in a sense, when researchers find something, it would be possible to ask the basic question about constraints for the existence of what it is they have found. In the context of research and investigation, it is possible to comprehend the constraints associated with the reality that underlies the concept and ideas behind the gadgets, devices, systems, and even inventions. Once these constraints are articulated, it may be possible to systematically relax these constraints individually, alternately, or in unison (e.g., by using new technology, a new environment, or new materials) and to adapt the existing device, gadget, or invention to do something that would be beneficial to society or that would solve a new set of problems. Such incremental and systematic investigations hardly qualify for basic research, although they may give rise to adaptations and even innovation.

The many steps that semiconductor engineers took during the migration toward LSI chips are good examples of adaptation. The size of the gate (the transistor) underwent changes of many orders of magnitude to improve speed and size and to reduce power consumption before complimentary metal oxide semiconductor (CMOS) technology was invented. Once again, the migration of the LSI chip to the VLSI chip and then to the ULSI chip is an instance of an adaptation or optimization.

On the other hand, some preliminary information about devices, gadgets, or systems is a prerequisite to adaptation or innovation. In addition, the boundary between adaptation and innovation is sometimes hard to discern. Many times, researchers do not disclose all of the circumstances surrounding inventions, and it becomes harder for others to understand the entire creative process that led to these inventions. In the context of research work, we suggest that mentors share their creative thoughts with their students in order to train them to progress beyond the individual perspective of either the mentor or the student.

2.6.4 The Recombination of Adaptations at Various Levels

Adaptations may be reengineered and improvised. They may be cascaded, permuted, and commuted, either individually or in groups. It is possible to perform an exhaustive search for all such re-combinations. In fact, with a little bit of luck, even simple genetic engineering has already developed opportunities for growing new crops. Yet, the approach is mechanical, computational and almost programmable. Sometimes even brainstorming leads to new solutions. In a sense, the intellectual activity required to invent a totally new solution can occur as a result of either an individual or collective process.

Mechanical as the process may seem, recognizing the connection between various activities and ideas can require Herculean mental efforts. Some innovations are steps toward a solution rather than actual inventions or breakthroughs. When a step becomes conceivable within the setting, the chain of adaptation leads to the innovation. The crux of the matter is the degree of obviousness or obscurity of the logical thought processes that lead to the innovation. Although patent examiners can become fussy about what is an invention or not an invention, experts in the academic environment can afford to be more flexible in this analysis.

From our observations, the development of an innovation or invention is an intensely personal experience. Such experiences are unpredictable and variable. Yet to some extent it can be fine-tuned and it can be articulated to the extent that the core concepts can be hurriedly put down on a piece of paper to be revisited, reexamined and given a through sanity check with the student–mentor team.

Many painters and poets undergo such experiences, though the language and rules of syntax are different. Yet a connection with the subconscious mind working in a different time scale is established for a short duration, and the most significant concept is communicated. At this stage of imagination, an image saved (on a piece of paper) is an image earned for future analysis and use. The ideas become the objects; the language of ideas becomes the mode of manifestation; the visual imagery or audible messages become the content. The assembly and cognition are abstract and fluid. The grammar of the concepts falls freely like a waterfall to reach and reveal the ultimate truth. Sometimes the core of the problem is obscured by mental reservations until the beauty and universality of the solution shine through.

In many cases, the rigor of the discipline starts after a flash of genius occurs. In some cases, such a flash of genius can also lead the student down a path of half truths. Mathematics, technology, and application programs are sometimes used to help the student to take the next few steps and avoid being sunk in the quicksand of doubt. Inner strength and perseverance can lead the student through this phase, and mentors frequently have the necessary experience to help guide the student.

Sometimes, the clues to a complex problem appear over an extended period of time, and the puzzle is resolved slowly. These clues may lead to a string of loosely connected inventions or patents. The legal departments of corporations or universities can provide the best suggestions regarding the documentation. It is often best that the logical chain of solutions be presented in book form even after the dissertation is complete.

Innovation and invention have vague boundaries. To some extent, the process of investigation alters the process under investigation, thanks to Heisenberg's uncertainty principle. The results of investigation could be positive or negative. The role of the mentor becomes obvious at this stage. On the one hand, the mentor can help to draw out the student's creative genius. On the other hand, the mentor could disregard what the student is really saying. In some cases, an unfriendly reception from a mentor can silence a student in much the same way a child's words can be suppressed by a hostile parental attitude.

2.6.5 The Creation of Novelty

Just as adaptations may be reengineered and improvised, innovations also may be adapted, reengineered, and improvised, either in part or in totality. If it is possible to reconstruct the early stages in the innovative processes (e.g., adaptation, reengineering, and improvisation) as a series of flowcharts, then it is possible to apply the constraint principle (Section 2.6.3) to each of the boxes within the flowcharts. It is also possible to reconstitute the flowchart or rearrange the local configuration of the boxes. Intuitively, some choices are more optimal than others, and rigorous examination of the most optimal choice becomes a demanding and scientific task. We reexamine these structural arrangements of the modules of creativity in Chapter 4.

The placement of critical path algorithms (CPM) techniques can also offer new flavor to the solution. All of the previously discussed manipulations within innovation, adaptation, reengineering, and improvisation become applicable here. It is our intent to warn the student against the mechanization of invention. We do believe that flashes of creativity often are many times faster and better than the results of years of toil. Yet we would like to point out that Mendel's laws of heredity are based on decades of lonesome work in a monastery! Many times, the creative and bold reach the scientific skies only to discover that the sky is not the limit.

The creation of something out of nothing is akin to the creation of mass out of energy (in the pre-Einstein era) or the creation of life from genetic code. One may blame the intellect for not providing the necessary connection, or one may leap into the vacuum and still land on something abstract or quite novel. The flight of the mind may lead to novelty that is far removed from the ordinary. If and when such novelty occurs, it may be rightfully labeled an invention. In many cases, a flash of inspiration brightens years of hard work. An ounce of intuition can outweigh a ton of logic.

The practice of creativity is an art rather than a science. Oddly enough, creativity can have as much of an effect on the evolution of science as it does on the evolution of the arts. It appears as if nature is challenging the intellect to abandon science to find truth. At times, the role of intuition almost replaces the role of logic. The syntax of ideas thus becomes the grammar of the language in which the ideas are themselves expressed and portrayed.

Suffice it say that we present examples of inventions in Chapter 3 rather than write any mathematical basis for their derivation. In a sense, we hope that readers examine the patents as presented and draw out the similarities that exist in their own work in order to create their own innovations and inventions in their own way. Since inventing is such a personal experience, we propose that the findings presented here be used to sharpen one's own skills. After all, re-creation is also a form of creation.

The technology of the future and the mind-set of the computer generation can only propel the creative process further. Such trends became evident over the last few decades when increasingly complex creative environments were assembled, defused, and dismantled quickly. For example, the space technology platforms of the 1950s and 1960s launched by the National Aeronautics and Space Administration (NASA) led to a flurry of new inventions. In addition, the packet-switching technologies launched by the NSF in the 1960s and 1970s led the information age of the 1990s. Today, it almost seems feasible to sow the seeds of creativity with a relatively large influx of funds and then manage creativity through long-term maneuvering rather than through the day-to-day management of scientists. Both the Federal government (e.g., NASA and NSF) and the universities have provided the fabric for newly created technologies and spinoff industries. These major efforts have provided society with microwave communications, satellites, and rocket propulsion, as well as wide area networks (WANs), the Internet, and global-communication capabilities.

2.6.6 Socially Significant Inventions

All patent-worthy inventions have some social effect, and all useless inventions do not deserve a patent. Most attorneys insist on knowing the potential use of an invention before they file for an application for a patent. The screening process is quite thorough. In a sense, of course, the inventor should be the first person to access the legitimacy of the concept and its social benefit. Most of patents presented in this book have had a social impact and, at the same time, have caused some social upheaval. The telephone, the television, the Internet, and intelligent networks are prime examples of inventions that have helped to take society on its evolutionary course. Flashes of genius thus can penetrate the social realm of the mind and crystallize as something concrete of benefit to society.

Inventions can have a negative influence, a positive influence, or a combination thereof. The invention of dynamite (Nobel) is as significant as the discovery of penicillin (Fleming). The electric chair (Edison) is as shocking as sculptures by Picasso. It is possible to classify inventions as having a positive impact or a negative impact. For instance, cigarettes, nerve gas, harmful drugs, etc., have had a substantial negative impact on society. In many cases, such products should not even be classified as inventions at all. They may even appear in nature just as neutral products.

For example, the invention of explosives led to the development of land mines. At times, scientists invent destructive tools. Ethically, inventors and scientists should resist any way to create negative inventions. The beauty of inventions lies in the benefits to society.

Creative thinking that leads to positive goals merits as much commendation as condemnation of the creative thinking for negative goals. It is the connection between creative thinking and social benefit that seems to produce the best inventions. Universities collectively bear the responsibility of actively barring their faculty and students from performing research that has a negative impact on society. For example, in the 1960s, research on LSD was banned by the federal government. Also, in some instances, universities have refused grants from the U.S. Army. Weapons research by any name can propagate human suffering. If the choice is between ethical human values and destructive innovations (e.g. nerve gas, bomber planes, missiles, and gun ships), then the balance swings toward inventions that have a positive impact on society.

In the same vein, mentors may have some responsibility to validate that the results of their research is firmly rooted in the rewards and benefits it brings to the community, society, and humankind. In many instances, no research is better than negative research and Mendel (a monk known for his work on heredity) excels over Nobel (a millionaire known for his invention of dynamite).

Inventions that have an impact on society are generally more than a patent. In many instances, the inventor considers the current social need and the prevalent level of technology without shortchanging either one. As we suggested earlier, invention is sometimes unpredictable. However, positive inventions that have some ethical values are always welcomed by a society. It requires much more of concerted effort to invent something that is noteworthy and that also has a positive impact on society. In some cases, some invention is better than no invention, but no invention is better than a negative invention.

A positive breakthrough invention perhaps achieves the highest honor in the hierarchy of invention. Some examples include Alexander Graham Bell's telephone, Thomas Alva Edison's lightbulb, the transistor, the stored program computer, magnetic resonance imaging, and masers and Silica glass for fiber-optic materials. Each of these contributions has had a profound impact on the evolution of modern society.

REFERENCES (C-2)

[2.1] March, James G., and Simon, Herbert A., *Organizations* 2d ed. Cambridge, MA, Blackwell Publishers, June 1993.

[2.2] Ballard, Dana Harry, and Brown, Christopher M. (Contributor), *Computer Vision.* Prentice-Hall, Englewood Cliffs, NJ, 1982.

[2.3] Li, E. Herbert (Editor), *Semiconductor Quantum Well Intermixing: Material Properties and Optoelectronic Applications (Optoelectronic Properties of Semiconductors and Superlattices)* 1st ed. Taylor & Francis, London, 2000. Also see Dutta, N. K., "III-V Device Technologies for Lightwave Applications." *AT&T Technical Journal,* February 1989, pp. 5–18.

[2.4] "Undersea Communication Technology." *AT&T Technical Journal,* Vol. 74, No. 1, February 1995, pp. 75–82. Also see "Erbium-Doped Fiber and Next Generation of Lightwave Systems," *AT&T Technical Journal,* No. 1, February–March 1992.

[2.5] Griffin, Peter B., Plummer, James D., and Deal, Michael D., *Silicon VLSI Technology: Fundamentals, Practice, and Modeling,* 1st ed. Prentice Hall, Upper Saddle River, NJ, 2000.

[2.6] Chang, C.-Y., Sze, S. M., *ULSI Devices.* Wiley-Interscience, New York, 2000.

[2.7] Antonakos, James L., *The Pentium Microprocessor,* 1st ed. Prentice Hall, Upper Saddle River, NJ, 1996.

[2.8] Born, Max, and Wolf, Emil, *Principles of Optics: Electromagnetic Theory of Propagation, Interference and Diffraction of Light,* 7th ed. Cambridge University Press, Cambridge, UK, 1999.

[2.9] Wolf, Wayne, *Modern VLSI Design: System-on-Chip Design,* 3d ed., Prentice Hall PTR, Upper Saddle River, NJ, 2002.

[2.10] Maslow, Abraham, *Motivation and Personality,* 2d ed., Harper & Row, New York, 1970. Also see Maslow, Abraham Harold, reproduction ed. *The Farther Reaches of Human Nature.* Arkana, New York, 1993. Also see Norwood, George, *Maslow's Hierarchy of Needs,* on the Internet at http://www.deepermind.com/20maslow.htm. (Accessed on March 24, 2004.)

[2.11] Bryce, Robert, and Ivins, Molly, *Pipe Dreams: Greed, Ego, and the Death of Enron,* 1st ed. Pubic Affairs, New York, 2002.

[2.12] Levitt, Arthur, and Dwyer, Paula, *Take On the Street: What Wall Street and Corporate America Don't Want You to Know,* 1st ed. Pantheon Books, Toronto, ON, 2002.

[2.13] Faraday, Michael, *Philosophical Magazine,* 1844, Vol. 24, p. 136. Also see Hammond, John W., *Men and Volts—The Story of General Electric.* J. B. Lippincott Company, Philadelphia, 1941.

[2.14] Einstein, Albert, *The Origin of the General Theory of Relativity.* Jackson, Wylie & Co., Glasgow, UK, 1933.

[2.15] Duffy, James P., *How to Earn an Advanced Degree without Going to Graduate School,* 2d ed., John Wiley & Sons, New York, 1994.

[2.16] Russell, Stuart J., and Norvig, Peter, *Artificial Intelligence: Modern Approach,* 1st ed., Prentice Hall, Englewood Cliffs, NJ, 1995.

[2.17] Moschytz G. S., and Ahamed, Syed V., "Transhybrid Loss with RC Balance Circuits for Primary Rate ISDN Transmission Systems." *IEEE J-SAC on High-speed Digital Subscriber Line* (special issue), Vol. 9, August 1991, pp. 951–959.

[2.18] Rosebury, Fred, *Handbook of Electron Tube and Vacuum Techniques,* reprint ed. American Vacuum Society Classics, Springer Verlag, 1993.

[2.19] Becker, P. C., Olsson, N. A., and Simpson, J. R., *Erbium-Doped Fiber Amplifiers,* 1st ed. Academic Press, San Diego, CA, 1999.

[2.20] Nussbaum, Allen, *Semiconductor Device Physics.* Prentice-Hall, Englewood Cliffs, NJ, 1962.

[2.21] O'Dell, Thomas Henry, *Ferromagnetodynamics: The Dynamics of Magnetic Bubbles, Domains, and Domain Walls.* Wiley, NY, 1981.

[2.22] Janesick, James R., *Scientific Charge-Coupled Devices.* SPIE PRESS Monograph Vol. PM83, Society of Photo-optical Instrumentation Engineers, Bellingham, WA, 2001.

[2.23] Holst, Gerald C., *CCD Arrays, Cameras, and Displays,* 2d ed. J C D Publications, Winter Park, FL, 1996.

[2.24] Cochran, Paul L., *Polyphase Induction Motors: Analysis, Design, and Application.* Electrical Engineering and Electronics Series 59, Marcel Dekker, New York, 1989.

[2.25] Dicksee, C.B., *High-Speed Compression-Ignition Engine.* Blackie Books, London, 1946.

[2.26] Stan Cornel (Editor), *Direct Injection Systems for Spark-Ignition and Compression-Ignition Engines.* Society of Automotive Engineers, Washington, DC, 2000.

[2.27] Ahamed, Syed V., and Lawrence, Victor B., *Design and Engineering of Intelligent Communication Systems.* Kluwer Academic Publishers, Boston, 1997.

[2.28] Choi, K. K., *The Physics of Quantum Well Infrared Photodetectors.* Series in Modern Condensed Matter Physics, Vol. 7. World Scientific, River Edge, NJ, 1997.

[2.29] Als-Nielsen, Jens, and McMorrow, Des, *Elements of Modern X-Ray Physics,* 1st ed. John Wiley & Sons, New York, 2001.

[2.30] Sarid, Dror, *Scanning Force Microscopy: With Applications to Electric, Magnetic and Atomic Forces,* revised ed. Oxford Series in Optical and Imaging Sciences, No. 5, Oxford University Press on Demand. Oxford University Press, New York, 1994.

[2.31] Bonnick, Sydney Lou, and Lewis, Lori Ann, *Bone Densitometry for Technologists.* Humana Press, Totowa, NJ, 2002.

[2.32] Foley, James D., van Dam, Andries, Feiner, Steven K., and Hughes, John F., *Computer Graphics: Principles and Practice in C*, 2d ed. Addison-Wesley Pub. Co., Reading, MA, 1995.

[2.33] Westbrook, Michael H., *The Electric Car: Development and Future of Battery, Hybrid and Fuel-Cell Cars*, 1st ed. Institution of Electrical Engineers, Warrendale, PA, 2001.

[2.34] Hunecke, Klaus, *Jet Engines: Fundamentals of Theory, Design and Operation*. Motorbooks International, Osceola, WI, 1997.

[2.35] Walter, Felix (director of studies), *Four Architectural Movement Studies for the Wheelchair and Ambulant Disabled*. Disabled Living Foundation, London, 1971.

[2.36] Khayyam, Omar, *The Rubaiyat*, in "The Internet Classics Archives," on the Internet at http://classics.mit.edu/Khayyam/rubaiyat.html. (Accessed on March 24, 2004.)

[2.37] Harrison, David M., *The Special Theory of Relativity*, on the Internet at http://www.upscale.utoronto.ca/GeneralInterest/Harrison/SpecRel/SepecRel.html. (Accessed on March 24, 2004.)

[2.38] Nielsen, Michael A., and Chuang, Isaac L., *Quantum Computation and Quantum Information*. Cambridge University Press, Cambridge, NY, 2000.

[2.39] Feitelson, Dror G., *Optical Computing, A Survey for Computer Scientists* reprint ed. MIT Press, Cambridge, MA, 1992.

[2.40] Kraus, John Daniel, and Fleisch Daniel A., *Electromagnetics*, 5th ed. McGraw-Hill Science/Engineering/Math, New York, 1999.

[2.41] Da Prato, Giuseppe, and Zabczyk, Jerzy, *Second Order Partial Differential Equations in Hilbert Spaces*, 1st ed. Cambridge University Press, Cambridge, Uk, 2002.

[2.42] Hatsopoulos, George N., *Principles of General Thermodynamics*. John Wiley and Sons, New York, 1965.

[2.43] Bankston, John, *Alexander Fleming and the Story of Penicillin*. Mitchell Lane Publishers, Inc., Bear, DE, 2002.

[2.44] Ludwig, Charles, *Michael Faraday, Father of Electronics*. Herald Press, Scottsdale, PA, 1988.

[2.45] Gauss, Carl Friedrich, *Disquisitiones Arithmaticae*. Tr. by Arthur A. Clarke, Yale University Press, New Haven, CT, 1965.

[2.46] Maxwell, James Clerk, *Treatise on Electricity and Magnetism*, 3d ed. Dover Publications, New York, 1991.

CHAPTER 3

CLASSIC BREAKTHROUGHS

Looking back at an invention is, in a sense, an injustice. While looking forward involves the projection of the thought process into the unknown, looking backward involves the rationalizing what is already known. Therefore, when we discuss the major inventions of the last century, we only inject our biases and modern sensibilities. For this reason, we do not directly discuss these inventions, but instead present the original patent document, which contains the invention as seen in the eyes of the inventor.

It is our hope and suggestion that readers attempt to "see" the invention as the inventor saw it, to duplicate the "rationality" if they see fit. We also challenge readers to use their own skills and mental capacities to systematize, analogize, conceptualize, rationalize, and generalize, as well as deduce and deduct concepts and relationships in the hope that new inventions or innovations may result. We also suggest reverse engineering and reengineering of the concepts embedded in these inventions in other disciplines. In following these steps at a conscious level, the development of a true invention may be rare; however, it is possible that an innovation may follow. The role of the subconscious and the mysterious mental process of channeling scientific truth, mathematical beauty, and social benefit into a realizable device, product, or a concept complements the conscious awareness of the problem at hand. The inventor is able to find a blissful balance between the known and the unknown, thus making each invention distinct and unique in its own right.

We have arranged these inventions in chronological order to allow readers to visualize how one major invention has influenced the inception of another. Conversely, subsequent inventions provided the technology needed to give previous inventions an even greater social and/or global effect. We have also alluded to the fact that major inventions are carried forward by succeeding inventions, thus making the original inventions even more beneficial to society. In a great many cases the secondary inventions only carry an incremental effect. Sometimes these intertwined chain of inventions follow the wave and ripple effects. We elaborate on these symbiotic and synergistic inventions in the following sections.

Symbiotic Inventions. The interdependence between major inventions such as cathode ray tubes (CRTs) (Section 3.4) and MRIs (Section 3.8) becomes evident if one visualizes the role of integration (e.g., Memory and Processor) in the semiconductor technology in the 1960s–80s. Damadian's MRI technology, developed in 1974, is seen as a direct beneficiary of Zworykin's 1938 CRT technology and his 1935 imaging techniques. Early magnetic-resonance imaging devices used updated versions of Zworykin's CRT. They also deployed the concepts that were the basis of processing signals and data documented for computer systems and stored program control in 1948 (Section 3.5), the transistor (Section 3.6), and the ensuing VLSI revolution and laser technology of the 1960s (Section 3.7). These technologies had already attained a certain amount of maturity by 1974, when the technology was incorporated in Damadian's MRI (1974) invention. The point we make is that Damadian's invention was not developed in isolation, but instead was profoundly influenced by past inventions. The underground water for the oasis in the desert is the intellectual and conceptual freedom that science and reason had brought to society in the twentieth century. Such freedom also exists today, but it is shrouded by the impact of computers, networks, and Internet environments. Moreover, the constant political micromanagement within the country, and then again within the institutions, curtail the conceptual freedom and clip the wings of unrestrained intellect.

Many earlier scientists and mathematicians have also acknowledged the importance of collaboration and appreciation for each other's contributions. Immanuel Kant, Friedrich Schelling, and Hans Christian Oersted [3.1] have jointly pondered the unification in nature that brings chemical affinity, static electricity, galvanism, and magnetism into one realm of thinking. They have stressed the need for analyses of prior principles, noting that this primary philosophic groundwork is necessary for the development of any practical generalizations. Such generalizations offer the starting points of other symbiotic contributions, inventions, and innovations. Maxwell's generalizations of the laws of electromagnetism are based on the work of numerous other thinkers and experimentalists. James Clerk Maxwell (1831–1879) [3.2] contributed significantly to the understanding of electricity and magnetism. Other notable scientists who have influenced Maxwell's theories are Michael Faraday (1791–1867) [3.3], Andre Marie Ampere (1775–1836) [3.4], Alessandro Volta (1745–1827) [3.5], Hans Christian Oersted (1777–1851) [3.1], and Carl Friedrich Gauss (1777–1855) [3.6]. Many other instances of symbiosis and synergy exist in more recent history. This remarkable and noteworthy confluence is documented by the eminent research of Bell (1862–1939) [3.7] and Edison (1847–1931) [3.8] in Sections 3.1 and 3.2; Shockley (1910–1989) [3.9], Bardeen (1908–1991) [3.10], and Brattain (1902–1987) [3.11] in Section 3.6; and Schawlow (1921–1999) [3.12] and Townes (1915–) [3.13] in Section 3.7).

The converse feedback effect may now be traced by looking at the relationship between previous and current inventions. Whereas the new crop of inventions developed from the 1970s through the 1990s substantially benefited society, older inventions are scrutinized in far greater detail in order to optimize the invention for use with modern technology, rather than to optimize the technology used to develop the invention. Newer and better inventions (e.g., innovations, improvisations, and reengineering of inventions) have resulted. The symbiosis has ensued in many disciplines, and experts in technology often work closely with inventors in order to capitalize on an idea that is genuine and beneficial.

The feed-forward effects of the semiconductor era (Section 3.6) on society are accelerated by the basic technologies of early communication concepts (Section 3.3), television systems (Section 3.4), and computer systems (Section 3.5). The reverse (i.e., feedback) effects are also equally evident. In fact, there are multiple combinations of feed-forward and feed-back effects, rather than a single rather simplistic feed-forward/feed-backward loop effect [i.e., the relationship of semiconductors (Section 3.6) to telecommunications (Section 3.3), television (Section 3.4), and computers (Section 3.5); the relationship of telecommunication concepts (Section 3.3) to semiconductors (Section 3.6), television (Section 3.4), and computers (Section 3.5)]. Tracing these effects to invent is not an invention, but rather an exhaustive mathematical process.

This symbiosis has occurred in other more recent instances. For example, the social effects of computer systems (Section 3.5) are carried on in the technologies of early communication concepts (Section 3.3) and television systems (Section 3.4). Likewise, the effects of Intelligent Networks (Section 3.10) are based on early communication concepts (Section 3.3), television systems (Section 3.4), and computer systems (Section 3.5). The greatest beneficiary of these cumulative effects is the Internet (Section 3.11), which is fueled by early communication concepts, television systems, computer systems, semiconductors (Section 3.6), and, to a large extent, the composition of glass and fiber-optics technology (Section 3.9) as global Internet signals are carried over fiber-optic information highways. Optical highways are likely to be greatly influenced by Intelligent Networks (Section 3.10) and to exploit a new breed of architecture, as the Internet becomes increasingly intelligent over the next decade or two.

Symbiotic and Synergistic Inventions. Symbiosis is a slower phenomenon that dominated earlier inventions. The act of extracting the symbols (or concepts) from one invention and then mixing them and mapping them with the symbols (or concepts) of another invention could produce a (downgraded) invention in its own right. Remote as it may sound, it is possible to do just this type of a conscious exercise to earn an invention. However, in comparison with the truly monumental achievements in history, such artificial invention bears the same mocking resemblance to true invention that artificial intelligence bears to natural intelligence.

The influence of the work of one inventor or artist upon the work of a subsequent or contemporary inventor or artist is frequently a premeditated act of intellectual collaboration based on appreciation and regard. Such collaboration could bring about competition and challenge [e.g., between Bell (Section 3.1) and Edison (Section 3.2) or between Marconi (Section 3.3.1) and Tesla (Section 3.3.2)]. The positive influence of prior inventors or co-inventors can become a conscious process of transmutation of inventions, where the invention in one area (e.g., vacuum tubes or solid-state amplification) is conceptually carried into another area (e.g., solid state to optical amplification). Whatever the reason, history shows that the symbiotic process seems to be slow, short lived, and unstructured.

More recently, in the era of high-speed network technology, the rapid feedback of inventions on the technology is essential for survival. To a large extent, inventions benefit not only from feedback, but also from the rate of synergy that can develop between related inventions and correlated technologies. For example, inventions in communication systems have facilitated feedback between inventors and scientists. The process is constantly improving. For example, on-line communication is becoming increasingly inexpensive as the result of digital subscriber lines and the Internet. Synergy is now as feasible as symbiosis. In fact, symbiosis can be accelerated by synergy, and synergy can be learned from selected symbiotic processes.

The shared effects among fiber optics, semiconductors, and the quantum technologies of the 1990s are also an example of both symbiosis and synergy that has facilitated global gigabit nets. Though the whole evolution of the technology and its implementation is quite recent, a platform of earlier inventions is nonetheless essential to the development of the technology. The role of accelerated inventions and innovations is obvious throughout the 1980s (e.g., VLSI technology), 1990s (e.g., network and fiber-optics technology), and into the present decade, though at a more moderate rate. Perhaps the saturation that curtails the duration of the creative phase in any discipline is a socioeconomic phenomenon.

External agencies (e.g., national laboratories, governmental agencies, professional societies, universities, and even mentors who work with students) can have two very distinct roles. First, they can provide the basis of symbiotic interdisciplinary interactions (e.g., between fiber-optic glass inventions and technologies and VLSI concepts and fabrications, between virtual reality and the recreational travel industry, between quantum computing and fiberonics, and between the aging processes of cells and the genetic codes carried by red blood cells). Second, they can provide a basis for synergistic and rapid feedback between the possible and the impossible, between concepts and the realization of any invention.

Major inventions sometimes follow a hierarchical order. The environment and the platform provided by lower-order inventions is usually a precursor to a contribution at the next higher level by uniquely convoluting the concepts incorporated in the inventions of the lower levels. The capacity to look back objectively and to recombine the principles of science to generate the next breakthrough are traits that can benefit many research students in a creative way. The order of influence could also be sideways rather than vertical. For example, the recombination of laser technology with magnetic disk storage technology has given rise to compact disc (CD) and video compact disc (VCD) [3.14] devices. Another example of such an invention (innovation) would be the derivation of Intelligent Network/2 from Intelligent Network/1 (see Section 3.10) or the deviation of Advanced Intelligent Networks from Intelligent Network/2 [3.15]. Other examples also exist. The extension of digital subscriber line (DSL) concepts to high-speed digital subscriber line (HDSL) concepts is also a classic example. For those readers with telecommunications and computing interests, the basic concepts of both subscriber lines are discussed in detail in [3.16].

Other researchers and scientists may have a different list of breakthrough patents and inventions, such as the Wright brothers' flying machines and their connection to today's modern spacecraft and medicine to cellular biology. We do not feel qualified to venture into these and other domains that are totally outside our area of expertise. For this reason, we confine ourselves to including patents that relate to telecommunications and computing, which are our specialties. These two disciplines have profoundly affected our lives, both directly and indirectly, over the last few decades. Mastery of these two disciplines provides a practical foundation for advanced students who are approaching the dissertation stage of their doctoral program.

As much as positive inventions have benefited society, we would like to note that catastrophes, wars, famines, and disasters have had many devastating influences on society. The development of such inventions [e.g., the atom bomb, intercontinental ballistic missiles (ICBMs), and poisonous agents within the biochemical industry] is sometimes encouraged by national governments and defense organizations. In the 16 major inventions presented, there is no reference to such negative inventions.

REFERENCES (C-3.0)

[3.1] Gregory, Frederick, *Oersted and the Discovery of Electromagnetism,* in "Episodes in Romantic Science," on the Internet at http://www.clas.ufl.edu/users/fgregory/oersted.htm. (Acessed on March 24, 2004.)

[3.2] Friedel, Robert D., *Lines and Waves: An Exhibit by the IEEE History Center (1981),* in "IEEE History Center," on the Internet at http://www.ieee.org/organizations/history_center/general_info/lines_menu.html. (Accessed on March 27,2004.) See also Maxwell, James Clerk, *A Treatise on Electricity and Magnetism,* 3d ed. Dover Publications, New York, 1991.

[3.3] Faraday, Michael, "A Speculation Touching Electric Conduction and the Nature of Matter." *Philosophical Magazine,* Vol. 204, No. 157, February 1844, pp. 136–144. Also see Hammond, John W., *Men and Volts—The Story of General Electric.* J. B. Lippincott Company, Philadelphia, 1941.

[3.4] Hoffmann, James R., *André-Marie Ampère.* Cambridge University Press, New York, 1996.

[3.5] Piuri, Vincenzo (Editor), IEEE Alessandro Volta Memorial Workshop on Low-Power Design: Proceedings in Como, Italy, March 4–5, 1999. IEEE Computer Society, Los Alimitos, CA, 1999.

[3.6] Phillips, George M., *Two millennia of Mathematics: From Archimedes to Gauss.* Springer Verlag, New York, 2000.

[3.7] Grosvenor, Edwin S., and Wesson, Morgan, Alexander Graham Bell: The Life and Times of the Man Who Invented the Telephone. Harry Abrams, New York, 1997.

[3.8] Josephson, Matthew, *Edison: A Biography,* reprint ed. John Wiley & Sons, New York, 1992. Also see Cramer, Carol (Editor), *Thomas Edison.* People Who Made History Series, Greenhaven Press, San Diego, CA, 2001.

[3.9] Shockley. William, *Electrons and Holes in Semiconductors, with Applications to Transistor Electronics.* R. E. Krieger Pub. Co., Huntington, NY, 1976. Also see *Transistor,* on the Internet at http://www.lucent.com/minds/transistor. (Accessed on March 28, 2004.)

[3.10] Hoddeson, Lillian, and Daitch, Vicki, *True Genius: The Life and Science of John Bardeen: The Only Winner of Two Nobel Prizes in Physics.* Joseph Henry Press, Washington, DC, 2002. Also see *Transistor,* on the Internet at http://www.lucent.com /minds/transistor. (Accessed on March 28, 2004.)

[3.11] The American Physical Society, "This Month in Physics History—November 17–December 23, 1947: Invention of the First Transistor." *APS News Online,* on the Internet at http://www.aps.org/apsnews/1100/110004.cfm. (Accessed on March 28, 2004.) Also see *Transistorized,* on the Internet at http:// www.pbs.org/transistor. (Accessed on March 28, 2004.) Also see *Transistor,* on the Internet at http://www.lucent.com/minds/transistor. (Accessed on March 28, 2004.)

[3.12] Schawlow, A. L., and Townes, C. H., "Infrared and Optical Masers." *Physical Review,* Vol. 112, December 1958, pp. 1940–1949. Also on the Internet at http://www.coseti.org/schawlow.htm. (Accessed on March 28, 2004.)

[3.13] Townes, C. H., and Miles, P. A. (Editors), *Quantum Electronics and Coherent Light.* Academic Press, New York, 1964. Also see Townes, C. H., and Schawlow, A. L., *Microwave Spectroscopy.* Dover Publications, New York, 1975.

[3.14] Lenk, J. D, Lenk's *Laser Handbook: Featuring CD, CDV, and CD-ROM Technology,* reprint ed. McGraw-Hill Professional, New York, 1991.

[3.15] Ahamed, Syed V., and Lawrence, Victor B., Intelligent Broadband Multimedia Networks: Generic Aspects and Architectures: Wireless, ISDN, Current and Future Intelligent Networks. Kluwer Academic Publishers, Boston, MA, 1997.

[3.16] Ahamed, Syed V., and Lawrence, Victor B., *Design and Engineering of Intelligent Communication Systems.* Kluwer Academic Publishers, Boston, MA, 1997.

Alexander Graham Bell (1847–1922)

Telephony

3.1 TELEPHONY

In the 1870s, long-distance communication was accomplished by telegraphy, a method of encoding the alphabet into a series of audible dots and dashes (Morse code) and transmitting it over wire. By listening to the string of audible dashes and dots, the listener was able to reconstruct the words. The message communicated could therefore be rewritten at a remote site. In the case of telephony, the electrical signals carrying the analog speech are transmitted over wire cables. The premise of telephony is the communication of speech-generated signals, thus eliminating the need for any intermediate Morse code.

The advantages of telephony claimed by Alexander Graham Bell in Patent No. 174,465 are based on the idea that voice rather than codes can be communicated bi-directionally and at a much faster rate. In a generic sense, this patent was the basis for most baseband telephone transmission systems for almost 50 years.

Like other patents of the late 1800s, this patent emphasizes the scientist's inventive talent and reinforces the results from his experimental observations. The scientific content in this patent is presented only as a means of explanation of the physical and electrical phenomena in the transmitter, the cable, and the receiver. The function of the microphone and the electromagnet to generate these undulating currents is clearly demonstrated by the experimental setup. The highly intuitive content of this patent is the highlight, as compared to the patents from the second half of the last century, which are discussed later in this chapter.

For the sake of our readers, we would like to point out the differences between the thinking and writing styles in Bell's patents and Nikola Tesla's patents. Bell's more closely structured and less intuitive pattern of reasoning and investigation (as described in Section 4.3.2) is distinct from Tesla's less structured and more intuitive pattern of reasoning and investigation (as described in Section 4.3.1). This concept of structured creativity gained momentum throughout the last century, and some recent patents are based on rigorous reason and strictly derived logical claims rather than on the inventions that should be the nucleus of a patent. The role of logic, induction, mathematics, and refined thinking starts to complement and supplement the central focus of thought or flash of genius of the inventor.

3.1.1. Improvement in Telegraphy (Bell, 1876)

UNITED STATES PATENT

March 7, 1876 174,465

Improvement in Telegraphy

ALEXANDER GRAHAM BELL OF SALEM, MASSACHUSETTS

Specification forming part of Letters Patent No. 174,465, dated March 7, 1876. Application filed February 14, 1876.

To all whom it may concern:

Be it known that I, ALEXANDER GRAHAM BELL, of Salem, Massachusetts, have invented certain new and useful Improvements in Telegraphy, of which the following is a specification:—

In Letters Patent granted to me April 6, 1875, No. 161,739, I have described a method of, and apparatus for, transmitting two or more telegraphic signals simultaneously along a single wire by the employment of transmitting instruments, each of which occasions a succession of electrical impulses differing in rate from the others; and of receiving instruments, each tuned to a pitch at which it will be put in vibration to produce its fundamental note by one only of the transmitting instruments; and of vibratory circuit-breakers operating to convert the vibratory movement of the receiving instrument into a permanent make or break (as the case may be) of a local circuit, in which is placed a Morse sounder, register, or other telegraphic apparatus. I have also therein described a form of autograph-telegraph based upon the action of the above-mentioned instruments.

In illustration of my method of multiple telegraphy I have shown in the patent aforesaid, as one form of transmitting instrument, an electro-magnet having a steel-spring armature, which is kept in vibration by the action of a local battery. This armature in vibrating makes and breaks the main circuit, producing an intermittent current upon the line wire. I have found, however, that upon this plan the limit to the number of signals that can be sent simultaneously over the same wire is very speedily reached; for, when a number of transmitting instruments, having different rates of vibration, are simultaneously making and breaking the same circuit, the effect on the main line is practically equivalent to one continuous current.

In a pending application for Letters Patent, filed in the United States Patent Office February 25, 1875, I have described two ways of producing the intermittent current—the one by actual make and break of contact, the other by alternately increasing and diminishing the intensity of the current without actually breaking the circuit. The current produced by the latter method I shall term, for distinction sake, a pulsatory current.

My present invention consists in the employment of a vibratory or undulatory current of electricity in contradistinction to a merely intermittent or pulsatory current, and of a method of, and apparatus for, producing electrical undulations upon the line-wire.

The distinction between an undulatory and a pulsatory current will be understood by considering that electrical pulsations are caused by sudden or instantaneous changes of intensity, and that electrical undulations result from gradual changes of intensity exactly analogous to the changes in the density of air occasioned by simple pendulous vibrations. The electrical movement, like the aerial motion, can be represented by a sinusoidal curve or by the resultant of several sinusoidal curves.

Intermittent or pulsatory and undulatory currents may be of two kinds, accordingly as the successive impulses have all the same polarity or are alternately positive and negative.

The advantages I claim to derive from the use of an undulatory current in place of merely intermittent one are, first, that a very much larger number of signals can he transmitted simultaneously on the same circuit; second, that a closed circuit and single main battery may be used; third, that communication in both directions is established without the necessity of special induction-coils; fourth, that cable despatches may be transmitted more rapidly than by means of intermittent current or by the methods at present in use; for, as it is unnecessary to discharge the cable before a new signal can he made, the lagging of cable signals is prevented; fifth, and that as the circuit is never broken a spark-arrester becomes unnecessary.

It has long been known that when a permanent magnet is caused to approach the pole of an electro-magnet a current of electricity is induced in the coils of the latter, and that when it is made to recede a current of opposite polarity to the first appears upon the wire. When, therefore, a permanent magnet is caused to vibrate in front of the pole of an electro-magnet an undulatory current of electricity is induced in the coils of the electro-magnet, the undulations of which correspond, in rapidity of succession, to the vibrations of the magnet, in polarity to the direction of its motion, and in intensity to the amplitude of its vibration.

That the difference between an undulatory and an intermittent current may be more clearly understood I shall describe the condition of the electrical current when the attempt is made to transmit two musical notes simultaneously—first upon the one plan and then upon the other. Let the interval between the two sounds be a major third; then their rates of vibration are in the ratio of 4 to 5. Now, when the intermittent current is used the circuit is made and broken four times by one transmitting-instrument in the same time that five makes and breaks are caused by the other. A and B, Figs. 1, 2, and 3, represent the intermittent currents produced, four impulses of B being made in the same time as five impulses of A. *c c c*, &c., show where and for how long time the circuit is made, and d d d, & c., indicate the duration of the breaks of the circuit. The line A and B shows the total effect upon the current when the transmitting-instruments for A and B are caused simultaneously to make and break the same circuit. The resultant effect depends very much upon the duration of the make relatively to the break. Fig. 1 the ratio is as 1 to 4; in Fig. 2, as 1 to 2; and in Fig. 3 the makes and breaks are of equal duration. The combined effect A and B, Fig. 3, is very nearly equivalent to a continuous current.

When many transmitting instruments of different rates of vibration are simultaneously making and breaking the same circuit the current upon the main line becomes for all practical purposes continuous.

Next, consider the effect when an undulatory current is employed. Electrical undulations, induced by the vibration of a body capable of inductive action, can be represented graphically, without error, by the same sinusoidal curve which expresses the vibration of the inducing body itself, and the effect of its vibration upon the air; for, as above stated, the rate of oscillation in the electrical body—that is, to the pitch of the sound produced. The intensity of the current varies with the amplitude of the vibration—that is, with the loudness of the sound; and the polarity of the current corresponds to the direction of the vibrating body—that is, to

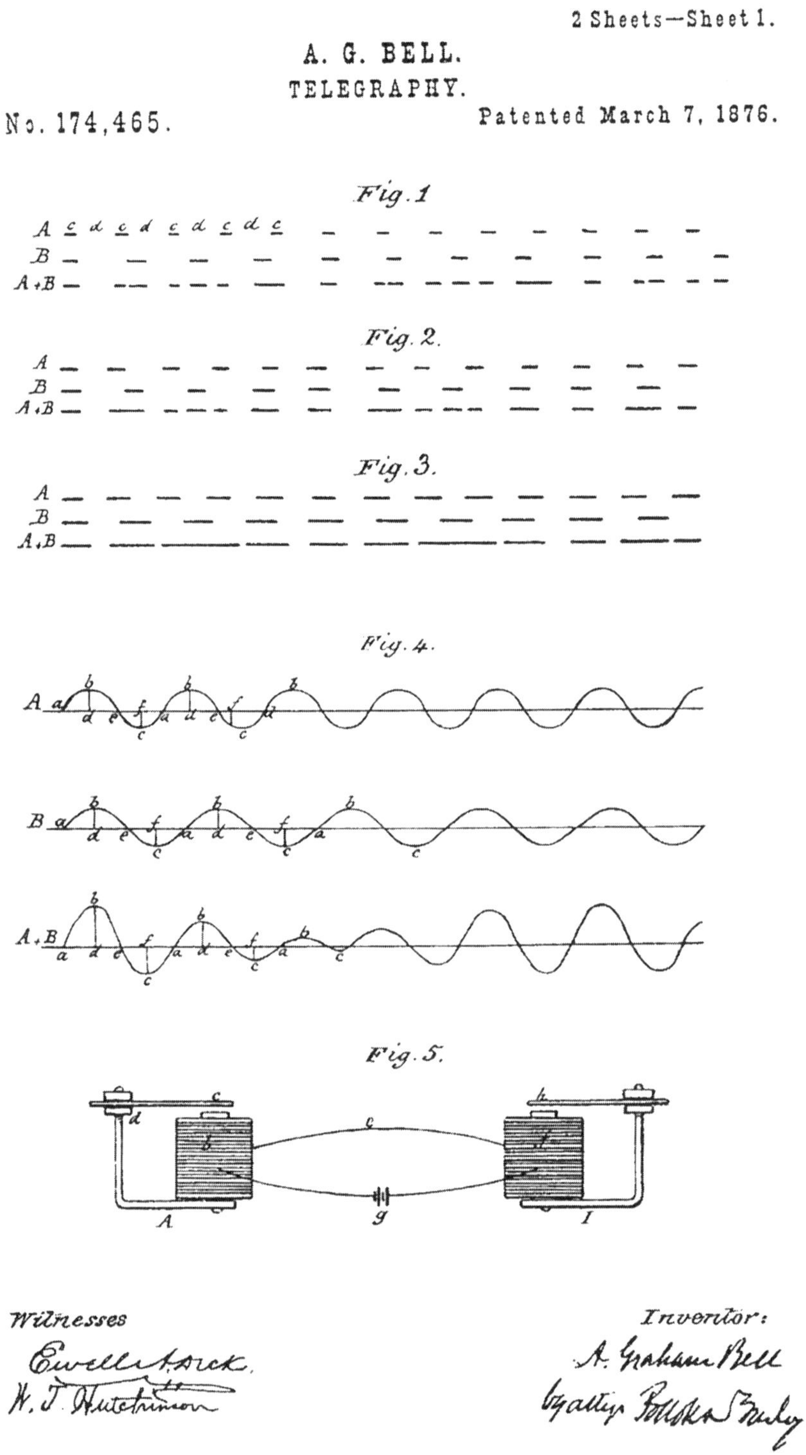
2 Sheets—Sheet 1.
A. G. BELL.
TELEGRAPHY.
No. 174,465.
Patented March 7, 1876.
Fig. 1
Fig. 2.
Fig. 3.
Fig. 4.
Fig. 5.
Witnesses
Inventor:
A. Graham Bell

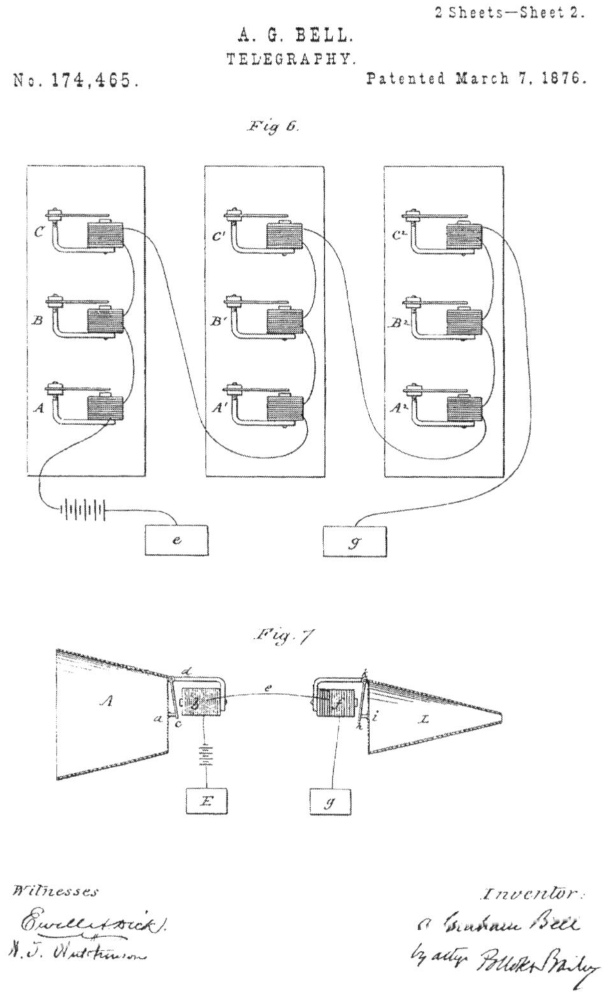

the condensations and rarefactions of air produced by the vibration. Hence, the sinusoidal curve A or B, Fig. 4, represents, graphically, the electrical undulations induced in a circuit by the vibration of a body capable of inductive action.

The horizontal line *a d e f*, &c., represents the zero of current. The elevation *b b b*, &c., indicate impulses of positive electricity. The depressions c c c, &c., show impulses of negative electricity. The vertical distance b d or c f of any portion of the curve from the zero line expresses the intensity of the positive or negative impulse at the part observed, and the horizontal distance a a indicates the duration of the electrical oscillation. The vibrations represented by the sinusoidal curves B and A, Fig. 4, are in the ratio aforesaid, of 4 to 5—that is, four oscillations of B are made in the same time as five oscillations of A.

The combined effect of A and B, when induced simultaneously on the same circuit, is expressed by the curve A + B, Fig. 4, which is the algebraic sum of the sinusoidal curves A and B. This curve A + B also indicates the actual motion of the air when the two musical notes considered are sounded simultaneously. Thus, when electrical undulations of different rates are simultaneously induced in the same circuit, an effect is produced exactly analogous to that occasioned in the air by the vibration of the inducing bodies. Hence, the co-existence upon a telegraphic circuit of electrical vibrations to different pitch is manifested, not by the peculiarities in the shapes of the electrical undulations, or, in other words, by the peculiarities in the shapes of the curves which represent those undulations.

There are many ways of producing undulatory currents of electricity, dependent for effect upon the vibrations or motions of bodies capable of inductive action. A few of the methods that may be employed I shall here specify. When a wire, through which a continuous current of electricity is passing, is caused to vibrate in the neighborhood of another wire, an undulatory current of electricity is induced in the latter. When a cylinder, upon which are arranged bar-magnets, is made to rotate in front of the pole of an electro-magnet, an undulatory current is induced in the coils of the electro-magnet.

Undulations are caused in a continuous voltaic current by the vibration of motion of bodies capable of inductive action; or by the vibration of the conducing-wire itself in the neighborhood of such bodies. Electrical undulations may also be caused by alternately increasing and diminishing the resistance of the circuit, or by alternately increasing and diminishing the power of the battery. The internal resistance of a battery is diminished by bringing the voltaic elements nearer together, and increased by placing them further apart. The reciprocal vibration of the elements of a battery, therefore, occasions an undulatory action in the voltaic current. The external resistance may also be varied. For instance, let mercury or some other liquid form part of a voltaic current, then the more deeply the conducting-wire is immersed in the mercury or other liquid, the less resistance does the liquid offer to the passage of the current. The vertical vibration of the elements of a battery in the liquid in which they are immersed produces an undulatory action in the current by alternately increasing and diminishing the power of the battery.

In illustration of the method of creating electrical undulations, I shall show and describe one form of apparatus for producing the effect. I prefer to employ for this purpose an electro-magnet A, Fig. 5, having a coil upon only one of its legs *b*. A steel-spring armature, *c*, is firmly clamped by one extremity to the uncovered leg *d* of the magnet, and its free end is allowed to project above the pole of the uncovered leg. The armature *c* can be set in vibration in a variety of ways, one of which is by wind, and, in vibrating, it produces a musical note of a certain definite pitch.

When the instrument A is placed in a voltaic circuit, *g b e f g*, the armature *c* becomes magnetic, and the polarity of its free end is opposed to that of the magnet underneath. So long as the armature *c* remains at rest, no effect is produced upon the voltaic current, but the moment it is set in vibration to produce its musical note a powerful induction action takes place, and electrical undulations traverse the circuit *g b e f g*. The vibratory current passing through the coil of the electro-magnet *f* causes vibration in its armature *h* when the armature *c h* of the two instruments A I are normally in unison with one another; but the armature *h* is unaffected by the passage of the undulatory current when the pitches of the two instruments are different.

A number of instruments may be placed upon a telegraphic circuit, instruments is set in vibration all the other instruments upon the circuit which are in unison with it respond, but those which have normally a different rate of vibration remain silent. Thus, if A, Fig. 6, is set in vibration, the armatures of A' and A^2 will vibrate also, but all the others will remain still. So, if B' is caused to emit its musical note the instruments B B^2 respond. They continue sounding so long as the mechanical vibration of B′ is continued, but become silent with the cessation of its motion. The duration of the sounds may be used to indicate the dot or dash of the Morse alphabet, and thus a telegraphic despatch may be indicated by alternately interrupting and renewing the sound. When two or more instruments of different pitch are simultaneously caused to vibrate, all the instruments of corresponding pitches upon the circuit are set in vibration, each responding to that one only of the transmitting instruments with which it is in unison. Thus the signals of A, Fig. 6, are repeated by A′ and A^2, but by no other instrument upon the circuit; the signals of B^2 by B and B′; and the signals of C′ by C and C^2—whether A, B^2, and C′ are successively or simultaneously caused to vibrate. Hence by these instruments two or more telegraphic signals or messages may be sent simultaneously over the same circuit without interfering with one another.

I desire here to remark that there are many other uses to which these instruments may be put, such as the simultaneous transmission of musical notes, differing in loudness as well as in pitch, and the telegraphic transmission of noises or sounds of any kind.

When the armature *c*, Fig. 5, is set in motion the armature *h* responds not only in pitch, but in loudness. Thus, when *c* vibrates with little amplitude, a very soft musical note proceeds from *h*; and when *c* vibrates forcibly the amplitude of the vibration of *h* is considerably increased, and the resulting sound becomes louder. So, if A and B, Fig. 6, are sounded simultaneously (A loudly and B softly), the instruments A′ and A^2 repeat loudly the signals of A, and B′ B^2 repeat softly those of B.

One of the ways in which the armature *c*, Fig. 5, may be set in motion has been stated above to be by wind. Another mode is shown in Fig. 7, whereby motions can be imparted to the armature by the human voice or by means of a musical instrument.

The armature *c*, Fig. 7, is fastened loosely by one extremity to the uncovered leg *d* of the electro-magnet *b*, and its other extremity is attached to the centre of a stretched membrane, *a*. A cone, A, is used to converge sound-vibrations upon the membrane. When a sound is uttered in the cone the membrane *a* is set in vibration, the armature *c* is forced to partake of the motion, and thus electrical undulations are created upon the circuit E *b e f g*. These undulations are similar in sound to the air vibrations caused by the sound that is, they are represented graphically by similar curves. The undulatory current passing through the electro-magnet *f* influences its armature *h* to copy the motions of the armature *c*. A similar sound to that uttered into A is the heard to proceed from L.

In this specification the three words "oscillation," "vibration," and "undulation," are used synonymously, and in contradistinction to the terms "intermittent" and "pulsatory." By the term "body capable of inductive action," I mean a body which, when in motion, produces dynamical electricity. I include in the category of bodies capable of inductive action—brass, copper, and other metals, as well as iron and steel.

Having described my invention, what I claim, and desire to secure by Letters Patent, is as follows:

1. A system of telegraphy in which the receiver is set in vibration by the employment of undulatory currents of electricity, substantially as set forth.
2. The combination, substantially as set forth, of a permanent magnet or other body capable of inductive action, with a closed circuit, so that the vibration of the one shall occasion electrical undulations in the other, or in itself, and this I claim, whether the

permanent magnet be set in vibration in the neighborhood of the conducting-wire forming the circuit, or whether the conducting-wire be set in vibration in the neighborhood of the permanent magnet, or whether the conducting wire and the permanent magnet both simultaneously be set in vibration in each other's neighborhood.

3. The method of producing undulations in a continuous voltaic current by the vibration of motion of bodies capable of inductive action, or by the vibration or motion of the conducting-wire itself, in the neighborhood of such bodies, as set forth.
4. The method of producing undulations in a continuous voltaic circuit by gradually increasing and diminishing the resistance of the circuit, or by gradually increasing and diminishing the power of the battery, as set forth.
5. The method of, and apparatus for, transmitting vocal or other sounds telegraphically, as herein described, by causing electrical undulations, similar in form to the vibrations of the air accompanying the said vocal or other sounds, substantially as set forth.

In testimony whereof I have hereunto signed my name this 20th day of January, A.D. 1876.

ALEX. GRAHAM BELL

Witnesses: Thomas E. Barry P. D. Richards

3.2 MOTION PICTURE CAMERA

Thomas Edison's contributions as an inventor are documented in his 1,093 U.S. patents initiated over a 63-year span (1868–1931). As a contemporary of Alexander Graham Bell, Edison had a healthy competition with Bell. Edison's numerous patents run through at least seven subjects: (a) 424 patents in electric light and power, (b) 199 patents in phonographs and sound recording, (c) 186 patents in telegraphy and telephony, (d) 147 patents in batteries, (e) 53 patents in mining and ore milling, (f) 49 patents in cement, and (g) 9 patents in motion pictures. It is our contention that the early inventors used their earlier inventions in their later inventions. In this vein, we suggest that readers use the patents of the early inventors to "blend" the documented innovative process within the trajectory of their own thinking.

During the time of Edison and Bell, many basic and fundamental inventions were made. Significant strides in the quantitative and algebraic formulation of electromagnetic* phenomena were also made during that time. Most of the early inventors did not demonstrate a clear mathematical insight into their patents. These inventions seem to be based more on intuition and experimental observations rather than on strict derivation and cause–effect relationships, although it was not absent. These inventions changed developments for the next two centuries. It is not necessary to scan all of Edison's numerous patents to realize that Edison was a determined inventor who developed largely through experimentation and insight.

Generally, any one particular device by itself is not sufficient to earn a reputation as a catalyst of scientific or engineering change. However, the scientist's sheer (almost dogmatic) persistence in building upon these individual devices to make systems that served society is what is most remarkable. This trait is

*Maxwell, James Clarke, *A Treatise on Electricity and Magnetism*. The original edition was published in 1873, based on the experimental work of Faraday, Ampère, and Gauss.

Thomas Edison (1847–1931)

Motion Picture Camera

also evident in the patents granted to Marconi and Tesla. The capacity of these scientists to build on their own early inventions to generate new inventions can be seen by studying both their early and their later patents. The number of patents can increase automatically by permeating and fusing the principle of one invention into the next, and so on. This process is purely a numbers game in deriving permutations and combinations of a very few rudimentary ideas. In the case of Edison, there are some noteworthy patents in spite of the very large number of fairly routine patents.

We have selected only one patent for Edison from his nine patents in motion pictures. It serves to illustrate that an entire industry was born from his seminal ideas for the kinetographic camera during the 1890s. It also served and satisfied the needs of society for a long time. Edison's first U.S. patent, "Electrographic Vote-Recorder," U.S. Patent No. 90,646, was executed in October 1868. Some of his early patents have enough merit to be qualified as "first" devices. Such examples include patents for transmitting (U.S. Patent No. 114,656, June 1870), recording (U.S. Patent No. 124,800, August 1871), and printing telegraphs (19 patents executed between January 1869 and April 1872). For perforated paper for telegraphs, Edison holds U.S. Patent No. 121,601, executed in August, 1871, and U.S. Patent No. 132,456, executed in March 1872. Edison's last patent "Holder for Article to be Electroplated," U.S. Patent No. 1,908,830, was executed in January 1931.

3.2.1 Kinetographic Camera (Edison, 1897)

UNITED STATES PATENT

August 31, 1897 — 589,168

Kinetographic Camera

THOMAS A. EDISON OF LLEWELLYN PARK, NEW JERSEY

Specification forming part of Letters Patent No. 589,168, August 31, 1897. Application filed August 24, 1891, Serial No. 403,534. (No model)

To all whom it may concern:

Be it known that I, THOMAS A. EDISON, a citizen of the United States, residing at Llewellyn Park, in the county of Essex and State of New Jersey, have invented a certain new and useful Improvement in Kinetoscopes, (Case No. 928,) of which the following is a specification.

The purpose I have in view is to produce pictures representing objects in motion throughout an extended period of time which may be utilized to exhibit the scene including such moving objects in a perfect and natural manner by means of a suitable exhibiting apparatus, such as that described in an application filed simultaneously herewith, (Patent No. 493,426, dated March 14, 1893.) I have found that it is possible to accomplish this end by means of photography.

In carrying out my invention I employ an apparatus for effecting by photography a representation suitable for reproduction of a scene including a moving object or objects comprising a means, such a single camera, for intermittently projecting at such rapid rate as to result in persistence of vision images of successive positions of the object or objects in motion as observed from a fixed and single point of view, a sensitized tape-like film, and a means for so moving the film as to cause the successive images to be received thereon separately and in single-line sequence. The movements of the tape-film may be continuous or intermittent, but the latter is preferable, and it is further preferable that the periods of rest of the film should be longer than the periods of movement.

By taking the photographs at a rate sufficiently high as to result in persistence of vision the developed photographs will, when brought successively into view by an exhibiting apparatus, reproduce the movements faithfully and naturally.

I have been able to take with a single camera and a tape-film as many as forty-six photographs per second, each having a size measured lengthwise of the tape of one inch, and I have also been able to hold the tape at rest for nine-tenths of the time; but I do not wish to limit the scope of my invention to this high rate of speed nor to this great disproportion between the periods of rest and the periods of motion, since with some subjects a speed as low as thirty pictures per second or even lower is sufficient, and while it is desirable to make the periods of rest as much longer than the periods of motion as possible any excess of the periods of rest over the periods of motion is advantageous.

In the accompanying drawings, forming a part hereof, Figure 1 is a plan view, with the top of the casing removed, of a form of apparatus which I have found highly useful for the taking of the photographs. Fig. 2 is a vertical longitudinal section on line *x x* in Fig. 1. Figs. 3 and 4 are enlarged views of the stop mechanism of the photographing apparatus. Figs. 5 and 6 are plan views of two different forms of shutters for the photographing apparatus, and Fig. 7 is a perspective view of a section of the tape-film with the photographs thereon.

Referring to the drawings, 3 indicates the transparent or translucent tape-film, which before the apparatus is put in operation is all coiled on a reel in the sheet-metal box or case 1, the free end being connected to an empty reel in the case 2. The film 3 is preferably of sufficient width to admit the taking of pictures one inch in diameter between the rows of holes 4, Figs. 2 and 7, arranged at regular intervals along the two edges of the film, and into which holes the teeth of the wheels 5, Figs. 1 and 2, enter for the purpose of positively advancing the film. When the film is narrow, it is not essential to use two rows of perforations and two feed-wheels, one feed-wheel being sufficient. Said wheels are mounted on a shaft 6, which carries a loose pulley 7 – that is, a pulley frictionally connected to its shaft and forming a yielding mechanical connection. This pulley is driven by a cord or belt 8 from a pulley 9 on the shaft 10, which shaft is driven by means of the beveled gears 11 12. The wheel 12 is preferably driven by an electric motor 13, which when the apparatus is in use is regulated to run at the desired uniform speed, being controlled by the centrifugal governor 14 and the

(No Model.) 3 Sheets—Sheet 1.

T. A. EDISON.
KINETOGRAPHIC CAMERA.

No. 589,168. Patented Aug. 31, 1897.

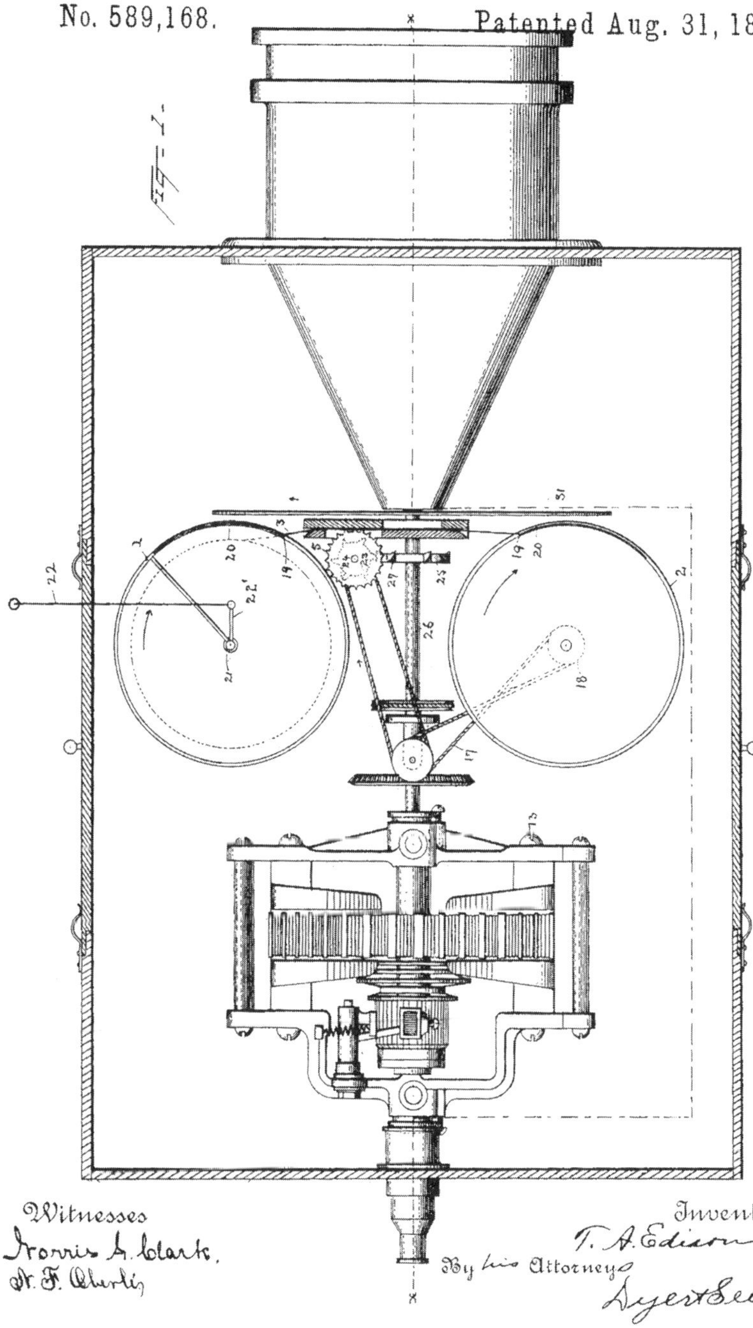

Fig. 1.

Witnesses
Norris G. Clark.
H. F. Oberlin

Inventor
T. A. Edison
By his Attorneys
Dyer & Seely

THE NORRIS PETERS CO., PHOTO-LITHO., WASHINGTON, D. C.

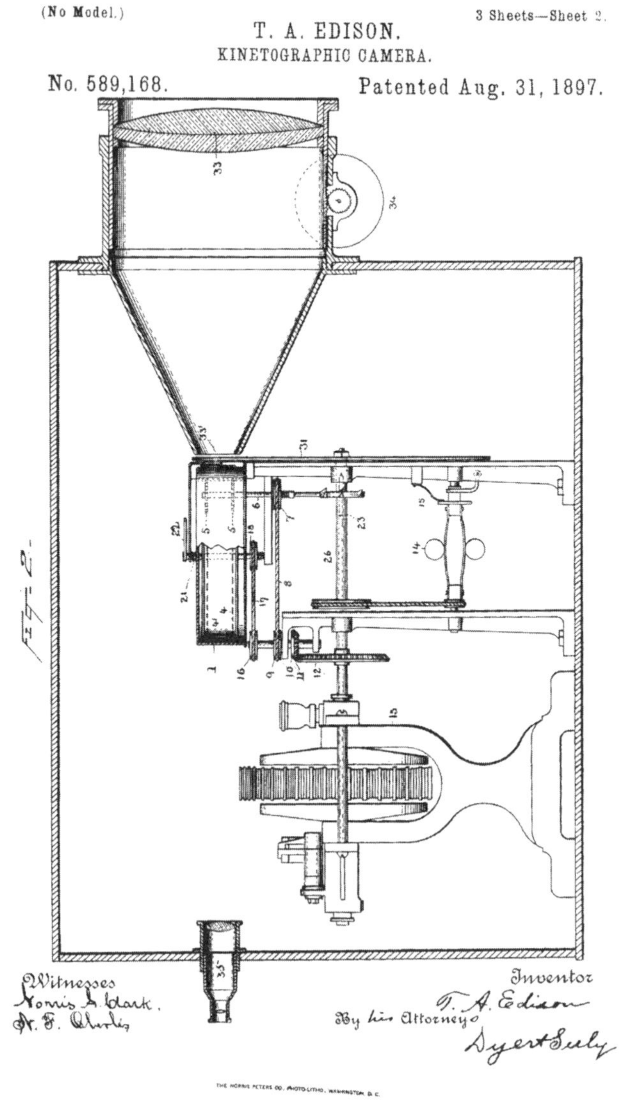

circuit-controller 15 in a well-known manner. On the shaft 10 is another pulley 16, which is connected by a cross-belt 17 to a pulley 18, also frictionally connected to its shaft, and which carries the reel to which the tape is connected in casing 2. The film passes from the casing 1 through a slit formed by the edge 19 and the sliding door 20, which is normally thrown forward by the spring 21, Fig. 2, with sufficient force to clamp the film and hold it from movement. When the door 20 is retracted by pulling on the rod or string 22, which is connected to the arm 22′, the film is liberated and allowed to advance. Film-case 2 is provided with a similar door, but the device for moving the door is not illustrated. This arrangement of the sliding door not only holds the film, but it tightly closes the casing, thus excluding

(No Model.) 3 Sheets—Sheet 3.

T. A. EDISON.

KINETOGRAPHIC CAMERA.

No. 589,168. Patented Aug. 31, 1897.

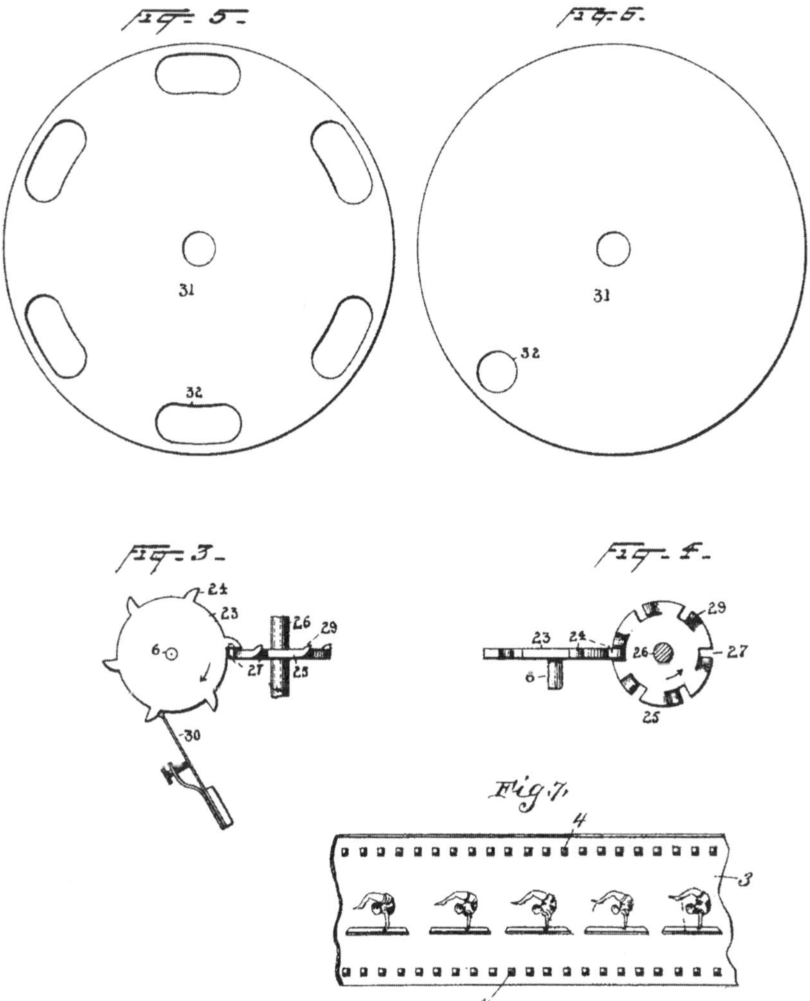

Witnesses
Norris A. Clark

Inventor
Thomas A. Edison
By his Attorneys Dyer & Seely

light and protecting the sensitive film. The casings or boxes 1 2 are removable, so that they, with the inclosed film, may be taken bodily from the apparatus. The shaft 6, heretofore referred to, is provided with a detent or stop-wheel 23, the form of which is most clearly shown in Figs. 3 and 4. The wheel 23 is provided with a number of projecting teeth 24, six being shown, which teeth are adapted to strike successively against the face of the cooperating detent or stop-wheel 25 on the shaft 26, which is the armature-shaft of the motor or a shaft which is constantly driven by the motor. The wheel 25 has a corresponding number of notches 27 at regular intervals around its periphery. These notches are of such size and shape that the teeth 24 can pass through them, and when the wheels 23 and 25 are rotated in the direction indicated by the arrows each tooth in succession will strike the face of wheel 23, thereby bringing the film absolutely to rest at the same moment that an opening in the shutter exposes the film and will then pass through a notch, allowing the tape-film to be moved forward another step while it is covered by the shutter. To avoid the danger of the wheel 25 moving so quickly that a tooth cannot enter the proper notch, a laterally-projecting tooth 29 is provided adjacent to each notch. When a tooth 29 strikes a tooth 24, the latter tooth will be guided by the tooth 29 into the adjacent notch 27.

30 is a detent spring or pawl to prevent backward movement of the wheel 23.

I prefer to so proportion the parts above described that the wheel 23 is at rest for nine-tenths of the time in order to give to the sensitized film as long an exposure as practicable and is moving forward one-tenth of the time, and said forward movement is made to take place thirty or more times per second, preferably at least as high as forty-six times per second, although the rapidity of movement or number of times per second may be regulated as desired to give satisfactory results. The longer interval of rest of the film insures a good impression of the object projected thereon and results in a picture having clean and sharp lines, since the film has sufficient time to become steady and overcome the vibration caused by the sudden and rapid motions of the feed mechanism. On the shaft 26 or on any suitable shaft driven by the motor is a revolving disk 31, serving as a shutter for alternately exposing and covering the sensitive film. This disk, which is continuously revolving, is provided with six or any other suitable number of apertures 32 at regular intervals around it near the edge, they being so arranged that one of the apertures passes directly between the camera-lens 33 and the film each time the film is brought to rest, the light-rays passing through the opening 33' and falling on the film half-way between the reels on which the film is wound.

34 is a device for adjusting the camera-lens toward or from the film, and 35 is a device by means of which the operator can focus the camera on the object to be photographed.

While I have described the use of an intermittently-moving film and a shutter having several openings so arranged as to expose the film at proper intervals, it would be possible and within my invention to use a continuously-moving film, and also to use a shutter with a single opening, as shown in Fig. 6, the shutter revolving at a high rate of speed.

To make the apparatus move the film continuously, it will only be necessary to omit the stop-wheels 23 and 25, which arrest the rotation of the shaft 6.

Although the operation has been partially indicated in the description of the apparatus it will now be set forth more in detail.

The apparatus is first charged with a sensitive tape-film several hundred or even thousands of feet long and the motor is set in operation. Since the spring 21 causes the door 20 to clamp the film, as already described, the loose pulleys 7-18 slip without pulling said film along, but when a moving object – for example, a man gesticulating – is placed in the field of the camera and the handle 22 is pulled the film is released and the pulleys operate to pull the same along. At the same time the reel in case 2 is rotated to wind up the film, thus transferring it from the reel in case 1 to the reel in case 2 . In the apparatus shown in Fig. 1 this movement is not continuous, but, as already indicated, is intermittent, the film advancing by very rapid steps, which are definitely and positively controlled by means of the peculiar detent or escapement described, and a photograph is taken after each step.

While I do not care to limit myself to any particular number of steps per second, there should be at least enough so that the eye of an observer cannot distinguish, or at least cannot clearly and positively distinguish, at a glance a difference in the position occupied by the object in the successive pictures, as illustrated in Fig. 7. A less speed in taking the pictures will cause a trembling or jerky appearance in the reproduced picture. When the movement of the object being photographed has ceased or the desired number of photographs has been obtained, the apparatus is stopped. The film is suitably treated for developing and fixing the pictures, when it is ready for use in an exhibiting apparatus.

What I claim is –

1. An apparatus for effecting by photography a representation, suitable for reproduction, of a scene including a moving object or objects, comprising a means for intermittently projecting at such rapid rate as to result in persistence of vision images of successive positions of the object or objects in motion, as observed from a fixed and single point of view, a sensitized tape-like film, and a means for so moving the film as to cause the successive images to be received thereon separately and in a single-line sequence.

2. An apparatus for taking photographs suitable for the exhibition of objects in motion, having in combination a single camera, and means for passing a sensitized tape-film at a high rate of speed across the lens of the camera and for exposing successive portions of the film in rapid succession, substantially as set forth.

3. An apparatus for taking photographs suitable for the exhibition of objects in motion, having in combination a single camera, and means for passing a sensitized tape film across the lens of the camera at a high rate of speed and with an intermittent motion, and for exposing successive portions of the film during the periods of rest, substantially as set forth.

4. An apparatus for taking photographs suitable for the exhibition of objects in motion, having in combination a single camera, and means for passing a sensitized tape-film across the lens at a high rate of speed and with an intermittent motion, and for exposing successive portions of the film during the periods of rest, the periods of rest being greater than the periods of motion, substantially as set forth.

5. An unbroken transparent or translucent tape-like photographic film having thereon equidistant photographs of successive positions of an object in motion, all taken from the same point of view, such photographs being arranged in a continuous straight-line sequence, unlimited in number save by the length of the film, substantially as described.

6. An unbroken transparent or translucent tape-like photographic film provided with perforated edges and having thereon equidistant photographs of successive positions of an object in motion, all taken from the same point of view, such photographs being arranged

in a continuous straight-line sequence, unlimited in number save by the length of the film, substantially as described.

This specification signed and witnessed this 31st day of July, 1891.

THOS. A. EDISON
Witnesses: John F. Randolph Frederick Ott

3.3 EARLY TELECOMMUNICATION CONCEPTS

Significant advances were made about 100 years ago in the field of wireless communication. The area was of such great importance at that time and the technological opportunities were so obvious that a large number of scientists (especially Marconi, Tesla, and others) wanted to claim inventions as their own. History has again repeated itself. In the area of computers, a large number of scientists (see Section 3.5) have made claims about the importance of their work. In the area of the Internet, a significant number of contributions have been made, even though the United States government [e.g., the Advanced Research Projects Agency (ARPA)] was the main funding source for the evolution of data networks to the Internet. Two exceptions (to this relationship between the number of scientists and the significance of the invention) are the invention of the transistor (Section 3.6) and invention of silica fiber-optic materials (Section 3.9). Even here, a number of scientists would like to stake a claim in the field.*

An abridged version of Marconi's first U.S. patent is included in Section 3.3.1. In this patent, Marconi demonstrated the possibilities of transmitting electrical signals through atmospheric media by picking up signals from a vertical wire on to a parabolic transmit antenna and then using the received energy to repeat [3.3.2] the signal. The tools, techniques, and apparatuses are consistent with the technology of that era. We include this patent to indicate the raw experimental ability and perseverance of these early scientists, which is also evident in Edison's exhaustive experiments with filament material for the lightbulb. Edison's broad genius and Marconi's more focused genius have earned them both a place in history. Both were persistent experimentalists who became renowned inventors. Today, this ability is translated into the software domain as the doctoral student tries to invent or discover new algorithms, such as Karmarkar's algorithm [3.3.1]. In this approach, the solution for very large linear-programming problems in polynomial time uses the interior-point method and is much faster than the traditional ellipsoid method. Other examples of efforts by Bell Laboratories scientists are the design of μ-Law [3.3.2] and A-Law companding [3.3.3] for speech in pulse code modulation (PCM) techniques, the C-message weighing function [3.3.2] for evaluating the effect of noise in speech channels, and so on. This talent is revisited many times as hardware, software, and firmware engineers design and optimize VLSI chips and then go on to find the right chip set for an application.

In Sections 3.3.2 and 3.3.3, we include two rather similar but basic patents by Tesla. Our comments are intended to educate our students rather than to criticize significant inventor of the past. These two patents carry concepts and ideas that are fundamental and original. The two patents are included in one section because the concepts overlap quite a bit. They are significant because they carry an idea for using a number of pulses to carry a signal from point A to point B in communications via microwave links [3.3.2 and 3.3.4]. Some aspects of balancing the signal and amplifying the received signals are also presented.

*It is not our intent to credit or discredit any scientists. We are simply stating the circumstances surrounding the invention and trying to provide some guidelines for readers and students.

Tesla had the ability to recycle his inventions in many applications. First, he was trying to send electrical power (not signals) over microwave media. He attempted this transmission by sending AC power at "high enough frequencies." Second, he resorted to sending "intelligence and signals" by causing a local receiver circuit to be triggered from one state to another in the same way that he was trying to send electrical power. In the patents included here, the use of his variable capacitors by using "powdery dielectric" is ingenious, though not practical.

The precision of the language and the characterization of the material properties as they are presented in the two classic patents by Schultz, et. al., for the fiber-optic material in Section 3.7 are indicative of a more mathematical approach to patents and inventions from Tesla (1901) to Schultz (1973). In a sense, Tesla had the ability to see through four or five components to make one more invention from his prior inventions without using any mathematical techniques. He was permuting the properties of the sub-components of a previous invention to create a new invention. In fact, Schultz did the same thing in his patents, but the reasoning relies deeply on the fundamentals of electromagnetic wave propagation. The rationality is proven by the symbols and numbers for the refractive indices of the silica-glass material for the core and cladding (Section 3.7). These remarks are feasible because we have the unfair advantage of hindsight, as well as an understanding of recent and refined mathematical techniques. However, the object is to learn from these great and classic inventions and to present them to our readers, both in retrospect and in perspective.

Tesla's patents do not provide experimental verification or analytical validation. For example, his concept about being able to carry power over free space in using generator systems would not have worked in 1901 or in 2001 unless the media was broken into a gaseous plasma state. However, this happens when lightning strikes, and the surge currents usually destroy the electrical apparatus rather than sustain stable power transfer. This conclusion is evident through the observation of thunderstorms and natural events, rather than through the use of circuit analysis or mathematics. However, in hindsight, we have the luxury to criticize.

Tesla's greatest contribution was his ability to grasp (or imagine) the notion of rotating electromagnetic fields for electric motors (a topic that is not discussed in this book). The invention of the induction motor alone is a large enough contribution to society. In fact, Tesla ventured into unknown territories to invent, (see Section 2.4.2) as we profess to our students. However, we believe that Bell (Section 3.1), Shockley (Section 3.6.1), and Bardeen and Brattain (Section 3.6.2) show greater flow for perfection and detail (see Section 2.4.1).

The ability to be a tireless experimentalist [fueled by unstructured creativity (Section 4.3.1)], as demonstrated by many earlier scientists, has led to inventions that have changed society. In these patents (3.3.1 to 3.3.3), there is very little mathematical or precise symbolic content. The lack of proof of the viability of the inventions is in stark contrast to recent patents of the modern era, which rely on precise mathematical and symbolic representation of the invention.

The innate mathematical ability that some people possess is sufficient to earn them an advanced degree in the areas of pure science, mathematics, or other disciplines. However, in our opinion, this innate ability is not, by itself, sufficient to allow them to acquire patents and then to make a blueprint for radically new devices. These devices (e.g., the airplane, the transistor, and the computer) help society to move forward. To this extent, the discipline of research (e.g., perfection, thoroughness, precision, and logic) and area of research (e.g., applied sciences, computers, and medicine) together generate the catalysts of change for society.

REFERENCES (C-3.3)

[3.3.1] Karmarkar, N., "A New Polynimial-Time Algorithm for Linear Programming." *Combinatorica*, 4, pp. 373–395. Also see Nazareth, J. L., *Differentiable Optimization and Equation Solving: A Treatise on Algorithmic Science and the Karmakar Revolution.* CMS Books in Mathematics 11, Springer Verlag, New York, 2003.

[3.3.2] Bell Telephone Laboratories, *Transmission Systems for Communications*, 5th ed. Bell Telephone Laboratories, Inc., Murray Hill, NJ, 1982.

[3.3.3] Flanagan James L., and Rabiner, Lawrence R. (Editors), *Speech Synthesis.* Dowden, Hutchinson & Ross, Stroudsburg, PA, 1973.

Guglielmo Marconi (1874–1937)

Transmitting Signals Through Air

3.3.1 Transmitting Electrical Signals (Marconi, 1897)

UNITED STATES PATENT

July 13, 1897 586,193

Transmitting Electrical Signals

GUGLIELMO MARCONI OF LONDON, ENGLAND

Specification forming part of Letters Patent No. 586,193 dated July 13, 1897. Application filed December 7, 1896. Serial No. 614,838. (No model.)

To all whom it may concern:

Be it known that I, GUGLIELMO MARCONI, student, a subject of the King of Italy, residing at 21 Burlington Road, London, in the county of Middlesex, England, have invented certain new and useful Improvements in Transmitting Electrical Impulses and Signals and in Apparatus Therefore, of which the following is a specification.

According to this invention electrical signals, actions, or manifestations are transmitted through the air, earth, or water by means of oscillations of high frequency, such as have been called the "Hertz rays" or "Hertz oscillations."

Usually all line-wires are dispensed with. At the transmitting-station I employ a Ruhmkorff coil, having in its primary circuit a Morse key or other signaling instrument and at its poles appliances for producing the desired oscillations. The Ruhmkorff coil may, however, be replaced by any other source of high-tension electricity. When working with large amounts of energy, it is, however, better to keep the coil or transformer constantly working for the time during which one is transmitting, and instead of interrupting the current of the primary interrupting the discharge of the secondary. In this case the contacts of the key should be immersed in oil, as otherwise, owing to the length of the spark, the current will continue to pass after the contacts have been separated. At the receiving-station, there is a local-battery circuit, containing any ordinary receiving instrument and an appliance for closing the circuit, the latter being actuated by the oscillations from the transmitting-station. When transmitting through the air and it is desired that the signal should only be sent in one direction, I place the oscillation-producer at the transmitting-station in the focus or focal line of a reflector directed to the receiving-station, and I place the circuit-closer at the receiving-station in a similar reflector directed towards the transmitting-station.

When transmitting signals through the earth, I connect one end of the oscillation-producer and one end of the circuit-closer to earth and the other ends to similar plates, preferably electrically tuned with each other in the air and insulated from earth.

Figure 1 is a diagrammatic front elevation of the instruments at the transmitting-station when signaling through the air, and Fig. 2 is a vertical section of the transmitter. Fig. 2^a is a longitudinal section of the oscillator to a larger scale. Fig. 3 shows a detail on a larger scale. Fig. 4 is a diagrammatic front elevation of the instruments at the receiving-station. Fig. 5 is a full-sized view of the receiver. Fig. 6 shows a modification of the tube *j*. Fig. 7 shows the detector. Fig. 8 is a full-sized view of the liquid-resistance. Figs. 9 and 10 show modifications of the arrangements at the transmitting-station. Fig. 11 shows a modification of the arrangements at the receiving-station.

Referring now to Fig. 1, *a* is a battery, and *b* an ordinary Morse key closing the circuit through the primary of a Ruhmkorff coil *c*. The terminals *c'* of the secondary circuit of the coil are connected to two metallic balls *d d*, fixed by heat or otherwise at the ends of tubes *d*, *d*, Fig. 2^a of insulating material, such as ebonite or vulcanite. *e e* are similar balls fixed in the other ends of the tubes *d*. The tubes *d* fit tightly in a similar tube d^2, having covers d^3, through which pass rods d^4, connecting the balls *d* to the conductors. One (or both) of the rods d^4 is connected to the ball *d* by a ball-and-socket joint and has a screw-thread upon it working in a nut in the cover d^3. By turning the rod therefore the distance of the balls *e* apart can be adjusted. d^5 are holes in the tube d^2, through which vaseline, oil, or like material is introduced into the space between the balls *e*.

The balls *d* and *e* are preferably of solid brass or copper, and the distance they should be apart depends on the quantity and electromotive force of the electricity employed, the effect increasing with the distance so long as the discharge passes freely. With a coil giving an ordinary eight-inch spark the distance between *e* and *e* should be from one twenty-fifth to one-thirtieth of an inch and the distance between *d* and *e* about one and a half inches. *f* is a cylindrical parabolic reflector made by bending a metallic sheet, preferably of brass or copper, to form and fixing it to metallic or wooden ribs *f*. Other conditions being equal the larger the balls the greater is the distance at which it is possible to communicate. I have generally used balls of solid brass of four inches diameter, giving oscillations of ten inches length of wave.

(No Model.) 3 Sheets—Sheet 1.

G. MARCONI.

TRANSMITTING ELECTRICAL SIGNALS

No. 586,193. Patented July 13, 1897.

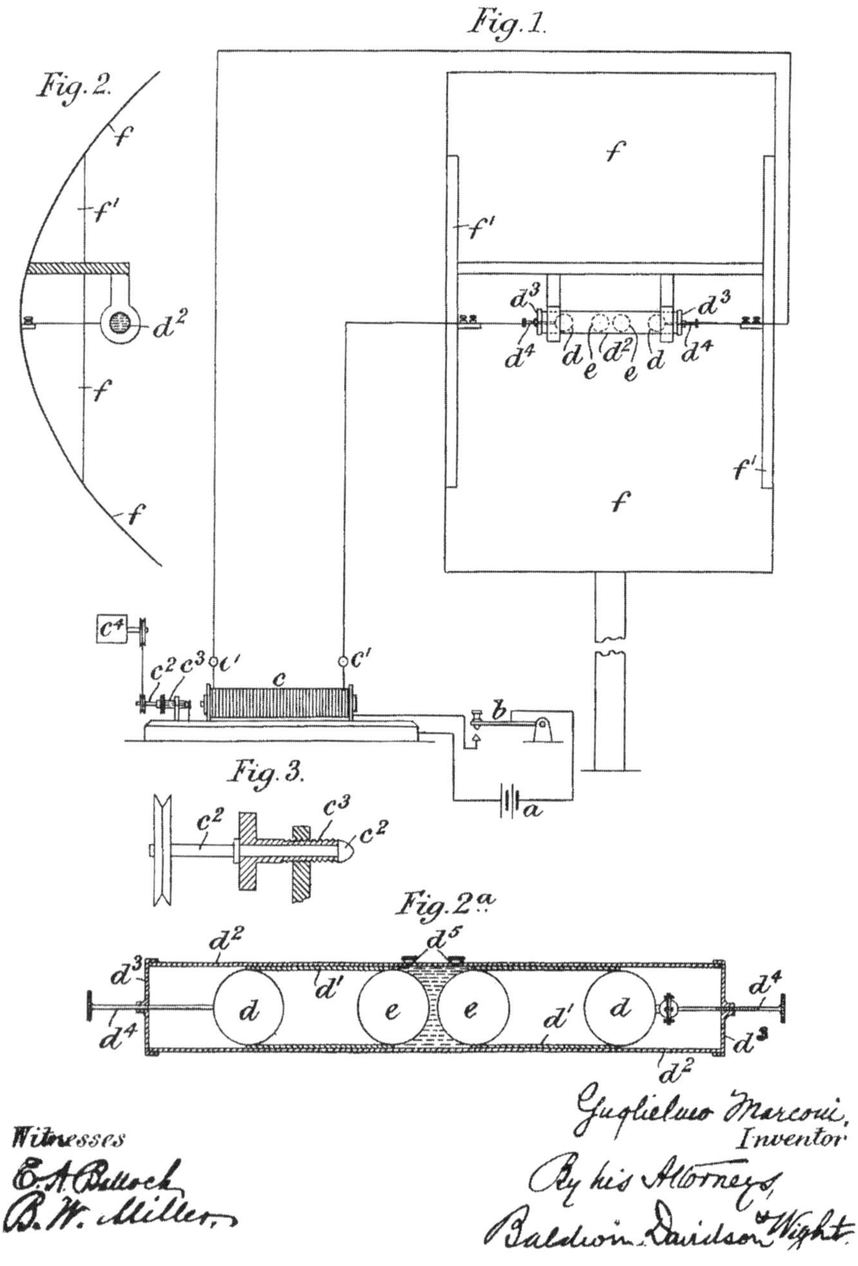

(No Model.) 3 Sheets—Sheet 2.

G. MARCONI.

TRANSMITTING ELECTRICAL SIGNALS.

No. 586,193. Patented July 13, 1897.

Fig. 4.

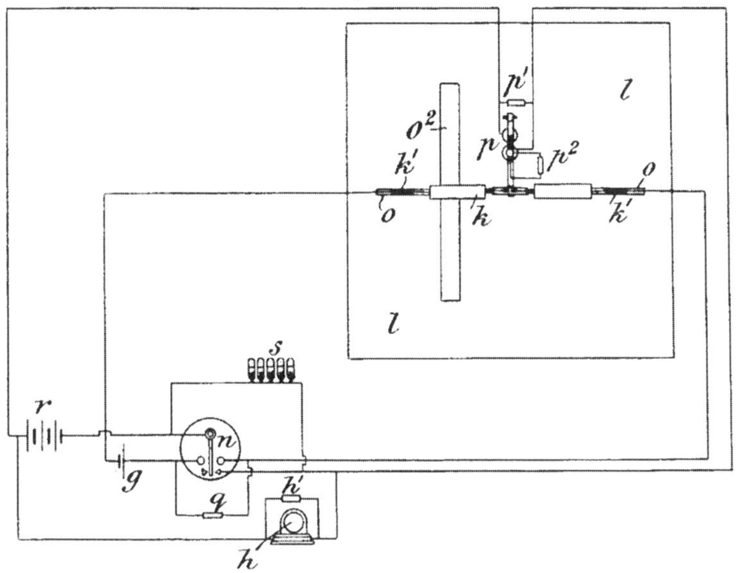

Fig. 5.

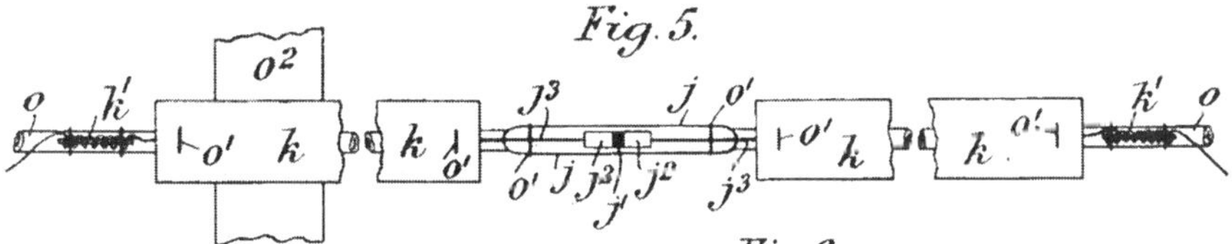

Fig. 6.

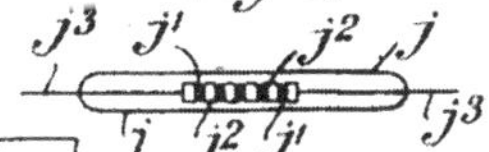

Fig. 7.

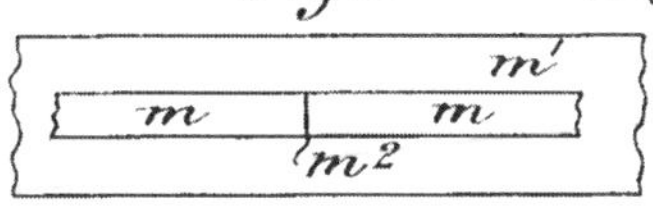

Fig. 8.

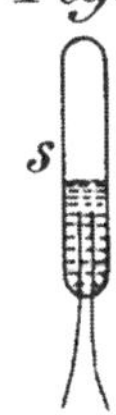

Witnesses
E. A. Ballock
B. W. Miller

Guglielmo Marconi,
Inventor
By his Attorneys
Baldwin, Davidson & Wight

(No Model.) 3 Sheets—Sheet 3.

G. MARCONI.

TRANSMITTING ELECTRICAL SIGNALS.

No. 586,193. Patented July 13, 1897.

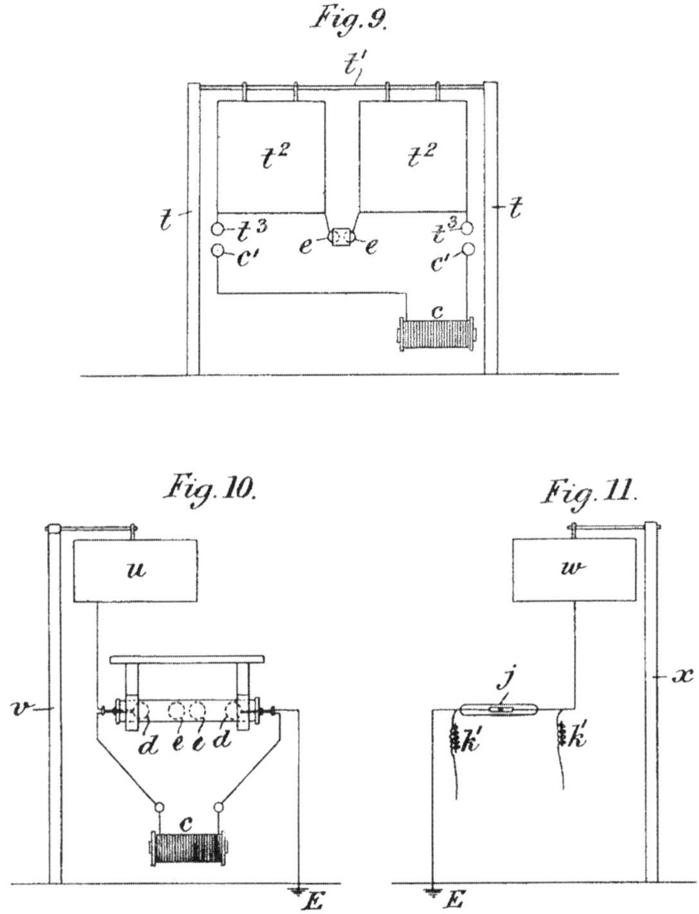

Witnesses
E. A. Bullock
B. W. Miller

Guglielmo Marconi
Inventor
By his Attorneys
Baldwin, Davidson & Wight

The reflectors applied to the receiver and transmitter ought to be preferably in length and opening the double at least of the length of wave emitted from the oscillator.

If a very powerful source of electricity giving a very long spark be employed, it is preferable to divide the spark gap between the central balls of the oscillator into several smaller gaps in series. This may be done by introducing between the big balls smaller ones, (of about half an inch diameter,) held in position by ebonite frames.

I find that the regularity, and power of the discharge of an ordinary Ruhmkorff coil with a trembler-break on its primary is greatly improved by causing one of the contacts of the vibrating break to revolve rapidly. I do this by having a revoluble central core c^2, Fig. 3, in the ordinary screw c^3, which is in communication with the platinum contacts. I cause the said central core with one of the platinum contacts attached to it to revolve by connecting it to a small electric motor c^4. This motor can be worked by the same circuit that works the coil, or if necessary, by a separate circuit. The connections are not shown in the drawings. By this means the platinums are kept smooth and any tendency to stick is removed. They last also much longer. At the receiving-station is a battery whose circuit includes an ordinary telegraphic instrument (or it may be a relay or other apparatus which it is desired to work from a distance) and a circuit-closer.

In Fig. 4, *g* is the battery, and *h* a telegraphic instrument on the derived circuit of a relay *n*.

The appliance I employ as a circuit-closer is shown full size at Fig. 5 and consists of a glass tube *j*, containing metallic powder or grains of metal j'; each end of the column of powder being connected to a metallic plate *k* of suitable length to cause the system to resonate electrically in unison with the electrical oscillations transmitted. The glass tube may be replaced in some cases by one of gutta-percha or like material. Two short pieces of thick silver wire j^2 of the same diameter as the internal diameter of the tube *j*, so as to fit tightly in it, are joined to two pieces of platinum wire j^3. The tube is closed and sealed-onto the platinum wires j^3 at both ends.

Many metals can be employed for producing the powder or filings j'; but I prefer to use a mixture of two or more different metals. I find hard nickel to be the best metal, and I prefer to add to the nickel filings about ten per cent of hard-silver filings, which increase greatly the sensitiveness of the tube to electric oscillations. By increasing the proportion of silver powder or grains the sensitiveness of the tube also increases; but it is better for ordinary work not to have a tube of too great sensitiveness, as it might be influenced by atmospheric or other electricity. The sensitiveness can also be increased by adding a very small amount of mercury to the filings and mixing up until the mercury is absorbed.

The mercury must not be in such a quantity as to clot or cake the filings. An almost imperceptible globule is sufficient for a tube. Instead of mixing the mercury with the powder one can obtain the same effects by slightly amalgamating the inner surfaces of the plugs which are to be in contact with the filings. Very little mercury must be used, just sufficient to brighten the surface of the metallic plugs without showing any free globules. The size of the tube and the distance between the two metallic stops may vary under certain limits. The greater the space allowed for the powder the larger and coarser ought to be the filings or grains.

I prefer to make my sensitive tubes of the following size: The tube *j* is one and one-half inches long and one-tenth or one-twelfth of an inch internal diameter. The length of the stops j^2 is about one-fifth of an inch, and the distance between the stops is about one-thirtieth of an inch. I find that the smaller the space between the stops in the tube the more sensitive it proves, but the space cannot under ordinary circumstances be excessively shortened without injuring the fidelity of the transmission.

The metallic powders ought not to be fine, but rather as coarse as can be produced by a large and rough file.

All the very fine powder ought to be removed by blowing or sifting.

The powder ought not to be compressed between the stops, but rather loose and in such a condition that when the tube is tapped the powder may be seen to move.

The tube must be sealed, but a vacuum inside it is not essential, except the slight vacuum which results from having heated it while sealing it. Care must also be taken not to heat the tube too much in the center when sealing it, as it would oxidize the surfaces of the silver stops and also the powder, which would diminish its sensitiveness. I use in sealing the tubes a hydrogen and air flame. A vacuum is, however, desirable, and I have used one of about one one-thousandth of an atmosphere, obtained by a mercury-pump. It is also necessary for the powder or grains to be dry and free from grease or dirt, and the files used in producing the same ought to be frequently washed and dried and used when warm.

If the tube has been well made, it should be sensitive to the induction of an ordinary electric bell when the same is working at one to two yards or more from the tube.

In order to keep the sensitive tube *j* in good working order, it is desirable, but not absolutely necessary, not to allow more than one milliampere to flow through it when active. If a stronger current is necessary, several tubes may be put in derivation between the tuned plates, but this arrangement is not quite as satisfactory as the single tube. It is necessary when using tubes of the type I have described not to insert in the circuit more than one cell of the Leclanche type, as a higher electromotive force than 1.5 volts is apt to pass a current through the tube even when no oscillations are transmitted. I can, however, construct tubes capable of working with a much higher electromotive force. Fig. 6 shows one of these tubes. In this tube instead of one space or gap filled with filings there are several spaces, separated by sections of tight-fitting silver wire. A tube thus constructed, observing also the rules of construction of my tubes in general, will work satisfactorily if the electromotive force of the battery in circuit with the tube is equal to 1.2 volts multiplied by the number of gaps. With this tube also it is well not to allow a current of more than one milliampere to pass.

The tube *j'* may be replaced by other forms of imperfect electrical contacts, but this is not desirable.

The plates *k* are of copper or aluminum or other metal, about half an inch or more broad, about one-fiftieth of an inch thick, and preferably of such a length as to be electrically tuned with the electric oscillations transmitted. The means I adopt for fixing the length of the plates is as follows: I stick a rectangular strip of tin-foil *m* (see Fig. 7) about twenty inches long (the length depends on the supposed length of wave that one is measuring) by means of a weak solution of gum onto a glass plate *m*'. Then by means of a very sharp penknife or point I cut across the middle of the tin-foil, leaving a mark of division m^2. If this detector is held in the proximity (four or five yards) and parallel with the axis of the oscillator in action, it will show little sparks at m^2. If the length of the pieces of tin-foil approximates to the length of wave emitted from the oscillator, the spark will take place between them at a certain distance from the transmitter, which is a maximum when they are of suitable length. By shortening or lengthening the strips, therefore, it is easy to find the length most appropriate to the length of wave emitted by the oscillator. It is desirable to try this detector in the focus or focal line of the reflector. The length so found is the proper length for the plates *k*, or rather these should be about half an inch shorter on account of the length of the sensitive tube *j*, connected between them.

l is a cylindrical parabolic reflector similar to that used at the transmitting-station.

The plates *k* may be in the form of tubes or even wires.

It is slightly advantageous for the focal distance of the reflector to be equal to one-fourth or three-fourths of the wave length of the oscillation transmitted.

When no oscillations are sent from the transmitting station, the tube *j* does not conduct the current and the local-battery circuit is broken, but when the powder or tube is influenced by the electrical oscillations from the transmitter it conducts and closes the circuit. I find, however, that when once started the powder in the tube continues to conduct even when the oscillations from the transmitter have ceased, but if it be shaken or tapped the circuit is broken. A tube well prepared will instantly interrupt the current passing through it at the slightest tap, provided it is inserted in a circuit in which there is little self-induction and small electromotive force, such as a single cell, and where the effects of self-induction have been removed by one of the methods which I will presently describe.

The two plates *k* communicate with the local circuit through two very small coils *k'*, which I will call "choking-coils," formed by winding a few inches of very thin and insulated copper wire around a bit of iron wire about an inch and a half long. The object of these choking-coils is to prevent the high-frequency oscillation induced across these plates by the transmitter from dissipating itself by running along the local-battery wires which might weaken its effect on the sensitive tube *j*. These choking-coils may, however, be sometimes replaced by simple thin wires. They may also be connected directly to the tube *j*. The local circuit in which the sensitive tube *j* is inserted contains a sensitive relay *n*, preferably wound to a resistance of about twelve hundred ohms. This resistance need not be necessarily that of the relay, but may be the sum of the resistance of the relay and another additional resistance. The relay ought to be one possessing small self-induction.

The plates *k*, tube *j*, and coils *k'* are fastened by means of wire stitches *o'* to a thin glass tube *o*, preferably not longer than twelve inches, firmly fixed at one end to a strong piece of timber o^2. This may be done by means of wood or ebonite grasping-screws.

[*A section of the patent has been removed for the sake of brevity.*]

What I claim is-

1. In a receiver for electrical oscillations the combination of an imperfect electrical contact, a circuit through the contact and means actuated by the circuit for shaking the contact.
2. In a receiver for electrical oscillations the combination of an imperfect electrical contact, metallic plates connected to it, a circuit through the contact and means actuated by the circuit for shaking the contact.
3. In a receiver for electrical oscillations the combination of an imperfect electrical contact, metallic plates connected to the contact, choking-coils connected to the contact, a circuit through the coils, and contact and means actuated by the circuit for shaking the contact.
4. In a receiver for electrical oscillations the combination of a tube containing metallic powder, a circuit through the powder and means actuated by the circuit for shaking the powder.
5. In a receiver for electrical oscillations the combination of a tube containing metallic powder, metallic plates connected to the powder, a circuit through the powder and means actuated by the circuit for shaking the powder.
6. In a receiver for electrical oscillations the combination of a tube containing metallic powder, metallic plates connected to the powder, choking-coils connected to the powder, a circuit through the coils and powder and means actuated by the circuit for shaking the powder.

7. In a receiver for electrical oscillations the combination of a tube containing a mixture of metallic powders, a circuit through the powder, and means actuated by the circuit for shaking the powder.
8. In a receiver for electrical oscillations the combination of a tube containing a mixture of metallic powder, metallic plates connected to the powder, a circuit through the powder and means actuated by the circuit for shaking the powder.
9. In a receiver for electrical oscillations the combination of a tube containing a mixture of metallic powders, metallic plates connected to the powder, choking-coils connected to the powder, a circuit through the coils, and powder and means actuated by the circuit for shaking the powder.
10. In a receiver for electrical oscillations the combination of a tube containing a mixture of metallic powder and mercury, a circuit through the powder and means actuated by the circuit for shaking the powder.

[*A section of the patent has been removed for the sake of brevity.*]

GUGLIELMO MARCONI
Witnesses: Wilfred Corpmael Fred C. Haries

Nikola Tesla (1856–1943)

Transmitting Signals Through Air

3.3.2 Transmitting from a Distance (Tesla, 1901)

UNITED STATES PATENT

November 5, 1901 685,955

Apparatus for Utilizing Effects Transmitted from a Distance to a Receiving Device Through Natural Media

NIKOLA TESLA OF NEW YORK, NEW YORK

Specification forming part of Letters Patent No. 685,955. Original application filed June 24, 1899, Serial No, 721,790.

Divided this application filed September 8, 1899. Renewed May 29, 1901, Serial No. 62,3 17. (No model)

To all whom it may concern:

Be it known that I, NIKOLA TESLA, a citizen of the United States, residing at the borough of Manhattan, in the city, county, and State of New York, have invented certain new and useful Improvements in Apparatus for Utilizing Effects Transmitted from a Distance to a Receiving Device Through the Natural Media, of which the following is specification, reference being had to the accompanying drawings, which form a part of the same.

This application is a division of an application filed by me June 24, 1899, Serial No. 721,790, in which a method of utilizing effects or disturbances transmitted through the natural media from a distant source is described and made the subject of the claims. The invention of my present application consists in the apparatus hereinafter described and claimed, by the use of which the method claimed in my said prior application may be practiced and by means of which results hitherto unattainable may be secured.

Several ways or methods of transmitting electrical disturbances through the natural media and utilizing them to operate distant receivers are now known and have been applied with more or less success for accomplishing a variety of useful results. One of these ways consists in producing by a suitable apparatus rays or radiations—that is, disturbances—which are propagated in straight lines through space, directing them upon a receiving or recording apparatus at a distance, and thereby bringing the latter into action. This method is the oldest and best known, and has been brought particularly into prominence in recent years through the investigations of Heinrich Hertz. Another method consists in passing a current through a circuit, preferably one in closing a very large area, inducing thereby in a similar circuit, situated at a distance, another current and affecting by the same in any convenient way a receiving device. Still another way, which has also been known for many years, is to pass in any suitable manner a current through a portion of the ground, as by connecting to two points of the same, preferably at a considerable distance from each other, the two terminals of a generator and to energize by a part of the current diffused through the earth a distant circuit, which is similarly arranged and grounded at two points widely apart and which is made to act upon a sensitive receiver. These various methods have their limitations, one, especially, which is common to all, being that the receiving circuit or instrument must be maintained in a definite position with respect to the transmitting apparatus, which often imposes great disadvantages upon the use of the apparatus.

In several applications filed by me and patents granted to me I have disclosed other methods of accomplishing results of this nature which may be briefly described as follows: In one system the potential of a point or region of the earth is varied by imparting to it intermittent or alternating electrifications through one of the terminals of a suitable source of electrical disturbances, which to heighten the effect has its other terminal connected to an insulated body, preferably of large surface and at an elevation. The electrifications communicated to the earth spread in all directions through the same, reaching a distant circuit, which generally has its terminals arranged and connected similarly to those of the transmitting source, and operates upon a highly-sensitive receive. Another method is based upon the fact that the atmospheric air, which behaves as an excellent insulator to currents generated by ordinary apparatus, becomes a conductor under the influence of currents of impulses of enormously high electromotive force which I have devised means for generating. By such means air strata, which are easily accessible, are rendered available for the production of many desired effects at

No. 685,955. Patented Nov. 5, 1901.

N. TESLA.

APPARATUS FOR UTILIZING EFFECTS TRANSMITTED FROM A DISTANCE TO A RECEIVING DEVICE THROUGH NATURAL MEDIA.

(Application filed Sept. 8, 1899. Renewed May 29, 1901.)

(No Model.)

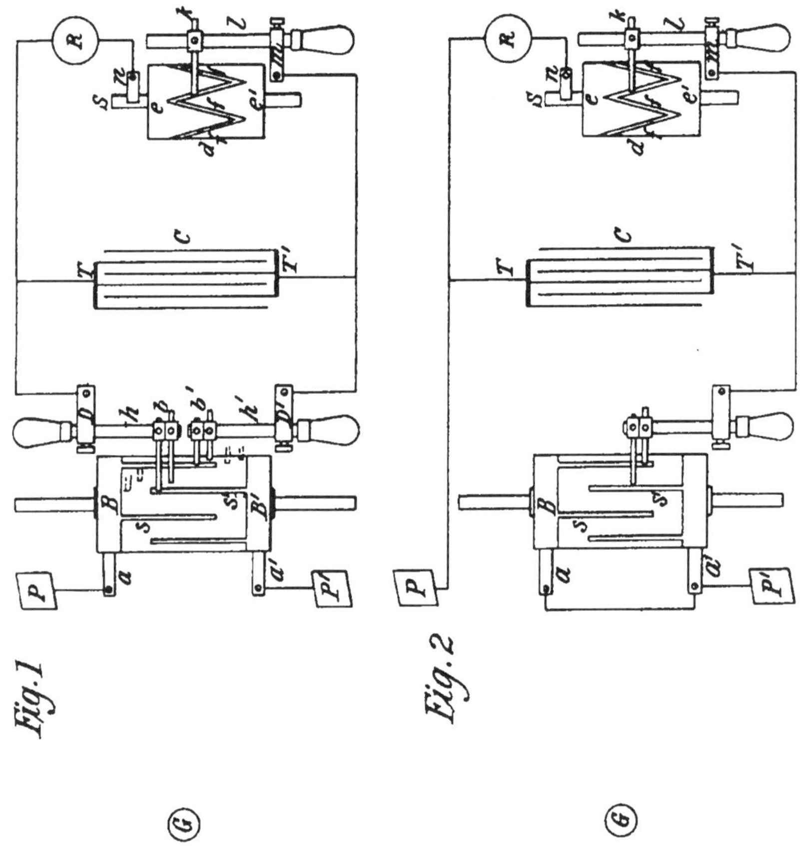

Witnesses:
G. B. Lsmo.
Hillary C. Messimer

Nikola Tesla, Inventor
by Ken. Page & Cooper.
Att'ys

distances, however great. This method, furthermore, allows advantage to be taken of many of those improvements which are practicable in the ordinary systems of transmission involving the use of a metallic conductor.

Obviously whatever method be employed it is desirable that the disturbances produced by the transmitting apparatus should be as powerful as possible, and by the use of certain forms of high-frequency apparatus which I have devised and which are now well known important practical advantages are in this respect secured. Furthermore, since in most cases the amount of energy conveyed to the distant circuit is but a minute fraction of the total energy emanating from the source, it is necessary for the attainment of the best results that whatever the character of the receiver and the nature of the disturbances as much as possible of tile energy conveyed should be made available for the operation of the receiver, and with this object in view I have heretofore, among other means, employed a receiving-circuit of high self-induction and very small resistance and of a period such as to vibrate in synchronism with the disturbances, whereby a number of separate impulses from the source were made to cooperate, thus magnifying the effect exerted upon and insuring the action of the receiving device. By these means decided advantages have been secured in many instances; but very often the improvement is either not applicable at all or if so the gain is very slight. Evidently when the source is one producing a continuous pressure or delivering impulses of long duration it is impracticable to magnify the effects in this manner, and when, on the other hand, it is one furnishing short impulses of extreme rapidity of succession the advantage obtained in this way is insignificant, owing to the radiation and the unavoidable frictional waste in the receiving-circuit. These losses reduce greatly both the intensity and the number of the cooperative impulses, and since the initial intensity of each of these is necessarily limited only an insignificant amount of energy is thus made available for a single operation of the receiver. As this amount is consequently dependent on the energy conveyed to the receiver by one single impulse, it is evidently necessary to employ either a very large and costly, and therefore objectionable transmitter, or else resort to the equally objectionable use of a receiving device too delicate and too easily deranged. Furthermore, the energy obtained through the cooperation of the impulses is in the form of extremely-rapid vibrations and because of this unsuitable for the operation of ordinary receivers, the more so as this form of energy imposes narrow restrictions in regard to the mode and time of its application to such devices. To overcome these and other limitations and disadvantages which have heretofore existed in such systems of transmission of signals or intelligence is the object of my invention, which comprises a novel form of apparatus for accomplishing these results.

The apparatus which is employed at the receiving, station, described in general terms, consists in the combination of a storage device included in a circuit connecting points at a distance from the source of the disturbances and between which a difference of potential is created by such disturbances, a receiving-circuit connected with the storage device, a receiver included in such receiving-circuit, and means for closing the receiving-circuit at any desired moment, and thereby causing the receiver to be operated by the energy with which the storage device has been charged.

The best form of apparatus for carrying out my invention of which I am now aware and the manner of using the same will be understood from the following description and the accompanying drawings, in which—

Figure 1 is a diagrammatic illustration of such apparatus, and Fig. 2 a modified form or arrangement of the same.

At any two points in the transmitting medium between which there exists or may be obtained in any manner, through the action of the disturbances or effects to be investigated or utilized a difference of electrical potential of any magnitude I arrange two plates or electrodes so that they may be oppositely charged through the agency of such effects or disturbances, and I connect these electrodes to the terminals of a highly-insulated condenser, generally of considerable capacity. To the condenser-terminals I also connect the receiver to be operated in series with a device of suitable construction which performs the function of periodically discharging the condenser through the receiver at and during such intervals of time as may be best suitable for the purpose contemplated. This device may merely consist of two stationary electrodes separated by a feeble dielectric layer of minute thickness, or it may comprise terminals one or more of which are movable and actuated by any suitable force and are adapted to be brought into and out of contact with each other in any convenient manner. It will now be readily seen that if the disturbances, of whatever nature they may be, cause definite amounts of electricity of the same sign to be conveyed to each of the plates or electrodes above mentioned either continuously or at intervals of time which are sufficiently long the condenser will be charged to a certain potential and an adequate amount of energy being thus stored during the time determined by the device effecting the discharge of the condenser the receiver will be periodically operated by the electrical energy so accumulated; but very often the character of tile impulses and the conditions of their use are such that without further provision not enough potential energy would be accumulated in the condenser to operate the receiving device. This is the case when, for example, each of the plates or terminals receives electricity of rapidly-changing sign or even when each receives electricity of the same sign, but only during periods which are short as compared with the intervals separating them. In such instances I resort to the use of a special device which I insert in the circuit between the plates and the condenser for the purpose of conveying to each of the terminals of the latter electrical charges of the proper quality and order of succession to enable the required amount of potential energy to be stored in the condenser.

There are a number of well-known devices, either without any moving parts or terminals or with elements reciprocated or rotated by the application of a suitable force, which offer a more ready passage to impulses of one sign or direction than to those of the other or permit only impulses of one kind or order of succession to traverse a path, and any of these or similar devices capable of fulfilling the requirements may be used in carrying my invention into practice. One such device of familiar construction which will serve to convey a clear understanding of this part of my invention and enable a person skilled in the art to apply the same is illustrated in the annexed drawings. It consists of a cylinder A, of insulating material, which is moved at a uniform rate of speed by clockwork or other suitable motive power and is provided with two metal rings B B′, upon which bear. brushes a and *a*′, which are connected, respectively, in the manner shown to the terminal plates P and P′, above referred to. From the rings B B′ extend narrow metallic segments *s* and *s*′, which by the rotation of the cylinder A are brought alternately into contact with double brushes *b* and *b*′, carried by and in contact with conducting-holders *h* and *h*′, which are adjustable longitudinally in the metallic supports D and D′, as shown. The latter are connected to the terminals T and T′ of a condenser C, and it should be understood that they are capable of angular displacement as ordinary brush-supports. The object of using two brushes as *b* and *b*′, in each of the holders *h* and *h*′ is to vary at will the duration of the electric contact of the plates P and P′ with the terminals T and T′, to which is connected a receiving-circuit, including a receiver R and a device *d* of the kind above referred to, which performs the duty of closing the receiving-circuit at predetermined intervals of time and discharging the stored energy through the receiver. In the present case this device consists of a cylinder made partly of conducting and partly of insulating material *e* and *e'*, respectively, which is rotated at the desired rate of speed by any suitable means. The conducting part *e* is in good electrical connection with the shaft S and is provided with tapering segments *f f*, upon which slides a brush *k*, supported on a conducting-rod *l*, capable of longitudinal adjustment in a metallic support *m*. Another brush *n* is arranged to bear upon the shaft S, and it will be seen that whenever one of the segments comes in contact with the brush *k* the circuit including the receiver R is completed and the condenser discharged through the same. By an adjustment of the speed of rotation of the cylinder *d* and a displacement of the brush *k* along the cylinder the circuit may be made to open and close in as rapid succession and remain open or closed during such intervals of time as may be desired. The plates P and P′ through which the electrifications are conveyed to the brushes *a* and *a'* may be at a considerable distance from each other and both in the ground Or both in the air or one in the ground and the other in the air, preferably at some height, or they may be connected to conductors extending to some distance Or to the terminals of any kind of apparatus supplying electrical energy which is obtained from the energy of the impulses or disturbances transmitted from a distance through the natural media.

In illustration of the operation of the devices described let it be assumed that alternating electrical impulses from a distant generator, as G, are transmitted through the earth and that it is desired to utilize these impulses in accordance with my method. This may be the case, for example, when such a generator is used for purposes of signaling in one of the ways before enumerated, as by having its terminals connected at two points of the earth distant from each other. In this ease the plates P and P′ are first connected to two properly-selected points of the earth, the speed of rotation of the cylinder A is varied until it is made to turn in synchronism with the alternate impulses of the generator, and, finally, the position of the brushes *b* and *b'* is adjusted by angular displacement, as usual, or in other ways, so that they are in contact with the segments *s* and *s'* during the periods when the impulses are at or near the maximum of their intensity. Only ordinary electrical skill and knowledge are required to make these adjustments, and a number of devices for effecting synchronous movement being well known and it being the chief object of my present application to set forth a novel apparatus embodying a general principle a detailed description of such devices is not considered necessary. I may state, however, that for practical purposes in the present case it is only necessary to shift the brushes back and forth until the maximum effect is secured. The above requirements being fulfilled, electrical charges of the same sign will be conveyed to each of the condenser-terminals as the cylinder A is rotated, and with each fresh impulse the condenser will be charged to a higher potential. The speed of rotation of the cylinder g being adjustable at will, the energy of any number of separate impulses may thus be accumulated in potential form and discharged through the receiver R upon the brush *k* coming in contact with one of the segments *f*. It will be of course understood that the capacity of the condenser should be such as to allow the storing of a much greater amount of energy than is required for the ordinary operation of the receiver. Since by this method a relatively great amount of energy and in a suitable form may be made available for the operation of a receiver, the latter need not be very sensitive; but of course when the impulses are very feeble, as when coming from a great distance or when it is desired to operate a receiver very rapidly, then any of the well-known devices capable of responding to very feeble influences may be used in this connection.

If instead of the alternating impulses short impulses of the same direction are conveyed to the plates P and P′, the apparatus described may still readily be used, and for this purpose it is merely necessary to shift the brushes *b* and *b′* into the position indicated by the dotted lines, while maintaining the same conditions in regard to synchronism as before, so that the succeeding impulses will be permitted to pass into the condenser, but prevented from returning to the ground or transmitting medium during the intervals between them, owing to the interruption during such intervals of the connections leading from the condenser-terminals to the plates.

Another way of using the apparatus with impulses of the same direction is to take off one pair of brushes, as *b*, disconnect the plate P from brush *a* and join it directly to the terminal T of the condenser, and to connect brush *a* with brush *a′*. When thus modified, the apparatus appears as shown in Fig. 2. Operated in this manner and assuming the speed of rotation of cylinder A to be the same, the apparatus will now be evidently adapted for a number of impulses per unit of time twice as great as in the preceding case. In all cases it is evidently important to adjust the duration of contact of segments *s* and *s′* with brushes *b b′* in the manner indicated.

When the apparatus I have described is used in connection with the transmission of signals or intelligence, it will of course be understood that the transmitter is operated in such a way as to produce disturbances or effects which are varied or intermitted in some arbitrary manner—for example, to produce longer and shorter successions of impulses, corresponding to the dashes and dots of the Morse alphabet—and the receiving device will respond to and indicate these variations or intermittences, since the storage device will be charged and discharged a number of times corresponding to the duration of the successions of impulses received.

Obviously the special appliances used in carrying out my invention may be varied in many ways without departing from the spirit of the same.

It is to be observed that it is the function of the cylinder A, with its brushes and connections, to render the electrical impulses coming from the plates P and P′ suitable for charging the condenser (assuming them to be unsuitable for this purpose in the form in which they are received) by rectifying them when they are originally alternating in direction or by selecting such parts of them as are suitable when all are not, and any other device performing this function will obviously answer the purpose. It is also evident that a device such as I have already referred to which offers a more ready passage to impulses of one sign or permits only impulses of the same sign to pass may also be used to perform this selective function in many cases when alternating impulses are received. When the impulses are long and all of the same direction, and even when they are alternating but sufficiently long in duration and sustained in electromotive force, the brushes *b* and *b′* may be adjusted so as to bear on the parts B B′ of the cylinder A, or the cylinder and its brushes may be omitted and the terminals of the condenser connected directly to the plates P and P′.

It will be seen that by the use of my invention results hitherto unattainable in utilizing disturbances or effects transmitted through natural media may be readily attained, since however great the distance of such transmission and however feeble or attenuated the impulses received enough energy may be accumulated from them by storing up the energy of succeeding impulses for a sufficient interval of time to render the sudden liberation of it highly effective in operating a receiver. In this way receivers of a variety of forms may be made to respond effectively to impulses too feeble to be detected or to be made to produce any sensible effect in any other way of which I am aware—a result of great value in scientific research as well as in various applications to practical use.

What I claim as my invention, and desire to secure by Letters Patent, is—

1. In an apparatus for utilizing electrical effects or disturbances transmitted through the natural media, the combination with a source of such effects or disturbances of a charging-circuit adapted to be energized by the action of such effects or disturbances, a storage device included in the charging-circuit and adapted to be charged thereby, a receiver, and means for causing the receiver to be operated by the energy accumulated in the storage device at arbitrary intervals of time, substantially as described.
2. In an apparatus for utilizing electrical effects or disturbances transmitted through the natural media, the combination with a source of such effects or disturbances of a charging-circuit adapted to be energized by the action of such effects or disturbances, a storage device included in the charging-circuit and adapted to be charged thereby, means for commutating, directing or selecting the current impulses in the charging-circuit, a receiving-circuit, and means for discharging the storage device through the receiving-circuit, substantially as described.
3. In an apparatus for utilizing electrical effects or disturbances transmitted through the natural media, the combination with a source of such effects or disturbances of a charging-circuit adapted to be energized by the action of such effects or disturbances, a condenser included in the charging-circuit and adapted to be charged thereby, means for commutating, directing or selecting the current impulses in the charging-circuit, a receiving-circuit, and means for discharging the condenser through the receiving-circuit, substantially as described.
4. In an apparatus for utilizing electrical effects or disturbances transmitted through the natural media, the combination with a source of such effects or disturbances of a charging-circuit adapted to be energized by the action of such effects or disturbances, a storage device included in the charging-circuit and adapted to be charged thereby, means for commutating, directing or selecting the current impulses in the charging-circuit so as to render them suitable for charging the storage device, a receiving-circuit, and means for discharging the storage device through the receiving-circuit, substantially as described.
5. In an apparatus for utilizing electrical effects or disturbances transmitted through the natural media, the combination with a source of such effects or disturbances of a charging-circuit adapted to be energized by the action of such effects or disturbances, a condenser included in the charging-circuit and adapted to be charged thereby, means for commutating, directing *or* selecting the current impulses in the charging-circuit so as to render them suitable for charging the condenser, a receiving-circuit, and means for discharging the condenser through the receiving-circuit, substantially as described.
6. In an apparatus for utilizing electrical effects or disturbances transmitted through the natural media, the combination with a source of such effects or disturbances of a charging-circuit adapted to be energized by the action of such effects or disturbances, a storage device included in the charging-circuit and adapted to be charged thereby, means for commutating, directing or selecting the current impulses in the charging-circuit so as to render them suitable for charging the storage device, a receiving-circuit, and means for discharging the storage device through the receiving-circuit at arbitrary intervals of time, substantially as described.
7. In an apparatus for utilizing electrical effects or disturbances transmitted to a distant receiving-station, the combination with a source of such effects or disturbances of a circuit distant from the source and adapted to have current impulses set up in it by the action of

the effects or disturbances, a storage device, means for commutating, directing or selecting the impulses and connecting the circuit with the storage device at succeeding intervals of time synchronizing with the impulses, a receiving-circuit, and means for periodically discharging the storage device through the receiving-circuit, substantially as described.

8. In an apparatus for utilizing electrical effects or disturbances transmitted to a distant receiving-station, the combination with a source of such effects or disturbances of a circuit distant from the source and adapted to have current impulses setup in it by the action of the effects or disturbances, a condenser, means for commutating, directing or selecting the impulses and connecting the circuit *with* the condenser at succeeding intervals of time synchronizing with the impulses, a receiving-circuit, and means for periodically discharging the condenser through the receiving-circuit, substantially as described.

9. In an apparatus for utilizing electrical effects or disturbances transmitted through the natural media, the combination with a source of such effects or disturbances of a circuit connecting points at a distance from the source between which a difference of potential is created by such effects *or disturbances,* a storage device included in such circuit and adapted to be charged with the energy supplied by the same, a receiving circuit connected with the storage device, a receiver included in such receiving-circuit, and means for closing the receiving-circuit and thereby causing the receiver to be operated by the energy accumulated in the storage device, substantially as described.

10. In an apparatus for utilizing electrical effects or disturbances transmitted through the natural media, the combination with a source of such effects or disturbances of a circuit at a distance from the source which is energized by such effects or disturbances, a storage device adapted to be charged with the energy Supplied by such circuit, means for connecting the storage device with the said circuit for periods of time predetermined as to succession and duration, a receiving-circuit connected with the storage device, a receiver included in such receiving-circuit, and means for closing the receiving-circuit and thereby causing the receiver to be operated by the energy accumulated in the storage device, substantially as described.

11. In an apparatus for utilizing electrical effects or disturbances transmitted through the natural media, the combination of a circuit connecting points at a distance from the source between which a difference of potential is created by such effects or disturbances, a storage device included in such circuit and adapted to be charged with the energy supplied by the same, a receiving-circuit, a receiver included in such circuit, and means for connecting the receiving-circuit with the storage device for periods of time predetermined as to succession and duration and thereby causing the receiver to be operated by the energy accumulated in the storage device, substantially as described.

12. In an apparatus for utilizing electrical effects or disturbances transmitted through the natural media, the combination of a circuit connecting points at a distance from the source between which a difference of potential is created by such effects or disturbances, a storage device adapted to be charged with the energy supplied by such circuit for succeeding and predetermined periods of time, a receiving-circuit, a receiver included in the receiving-circuit, and means for connecting the receiving-circuit with the storage device for periods of time predetermined as to succession and duration and thereby causing

the receiver to be operated by tile energy accumulated in the storage device, substantially as described.

13. In an apparatus for utilizing electrical effects or disturbances transmitted through tile natural media, the combination of a circuit connecting points at a distance from the source, between which a difference of potential is created by such effects or disturbances, a condenser included in such circuit and adapted to be charged by the current in the same, a receiving-circuit connected with tile eon-denser, a receiver included in such receiving-circuit, and a device adapted to close the receiving-circuit at arbitrary intervals of time and I hereby cause the receiver to be operated by the electrical energy accumulated in the condenser, substantially as described.

14. In an apparatus for utilizing electrical effects or disturbances transmitted through the natural media, the combination of a charging-circuit distant from the source and energized by the effects or disturbances, a storage device included in the charging-circuit, means included in the charging-circuit and acting in synchronism with the impulses therein for commutating, directing or selecting the impulses, a receiving-circuit and means for periodically discharging the storage device through the receiving-circuit, substantially as described.

15. In an apparatus for utilizing electrical effects or disturbances transmitted through the natural media, the combination of a charging-circuit distant from the source and energized by the effects or disturbances, a eon-denser included in the charging-circuit, means included in the charging-circuit and acting in synchronism with the impulses therein for commutating, directing or selecting the impulses, a receiving-circuit and means for periodically discharging the condenser through the receiving-circuit, substantially as described.

16. In an apparatus for transmitting signals or intelligence through the natural media from a sending-station to a distant point, the combination of a generator or transmitter adapted to produce arbitrarily varied or intermitted electrical disturbances or effects in the natural media, a charging-circuit at the distant point adapted to receive corresponding electrical impulses or effects from the disturbances or effects so produced, a storage device included in the charging-circuit, means included in the charging-circuit and acting in synchronism with the impulses therein for commutating, directing or selecting the impulses so as to render them suitable for charging the storage device, a receiving-circuit and means for periodically discharging the storage device through the receiving-circuit, substantially as described.

17. In an apparatus for transmitting signals or intelligence through the natural media from a sending-station to a distant point, the combination of a generator or transmitter adapted to produce arbitrarily varied or intermitted electrical disturbances or effects in the natural media, a charging-circuit at the distant point adapted to receive corresponding electrical impulses or effects from the disturbances or effects so produced, a condenser included in the charging-circuit, moans included in the charging-circuit and acting ill synchronism with the impulses therein for commutating, directing or selecting the impulses so as to render them suitable for charging the condenser, a receiving-circuit and means for periodically discharging the condenser through the receiving-circuit, substantially as described.

18. In an apparatus for transmitting signals or intelligence through the natural media from a sending-station to a distant point, the combination of a generator or transmitter, adapted to produce arbitrarily varied or intermitted electrical disturbances or effects in the natural media, a circuit at the distant point adapted to receive corresponding electrical impulses or disturbances from the disturbances or effects so transmitted, a storage device in-eluded in such circuit and adapted to be charged thereby, a receiving-circuit connected with the storage device, a receiver included in the receiving-circuit and a device for closing the receiving-circuit at arbitrary intervals of time and thereby causing the receiver to be operated by the energy accumulated in the storage device, substantially as described.

3.3.3 Transmitting Through Natural Media (Tesla, 1901)

UNITED STATES PATENT

November 5, 1901 685,956

Apparatus for Utilizing Effects Transmitted Through Natural Media

NIKOLA TESLA OF NEW YORK, NEW YORK

Specification forming part of Letter Patent No, 685,956, dated November 5, 1901.

Original application filed August 1, 1899, Serial No. 725,749. Divided and this application filed November 2,1899. Renewed May 29, 1901. Serial No. 62,318. (No model.)

To all whom it may concern:

Be it known that I, NIKOLA TESLA, a citizen of the United States, residing at New York city, in the county and State of Now York, have invented a now and Useful Improvement in Apparatus for Utilizing Effects Transmitted from a Distance to a Receiving Device Through the Natural Media, of which the following is a specification, reference being had to the accompanying drawings, which forth a part of tile same.

The subject of my present invention is an improvement in the art of utilizing effects transmitted from a distance to a receiving device through the natural media; and it consists in the novel apparatus hereinafter described.

This application is a division of one filed by me August 1, 1899, Serial No. 725,749, and based upon and claiming the method herein described and which may be practiced by the use of the apparatus forming the subject of this application.

My invention is particularly useful in connection with methods and apparatus for operating distant receiving devices by means of electrical disturbances produced by proper transmitters and conveyed to such receiving devices through thc natural media; but it obviously has a wider range of applicability and may be employed, for example, in the investigation or utilization of terrestrial, solar, or other disturbances produced by natural causes.

Several ways or methods of transmitting electrical disturbances through the natural media and utilizing them to operate distant receivers are now known and have been applied with more or less success for accomplishing a variety of useful results. One of these ways consists

No. 685,956. **Patented Nov. 5, 1901.**

N. TESLA.

APPARATUS FOR UTILIZING EFFECTS TRANSMITTED THROUGH NATURAL MEDIA.

(Application filed Nov. 2, 1899. Renewed May 29, 1901.)

(No Model.)

Fig. 1

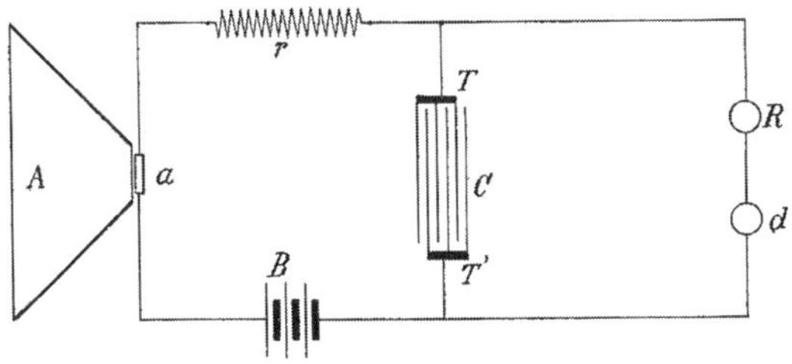

Fig. 2

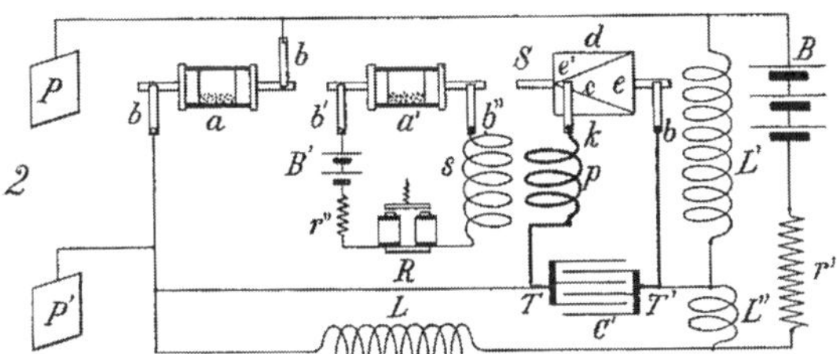

Fig. 3

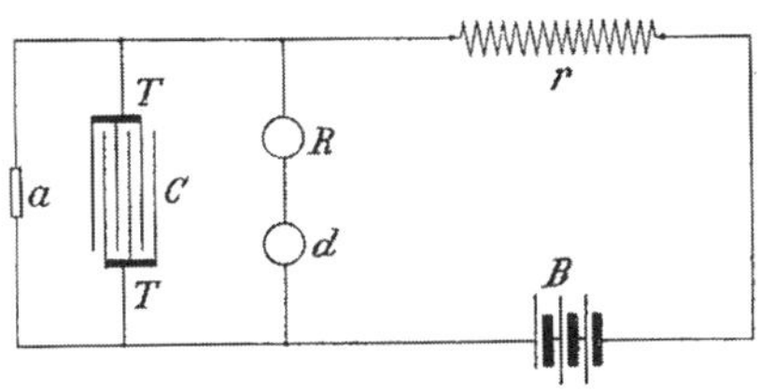

Fig. 4

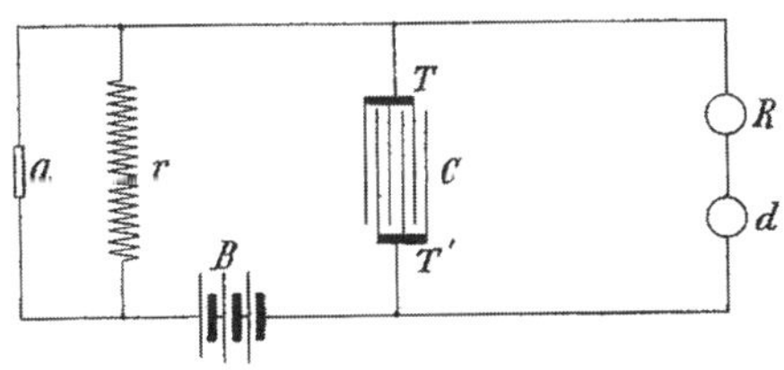

Fig. 5

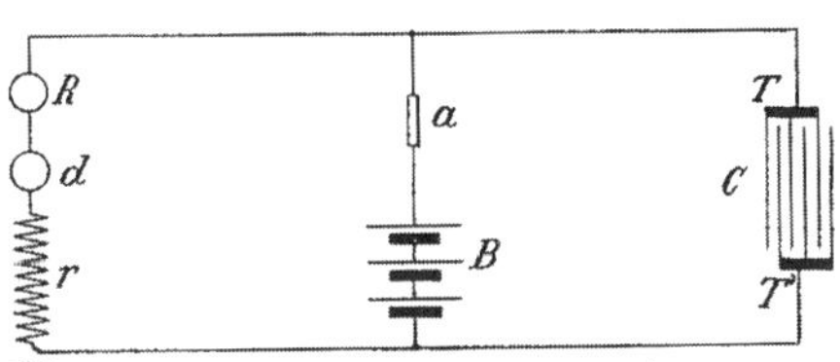

Witnesses:
Raphaël Netter
M. Lawson Dyer.

Inventor
Nikola Tesla
by Kerr, Page & Cooper, Attorneys.

THE NORRIS PETERS CO., PHOTO-LITHO., WASHINGTON, D. C.

in producing by a suitable apparatus rays or radiations—that is, disturbances—which are propagated in straight lines through space, directing them upon a receiving or recording apparatus at a distance, and thereby brining tile latter into action. This method is the oldest and best known and has been brought particularly into prominence in recent years through the investigations of Heinrich Hertz. Another method consists in passing a current through a circuit, preferably one in closing a very large area, inducing thereby in a similar circuit situated at a distance another current and affecting by the same in any convenient way a receiving device. Still another way, which has also been known for many years, is to pass in any suitable manner a current through a portion of the ground, as by connecting to two points of the same, preferably at a considerable distance from each other, the two terminals of a generator and to energize by a part of the current diffused through the earth a distant circuit, which is similarly arranged and grounded at two points widely apart and which is made to act upon a sensitive receiver. These various methods have their limitations, one especially, which is common to all, being that the receiving circuit or instrument must, be maintained in a definite position with respect to the transmitting apparatus, which often imposes great disadvantages upon the use of the apparatus.

In several applications filed by me and patents granted to me I have disclosed other methods of accomplishing results of this nature, which may be briefly described as follows: In one system the potential of about or region of the earth is varied by imparting to it intermittent or alternating electrifications through one of the terminals of a suitable source of electrical disturbances, which, to heighten the effect, has its other terminal connected to an insulated body, preferably of large surface and at an elevation. The electrifications communicated to the earth spread in all directions through the same, reaching a distant circuit, which generally has its terminals arranged and connected similarly to those of the transmitting source and operates upon a highly-sensitive receiver. Another method is based upon the fact that the atmospheric air, which behaves as an excellent insulator to currents generated by ordinary apparatus, becomes a conductor under the influence of currents or impulses of enormously high electromotive force, which I have devised means for generating. By such means air strata, which are easily accessible, are rendered available for the production of many desired effects at distances however great. This method, furthermore, allows advantage to be taken of many of those improvements which are practicable in the ordinary systems of transmission involving the use of a metallic conductor.

Obviously whatever method be employed it is desirable that the disturbances produced by the transmitting apparatus should be as powerful as possible, and by the use of certain forms of high-frequency apparatus, which I have devised and which are now well known, important practical advantages arc in this respect secured. Furthermore, since in most cases the amount of energy conveyed to the distant circuit is but a minute fraction of the total energy emanating from the source it is necessary for the attainment of the best results that whatever the character of the receiver and the nature of the disturbances as much as possible of the energy conveyed should be made available for the operation of the receiver, and with this object in view I have heretofore, among Other means, employed a receiving-circuit of high self-induction and very small resistance and of a period such as to vibrate in synchronism with the disturbances, whereby a number of separate in-pulses from the source were made to cooperate, thus magnifying the effect exerted upon and insuring the action of the receiving device. By these means decided ad vantages have been secured in many instances; but very often the improvement is either not applicable at all, m', if so, the gain is very slight. Evidently when the source is one producing a continuous pressure or delivering impulses of long duration it is impracticable to magnify the effects in this manner, and when, on the other hand, it is one furnishing short impulses of extreme rapidity of succession the advantage obtained in this way is insignificant, owing to the radiation and the unavoidable frictional waste in the receiving-circuit. These losses reduce greatly both the intensity and the number of the cooperative impulses, and since the initial intensity of each of these is necessarily limited only an insignificant amount of energy is thus made available for a single operation of the receiver. As this amount is consequently dependent on the energy conveyed to the receiver by one single impulse, it is evidently necessary to employ either a very large and costly, and therefore objectionable, transmitter, or else to resort to she equally objectionable use of a receiving device too delicate and too easily deranged. Furthermore, the energy obtained through the cooperation of the impulses is in the form of extremely rapid vibrations and because of this unsuitable for the operation of ordinary receivers, the more so as this form of energy imposes narrow restrictions in regard to the mode and time of its application to such devices. To overcome these and other limitations and disadvantages that have heretofore existed in such systems of transmission of signals or intelligence and to render possible an investigation of impulses or disturbances propagated through the natural media from any kind of source and their practical utilization for any purpose to which they are applicable, I have devised a novel method, which I have described in a pending application, filed June 24, 1899, Serial No. 721,790, and which, broadly stated, consists in effecting during any desired time interval a storage of energy derived from such impulses and utilizing the potential energy so obtained for operating a receiving device.

My present invention is intended for the same general purposes, and it comprises another apparatus by means of which similar results may be obtained.

The chief feature which distinguishes the method of my present from that of my former invention, just referred to, is that the energy stored is not, as in the former instance, obtained from the energy of the disturbances or effects transmitted from a distance, but from an independent source.

Expressed generally, the present method consists in charging a storage device with energy from an independent source, controlling the charging of said device by the action of the effects or disturbances transmitted through the natural media, and coincidently using the stored energy for operating a receiving device.

A great variety of disturbances, produced either by suitably-constructed transmitters or by natural causes, are at present known to be propagated through the natural media, and there are also a variety of means or devices enabling energy to be stored, and in view of this I wish to say that I consider the utilization of any such disturbances and the employment of any of these means as within the scope of my present invention so long as the use of the general method herein before stated is involved.

The best way of carrying out my invention which I at present know is to store electrical energy obtained from a suitable electrical generator in a condenser and to control the storage or the application of this energy by means of a sensitive, device acted upon by the effects or disturbances, and thereby cause the operation of the receiver.

In the practical application of this method I usually proceed as follows: At any point where I desire to investigate or to utilize for any purpose effects or disturbances propagated through the natural media from any kind of source I provide a suitable generator of electricity—as, for example, a battery and a condenser—which I connect to the poles of the generator in series with a sensitive device capable of being modified in its electrical resistance or other property by the action of the disturbances emitted from the source. To the terminals of the condenser I connect the receiver which is to be operated in series with another device of suitable construction, which performs the function of periodically discharging the condenser through the receiver at and during such intervals of time as may be best suitable for the purpose contemplated. This latter device may merely consist of two stationary electrodes separated by a feeble dielectric layer of minute thickness, but sufficient to greatly reduce or practically interrupt the current in the circuit under normal conditions, or it may comprise terminals one or more of which are movable and actuated by any suitable force and are adapted to be brought into and out of contact with each other in any convenient manner. The sensitive device maybe any of the many devices of this kind which are known, to be affected by the disturbances, impulses, or effects propagated through the media, and it may be of such a character that normally—that is, when not acted upon—it entirely prevents the passage of electricity from the generator to the condenser, or it way be such that it allows a gradual leaking through of the current and a charging of the condenser at a slow rate. In any case it will be seen that if the disturbances, of whatever nature they may be, cause an appreciable diminution in the electrical resistance of the sensitive device the current from the battery will pass more readily into the condenser, which will be charged at a more rapid rate, and consequently each of its discharges through the receiver, periodically effected by the special device before referred to which performs this function, will be stronger than normally—that is, when the sensitive device is not acted upon by the disturbances. Evidently then if the receiver be so adjusted that it does not respond to the comparatively feeble normal discharges of the condenser, if they should occur, but only to those stronger ones which take place upon the diminution of the resistance of the sensitive device it will be operated only when this device is acted upon by the disturbances, thus making it possible to investigate and to utilize the latter for any desired purpose.

The general principle underlying my invention and the operation of the various devices used will be clearly understood by reference to the accompanying drawings, in which—

Figure 1 is a diagram illustrating a typical arrangement of apparatus which may be used in carrying my method into practice, and Figs. 2, 3, 4, and 5 similar diagrams of modified arrangements of apparatus for the same purpose.

In Fig. 1, C is a condenser, to the terminals T and T′ of which is connected a charging-circuit, including a battery B, a sensitive device *a*, and a resistance *r*, all connected in series, as illustrated. The battery should be preferably of very constant electromotive force and or an intensity carefully determined to secure the best results. The resistance *r*, which may be a frictional or an inductive one, is not absolutely necessary; but it is of advantage to use it in order to facilitate adjustment, and for this purpose it may be made variable in any convenient and preferably continuous manner. Assuming that the disturbances which are to be investigated or utilized for some practical end are rays identical with or resembling those of ordinary light, the sensitive device a may be a selenium cell properly prepared, so as to be highly susceptible to the influence of the rays, the action of which should be intensified by the use of a reflector A. (Shown in the drawings.) It is well known that when cells of this kind are exposed to such rays of greatly-varying intensity they undergo corresponding modifications of their electrical resistance; but in the ways they have been heretofore used they have been of very limited utility.

In addition to the circuit including the sensitive device or cell *a* another circuit is provided, which is likewise connected to the terminals T T′ of the condenser. This circuit, which may be called the "receiving-circuit," includes the receiver R and in series with it a device *d*, before referred to, which performs the duty of periodically discharging the condenser through the receiver. It will be noted that, as shown in Fig. 1, the receiving-circuit is in permanent connection with the battery and condenser-terminal T, and it should be stated that it is sometimes desirable to entirely insulate the receiving-circuit at all times, except the moments when the device *d* operates to discharge the condenser, thus preventing any disturbing influence which might otherwise be caused in this circuit by the battery or the condenser during the period when the receiver should not be acted upon. In such a case two devices, as *d*, may be used, one in each connection from the condenser to the receiving-circuit, or else one single device of this kind but of a suitably-modified construction, so that it will make and break simultaneously and at proper intervals of time both of the connections of this circuit with the condenser T and T′.

From the foregoing the operation of the apparatus, as illustrated in Fig. 1, will be at once understood. Normally—that is, when it is not influenced by the rays at all or very slightly—the cell *a*, being of a comparatively high resistance, permits only a relatively feeble current to pass from the battery into the condenser, and hence the latter is charged at too slow a rate to accumulate during the time interval between two succeeding operations of the device *d* sufficient energy to operate the receiver or, generally speaking, to produce the required change in the receiving-circuit. This condition is readily secured by a proper selection and adjustment of the various devices described, so that the receiver will remain unresponsive to the feeble discharges of the condenser which may take place when the cell *a* is acted upon but slightly or not at all by the rays or disturbances; but if now new rays are permitted to fall upon the cell or if the intensity of those already acting upon it be increased by any cause then its resistance will be diminished and the condenser will be charged by the battery at a more rapid rate, enabling sufficient potential energy to be stored in the condenser during the period of inaction of the device *d* to operate the receiver or to bring about any desired change in the receiving-circuit when the device *d* acts. If the rays acting upon the cell or sensitive device *a* are varied or intermitted in any arbitrary manner, as when transmitting intelligence in the usual way from a distant station by means of short and long signals, the apparatus may readily be made to record or to enable an operator to read the message, since the receiver—supposing it to be an ordinary magnetic relay for example—will be operated by each signal from the sending-station a certain number of times, having some relation to the duration of each signal. It will be readily seen, however, that if the rays are varied in any other way, as by impressing upon them changes in intensity, the succeeding condenser discharges will undergo corresponding changes in intensity, which may be indicated or recorded by a suitable receiver and distinguished irrespectively of duration.

With reference to Fig. 1 it may be useful to state that the electrical connections of the various devices illustrated may be made in many different ways. For instance, the sensitive device instead of being in series, as shown, may be in a shunt to the condenser, this modification being illustrated in Fig. 3, in which the devices already described are indicated by similar letters to correspond with those of Fig. 1. In this case it will be observed that the condenser, which is being charged from the battery B through the resistance *r*, preferably inductive and properly related to the capacity of the condenser, will store less energy when the sensitive device *a* is energized by the rays, and its resistance thereby diminished. The adjustment of the various instruments may then be such that the receiver will be operated only when the rays are diminished in intensity or interrupted and entirely prevented from falling upon the sensitive cell, or the sensitive device may be placed, as shown in Fig. 4, in a shunt to the resistance *r* or inserted in any suitable way in the circuit containing the receiver—for example, as illustrated in Fig. 5—in both of which figures the various devices are lettered to correspond with those in Fig. 1, so that the figures become self explanatory. Again, the several instruments may be connected in the manner of a Wheatstone bridge, as will be hereinafter explained with reference to Fig. 2, or otherwise connected or related; but in each case the sensitive device will have the same duty to perform—that is, to control the energy stored and utilized in some suitable way for causing the operation of the receiver in correspondence with the intermittances or variations of the effects or disturbances—and in each instance by a judicious selection of the devices and careful adjustment the advantages of my method may be more or less completely secured. I find it preferable, however, to follow the plan which I have illustrated and described.

It will be observed that the condenser is an important element in the combination. I have shown that by reason of its unique properties it greatly adds to the efficacy of this method. It allows the energy accumulated in it to be discharged instantaneously, and therefore in a highly effective manner. It magnifies in a large degree the current supplied from the battery, and owing to these features it permits energy to be stored and discharged at practically any rate desired, and thereby makes it possible to obtain in the receiving-circuit very great changes of the current strength by impressing upon the battery-current very small variations. Other means of storage possessing these characteristics to a useful degree may be employed without departing from the broad spirit of my invention; but I prefer to use a condenser, since in these respects it excels any other storage device of which I have knowledge.

In Fig. 2 a modified arrangement of apparatus is illustrated which is particularly adapted for the investigation and utilization of very feeble impulses or disturbances, such as may be used in conveying signals or producing other desired effects at very great distances. In this case the energy stored in the condenser is passed through the primary o~ a transformer, the secondary circuit of which contains the receiver, and in order to render the apparatus still more suitable for use in detecting feeble impulses, in addition to the sensitive device which is acted upon by the impulses, another such device is included in the secondary circuit of the transformer. The schema of connections is in the main that of a Wheatstone bridge, the four branches of which are formed by the sensitive device *a* and resistances L, L′, and L″, all of which should be preferably inductive and also adjustable in a continuous manner, or at least by very small steps. The condenser C′, which is generally made of considerable capacity, is connected to two opposite points of the bridge, while a battery B, in series with a continuously adjustable non-inductive resistance *r*′, is connected to the other pair of opposite points, as usual. The four resistances included in the branches of the bridge namely, *a*, L, L′, and L″—are of a suitable size and so proportioned that under normal conditions—that is, when the device *a* is not influenced at all or only slightly by the disturbances—there will be no difference of potential, or, in any case, the minimum of the same at the terminals T and T′ of the condenser. It is assumed in the present instance that the disturbances to be investigated or utilized are such as will produce a difference of electric potential, however small, between two points or regions in the natural media, as the earth, the water, or the air, and in order to apply this potential difference effectively to the sensitive device a the terminals of the same are connected to two plates P and P′, which should be of as large a surface as practicable and so located in the media that the largest possible difference of potential will be produced by the disturbances between the terminals of the sensitive device. This device is in the present case one of familiar construction, consisting of an insulating-tube, which is indicated by the heavy lines in the drawings and. which has its ends closed tightly by two conducting-plugs with reduced extensions, upon which bear two brushes *b b*, through which the currents are conveyed to the device. Tile tubular space between the plugs is partially filled with a conducting sensitive powder, as indicated, the proper amount of the same and the size of its grains being determined and adjusted beforehand by experiment. This tube I rotate by clockwork or other means at a uniform and suitable rate of speed, and under these conditions I find that this device behaves toward disturbances of the kind before assumed in a manner similar to that of a stationary cell of celenium toward rays of light. Its electrical resistance is diminished when it is acted upon by the disturbances and is automatically restored upon the cessation of their influence. It is of advantage to employ round grains of powder in the tube, and in any event it is important that they should be of as uniform size and shape as possible and that provision should be made for maintaining an unchanging and very dry atmosphere in the tube. To the terminals T and T′ of the condenser C′ is connected a coil *p* usually consisting of a few turns of a conductor of very small resistance, which is the primary of the transformer before referred to, in series with a device *d*, which effects the discharge of the condenser through the coil *p* at predetermined intervals of time. In the present case this device consists of a cylinder made partly of conducting and partly of insulating material *e* and *e′*, respectively, which is rotated at the desired-rate of speed by any suitable means. The conducting part *e* is in good electrical connection with shaft S and is provided with tapering segments, as *f*, upon which slides a brush *k*, which should preferably be capable of longitudinal adjustment along the cylinder. Another brush *b′*, which is connected to the condenser-terminal T′, being arranged to bear upon the shaft S, it will be seen that whenever the brush *k* comes in contact with a conducting-segment *e* the circuit including the primary *p* will be completed and the condenser, if energized, discharged through the same. By an adjustment of the speed of rotation of the cylinder and a displacement of the brush *k* along the axis of the same the circuit may be made to open and close in as rapid succession and remain open or closed during such intervals of time as may be desired. In inductive relation to the primary *p* is a secondary coil *s*, usually of much thinner wire and of many more turns than the former, to which are connected in a series a receiver R, illustrated as an ordinary magnetic relay, a continuously-adjustable non inductive resistance *r″*, a battery B′ of a properly-determined and very constant electromotive force, and finally a sensitive device *a′* of the same or similar construction as *a*, which is likewise rotated at a uniform speed and which, with its brushes *b′ b″*, closes the secondary circuit. The electromotive force of the battery B' is so graduated by means of the adjustable resistance *r″* that the dielectric layers in the sensitive device *a′* are strained very nearly to the point of breaking down and give way upon a slight increase of the electrical pressure on the terminals of the device. It will of course be understood that the resistance *r″* is used mainly because of convenience and that it may be dispensed with, in which case the adjustment may be effected in many other ways, as by determining the proper amount or coarseness of the sensitive powder or by varying the distance apart of the metallic plugs in the ends of the tube. The same may be said of the resistance *r′*, which is in series with the battery B and serves to graduate the force of the latter, so that the dielectric layers of the sensitive device *a* are subjected to a similar strain and maintained in a state of delicate poise.

The various instruments being connected and adjusted in the manner described, it will now be readily seen from the foregoing that under normal conditions, the device a being unaffected by the disturbances, or practically so, and there, being no or only a very insignificant amount of energy stored in the condenser, the periodical closure of the primary circuit of the transformer through the operation of the device *d* will have no appreciable effect upon the primary coil *p*, and hence no currents will be generated in the secondary coil *s*, at least not such as would disturb the state of delicate balance existing in the secondary circuit including the receiver, and therefore the latter wilt not be actuated by the battery B′; but when, owing to the disturbances or impulses propagated through the media from a distant source, an additional electromotive force, however small, is created between the terminals of the device *a* the dielectric layers in the same, unable to support the increased strain, give way and allow the current of the battery B to pass through, thus causing a difference of potential at the terminals T and T′ of the condenser. A sufficient amount of energy being now stored in this instrument during the time interval between each two succeeding operations of the device *d*, each closure of the primary circuit by the latter results in the passage of a sudden current impulse through the coil *p*, which induces a corresponding current of relatively high electromotive force in the secondary coil *s*. Owing to this the dielectric in the device *a*′ gives way, and the current of the battery B′ being allowed to pass the receiver R is operated, but only for a moment, since by the rotation of the devices a, *a*′, and *d*, which may be all driven from the same shaft, the original conditions are restored, assuming, of course, that the electromotive force set up by the disturbances at the terminals of the sensitive device *a* is only momentary or of a duration not longer than the time of closure of the primary circuit; otherwise the receiver will be actuated a number of times and so long as the influence of the disturbances upon the device a continues. In order to render the discharged energy of the condenser more effective in causing the operation of the receiver, the resistance of the primary circuit should be very small and the secondary coil *s* should have a number of turns many times greater than that of the primary coil *p*. It will be noted that since the condenser under the above assumptions is always charged in the stone direction the strongest current impulse in the secondary coil, which is induced at the moment when the brush *k* comes in contact with segment *e*, is also of unchanging, direction, and for the attainment of the best results It is necessary to connect the secondary coil so that the electromotive force of this impulse will be added to that of the battery and will momentarily strengthen the same. However, under certain conditions, which are well understood by those skilled in the art, the devices will operate whichever way the secondary be connected. It is preferable to make the inductive resistances L and L' relatively large, as they are in a shunt to the device *a* and might, if made too small, impair its sensitiveness. On the other hand, the resistance L″ should not be too large and should be related to the capacity of the condenser arid the number of makes and breaks effected by the device *d* in well-known ways. Similar considerations apply, of course, to the circuits including the primary *p* and secondary *s*, respectively.

By carefully observing well-known rules of scientific design and adjustment of the instruments the apparatus may be made extremely sensitive and capable of responding to the feeblest influences, thus making it possible to utilize impulses or disturbances transmitted from very great distances and too feeble to be detected or utilized in any of the ways heretofore known, and on this account the method here described lends itself to many scientific, and practical uses of great value.

Obviously the character of the devices and the manner in which they are connected or related may be greatly varied without departing from the spirit of my invention.

What I claim as new, and desire to secure by Letters Patent, is—

1. In an apparatus for utilizing effects or disturbances transmitted through the natural media from a distant source, the combination of an electrical storage device, a charging-circuit

connected therewith and including a device sensitive to the action of the effects or disturbances and determining under their control the flow of current in the charging-circuit, a receiving-circuit including a receiver, and means for periodically discharging the storage device through the receiving-circuit, substantially as described.

2. In an apparatus for utilizing effects or disturbances transmitted through the natural media from a distant source, the combination of a condenser, a charging-circuit connected therewith and including a source of electricity and device sensitive to the action of the effects or disturbances and determining under their control the flow of current in the charging-circuit, a receiving-circuit including a receiver, and means for periodically discharging the condenser through the receiving-circuit, substantially as described.

3. In an apparatus for utilizing effects or disturbances transmitted through the natural media from a distant source, the combination of a circuit including a source, of electricity, a storage device adapted to be charged thereby and a device normally of very high resistance but adapted to have its resistance reduced when acted upon by the effects or disturbances, with a receiving-circuit connected with the storage device and including a receiver and a device adapted to open and close the receiving circuit, at predetermined intervals of time, substantially as described.

4. In an apparatus for utilizing effects or disturbances transmitted through the natural media from a distant source, the combination of a circuit including a source of electricity, a condenser and a device normally of very high resistance but adapted to have its resistance reduced when acted upon by the effects or disturbances, with a receiving-circuit connected with the condenser and including a receiver and a device adapted to open and close the receiving-circuit at predetermined intervals of time, substantially as described.

5. In an apparatus for utilizing effects or disturbances transmitted from a distant source, the combination of a circuit including a sourer of electricity, a storage device adapted to be charged thereby and a device, normally of very high resistance but adapted to have its resistance reduced when acted upon by the effects or disturbances, with a receiving-circuit connected with the storage device and including the primary of a transformer and a device adapted to open and close such second circuit at predetermined intervals of time, and a receiver included in the secondary of the transformer, substantially as described.

6. In an apparatus for utilizing effects or disturbances transmitted from a distant source, the combination of an electrical storage device, a charging-circuit connected therewith and including a device sensitive to the action of the effects or disturbances and determining under their control the flow of the current in the charging-circuit, and a receiving-circuit supplied with energy from the storage device and including a receiver and a device sensitive to electrical variations in the receiving-circuit, substantially as described.

7. In an apparatus for utilizing effects or disturbances transmitted through the natural media from a distant source, the combination of a condenser, a charging-circuit connected therewith and including a device sensitive to the action of the effects or disturbances and determining under their control the flow of the current in the charging-circuit, and a receiving-circuit supplied with energy from the condenser and including a receiver and a device sensitive to electrical variations in the receiving-circuit, substantially as described.

8. In an apparatus for utilizing effects or disturbances transmitted through the natural media from a distant source, the combination of a circuit, an independent local source of electricity

included therein; a storage device connected with the said circuit and adapted to receive energy from the said source, a device normally of very high resistance, but adapted to have its resistance reduced when acted upon by the effects or disturbances, a receiving-circuit connected with the storage device, a transformer, the primary of which is included in said receiving-circuit, a device adapted to open and close the receiving-circuit at predetermined intervals of time, a receiver, and a device, normally of very high resistance, but adapted to have its resistance reduced when acted upon by the effects or disturbances, and included in the secondary circuit of the transformer, as set forth.

9. In an apparatus for utilizing effects or disturbances transmitted from a distant source, the combination with a storage device and an independent source of energy for charging the same, of a receiving-circuit connected with the storage device, a device sensitive to the effects or disturbances and determining under their control the flow of current in the receiving-circuit, substantially as set forth.

10. In an apparatus for utilizing effects or disturbances transmitted through the natural media from a distant source, the combination with a storage device and an independent source of energy for charging the same, of a receiving-circuit connected with the storage device, a device sensitive to the effects or disturbances and adapted to have its resistance varied, by the action thereon of such effects or disturbances and determining under their control the flow of current in the receiving-circuit, substantially as set forth.

11. In an apparatus for utilizing effects or disturbances transmitted from a distant source, the combination of a storage device, a battery or similar independent source of energy for charging the same, a sensitive device adapted to have its resistance varied under the influence of the effects or disturbances, a receiver adapted to be operated by the discharge of the storage device, the sensitive device being in one branch of a Wheatstone bridge, the storage device in one of the cross connections between two opposite points of the bridge, and the battery in the other cross connection, and resistances L, L′, and L″ in the three remaining branches of the bridge, as set forth.

12. In an apparatus for utilizing effects or disturbances transmitted through the natural media from a distant source, the combination of a storage device, a battery or like independent source of energy for charging the same, a sensitive device adapted to have its resistance varied Under the influence of the effects or disturbances, a circuit connected with the terminals of the storage device, a transformer having its primary in said circuit and a receiver in the secondary circuit of the transformer, the sensitive device being in one branch of a Wheatstone bridge, the storage device in one of the cross connections between opposite points of the bridge and the battery in the other cross connection, and resistances L, L′, and L″ in the three remaining branches of the bridge, as set forth.

NIKOLA TESLA.

WITNESSES: E. A. Subiderlin, D. D. Lord.

3.4 TELEVISION SYSTEMS

The stage for television was set after the successful launching of the telephone by Bell. The telephone converts voice and other sound signals into a form that can be transmitted to remote locations and that can receive and reconvert waves into sound signals. The capacity to reproduce the effect of movement by a series of still pictures was patented by Edison in 1897 (see Section 3.2). The major issue that is evident in these television patents is the ability to capture the light and color content of pictures (in small segments, or pixels as they are now called) as electrical signals and to resynchronize the rate of delivery at the receiver. Aspects of both the transmitter and the receiver are covered in the three patents.

During the three decades between 1900 and 1930, many experiments were designed and verified to carry electrical signals farther and farther for use and deployment at higher frequencies. During this time, a genuine understanding of the principles of electricity and magnetism was also evolving. The educated guesswork, verified by well-designed experiments, led to the early television systems and the ensuing patents. In that era, purely mathematical formulations would not have led to the devices or to the patents. Purely experimental rigs would not prove the validity of the novel technological process behind the invention nor would they have proved the concept to derive other patents from this patent. A true blending of the imagination of the scientists and the skill of the experimentalists (two aspects of the same personality) has been demonstrated in these patents, thus forming the basis on which the patents are presented in this section.

The features of the Farnsworth 1930 patent (Section 3.4.1) and the two Zworykin 1935 and 1938 patents (Sections 3.4.2 and 3.4.3) are noteworthy and indicative of two major trends during the 1930s. First, the rigor and expertise in proving the validity of the patent are remarkable (compared with the Marconi or Tesla patents), and second, the technical complexity on which the device is based shows greater reason than intuition. These two features have become increasingly dominant in the later patents of the last century. In fact, the emphasis has shifted to the concept, the symbolic and mathematical representation, and the proof of functionality rather than a device to "see" it functioning in the particular way it is claimed. These two features are valid and reinforce the notion that some amount of logic and derivation is involved with the process of invention rather than simply a flash of true genius. However, a blend of the two is essential to produce a sequence of significant inventions. This blending in the seminal patents has led to scores of patents in the area of computing, fiber optics, masers and lasers.

We strongly recommend to our readers that they look for the growing rigor and basic concepts that are evident in these three patents, rather than simply looking at a detailed description of novel devices. This is a major shift in the attitude of the inventors of the 1930s and 1940s. In reality, the device is the basis of the patent, and the principle is the basis of the device. In these three patents, it is noteworthy that the inventors' imagination and creativity are focused on both aspects (concept and device) necessary for a seminal patent.

Philo T. Farnsworth (1906–1971)

Early Television Systems

3.4.1 Television Systems (Farnsworth, 1930)

UNITED STATES PATENT

August 26, 1930 1,773,980

Television System

PHILO T. FARNSWORTH, OF BERKELEY, CALIFORNIA, ASSIGNOR, BY MESNE ASSIGNMENTS, TO TELEVISION LABORATORIES, INC., OF SAN FRANCISCO, CALIFORNIA, A CORPORATION OF CALIFORNIA

Application filed January 7, 1927, Serial No. 159,540.

This invention relates to a television apparatus and process, that is, it is directed to an apparatus and process for the instantaneous transmission of a scene or moving image of an object located at a distance in which the transmission is by electricity.

Heretofore attempts have been made to transmit an image of an object by electricity so that the image of the object will instantaneously appear at a distance. These prior attempts at television have generally embodied an apparatus and method in which each particular elementary area of the image of the object is successively converted into an electrical current, the intensity of which is proportional to the intensity of the light at that particular elementary area; all the elementary areas of the image being covered in that fraction of a second during which the eye will retain a picture, hereafter referred to as the optical period. This is followed by a transmission of such current and a conversion of such current to light corresponding in intensity to the intensities of the light of the individual areas of the original image; the reconversion process likewise being performed within the optical period so that, by a proper coordination of the developed light, an image of the object to be transmitted appears as instantly formed at the receiving end of the apparatus and method.

Aug. 26, 1930. P. T. FARNSWORTH **1,773,980**

TELEVISION SYSTEM

Filed Jan. 7, 1927 4 Sheets-Sheet 1

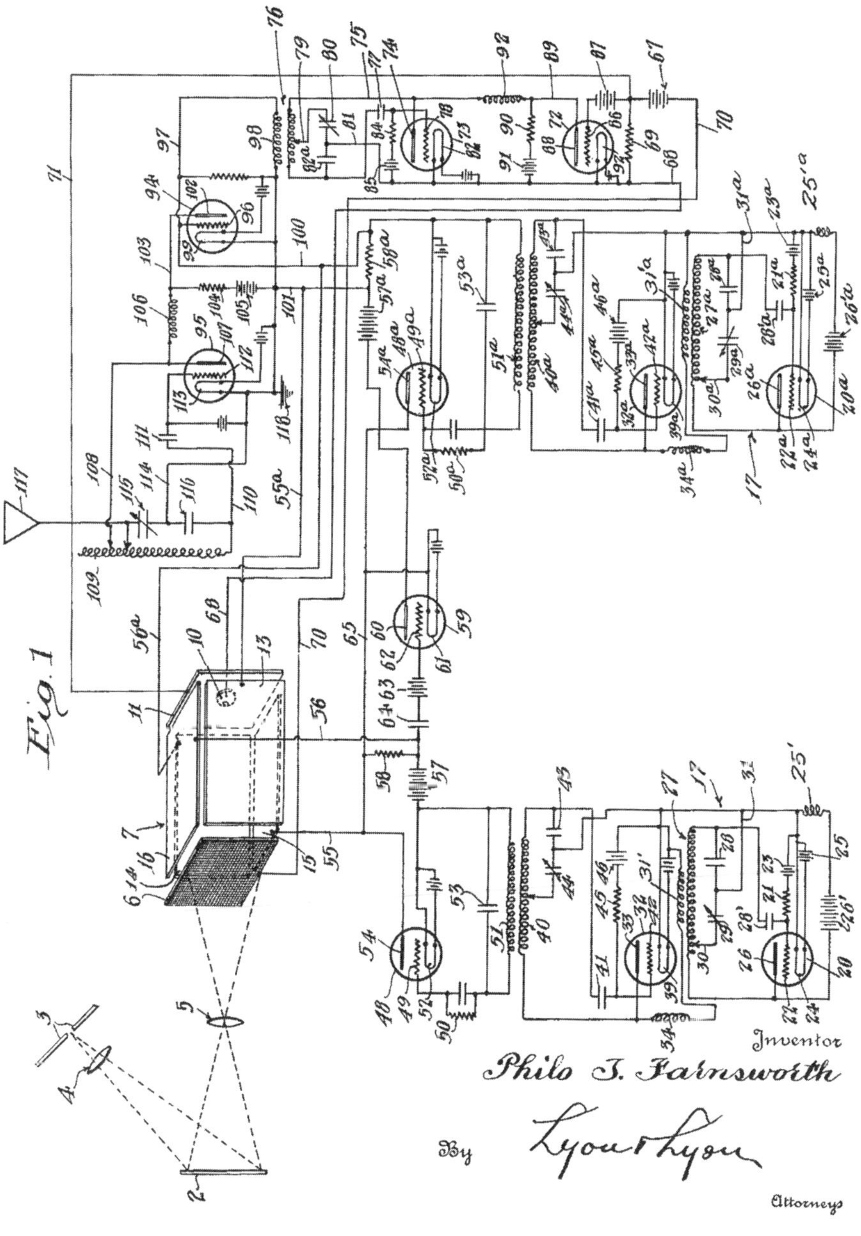

Fig. 1

Inventor

Philo T. Farnsworth

By *Lyon & Lyon*

Attorneys

Aug. 26, 1930. P. T. FARNSWORTH **1,773,980**

TELEVISION SYSTEM

Filed Jan. 7, 1927 4 Sheets-Sheet 2

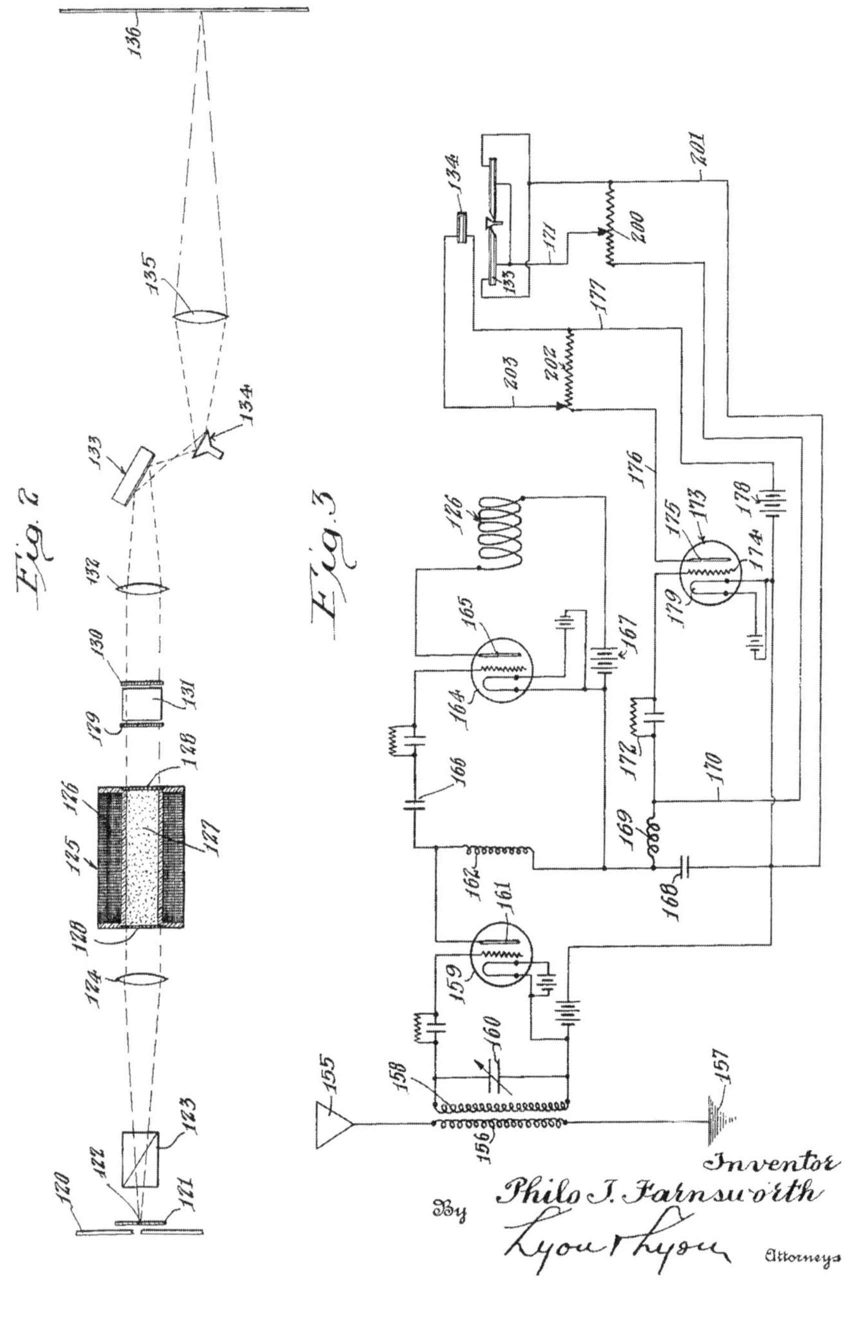

Aug. 26, 1930. P. T. FARNSWORTH **1,773,980**

TELEVISION SYSTEM

Filed Jan. 7, 1927 4 Sheets-Sheet 3

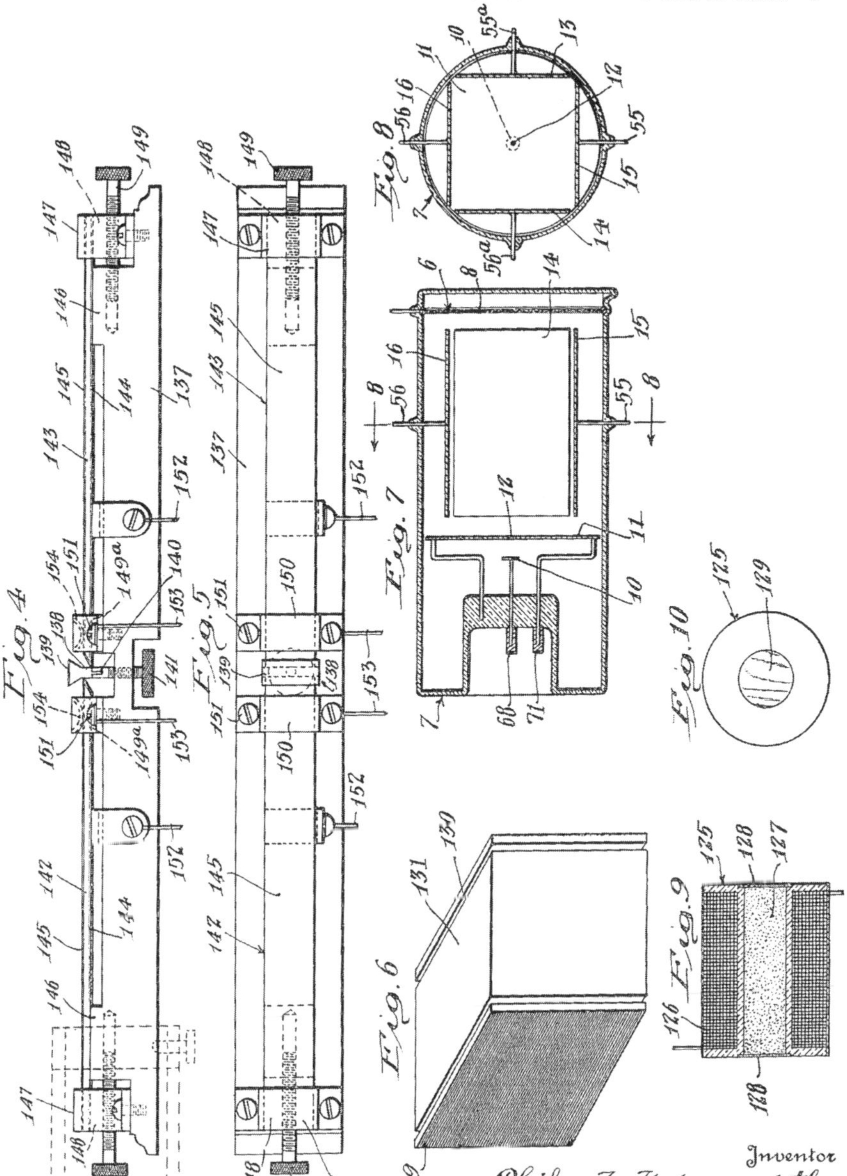

Inventor

Philo T. Farnsworth

By *Lyon & Lyon*

Attorneys

Aug. 26, 1930. P. T. FARNSWORTH **1,773,980**

TELEVISION SYSTEM

Filed Jan. 7, 1927 4 Sheets-Sheet 4

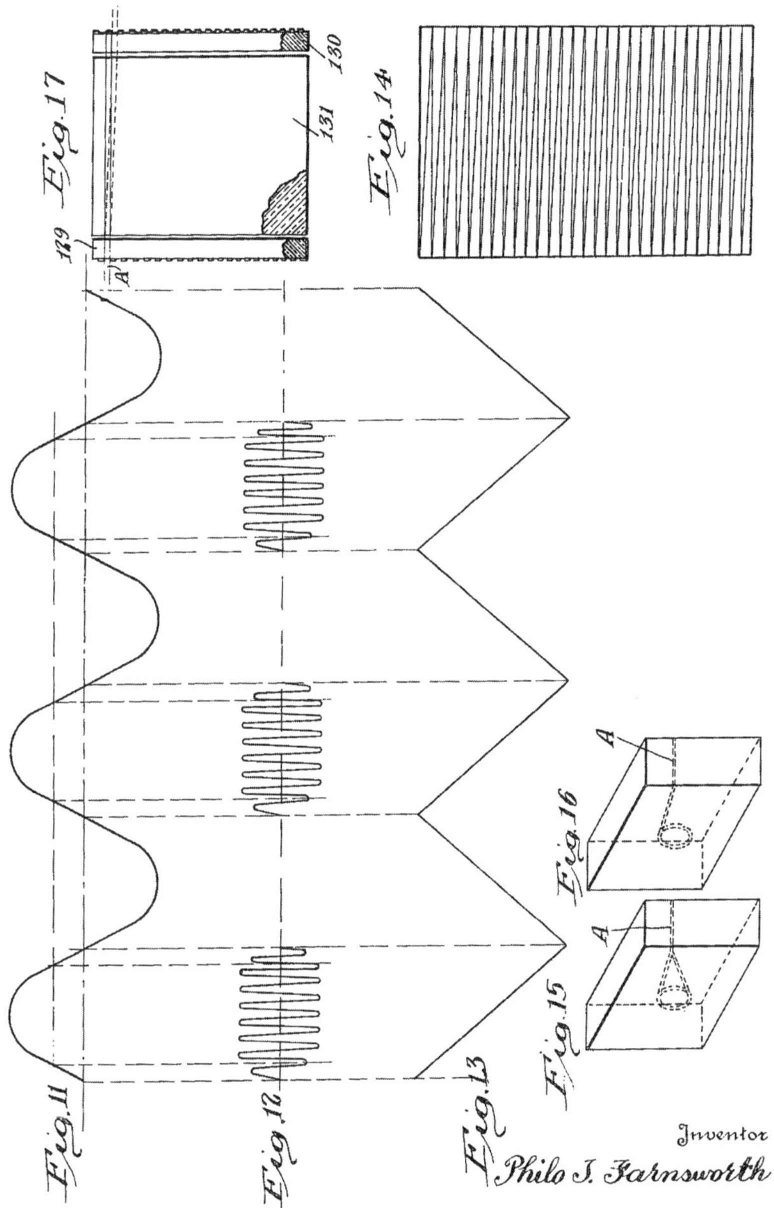

Inventor

Philo T. Farnsworth

By *Lyon & Lyon*

Attorneys

The time during which the human eye will retain a picture is of such short duration that the conversion of the light shades of the original image of the object to electricity and the reconversion of said electricity to light and the proper coordination of such light must be performed at a very tremendous speed. All prior attempts at television have attempted to employ some mechanically moving part for dissecting the image of the original object during the process of forming an electrical current which varies in intensity in accordance with the light shades of the respective elementary areas of the image. None of these prior attempts at television have proven successful. They have resulted at best in the production of a crude moving silhouette of the object to be transmitted. This has generally been due to the fact that the mechanically moving parts of the prior apparatus have not been able to travel at the necessary speed requirements with the synchronism required in a television apparatus.

An object of the present invention is to provide a method and apparatus for television, which is adapted to transmit electrically a true moving image in full light shades of the object to be transmitted.

Another object of the present invention is to provide a method and apparatus for television in which the conversion and dissecting of the light shades of the object to be transmitted, to electricity and the reconversion of such electricity to form an image is accomplished in the following manner:

In the process and apparatus of the present invention, light from all portions of the object whose image is to be transmitted, is focused at one time upon a light sensitive plate of a photo-electrical cell to thereby develop an electronic discharge from said plate, in which each portion of the cross section of such electronic discharge will correspond in electrical intensity with the intensity of light imposed on that portion of the sensitive plate from which the electrical discharge originated. Such a discharge is herein termed an electrical image. An electrical shutter is then interposed between said sensitive plate and the anode of the photo-electrical cell, the shutter having a small aperture therein so that there can be received upon said anode at one instant, only the electrons which originate from one elementary area of the light sensitive plate. There is then imposed upon the electrical discharge a plurality of electrical potentials of different frequencies for causing the electrical discharge to bend in two directions, whereby the electrons from each elementary portion of the sensitive plate are successively directed through said shutter, this action taking place so as to completely cover the area of the sensitive plate within the optical period. The scene to be transmitted is thus analyzed or dissected to produce an electrical current or "light" current having variations in intensity in accordance with the light shades of the object to be transmitted and this is accomplished within the optical period without the necessity of employing any mechanically moving parts.

The produced electrical current or "light" current may be transmitted to the receiving end of the apparatus by either wires or may be superimposed upon a wireless carrier wave. There is also transmitted at the same time and preferably superimposed upon the same carrier wave, the two electric potentials of different frequencies which are employed in analyzing the image so that such currents may be employed to synchronize the receiving apparatus and process.

At the receiving end of the apparatus and process, the "light" current is reconverted to light and the light coordinated to form an image of the object transmitted in accordance with the following apparatus and process.

Preferably a constant source of light is utilized which is directed, first, through a polarizing prism and hence through an apparatus or means by which the plane of polarization of the light may be rotated by the "light" current. In this manner an instantaneous response to the variations of such light current is obtained in the rotation of the plane of polarization of the light. The light is then directed through a suitable screen capable of shutting off the light in accordance with the rotation of its plane of polarization. In this manner, a beam of light is developed fluctuating in intensity to the variations of intensity of the "light" current transmitted without the necessity of employing any mechanically moving parts. This said beam of light is then projected by means of two cooperating oscillographs upon the screen where the image is to be transmitted, said oscillographs being operated by the synchronizing frequencies transmitted with the "light" current to correctly coordinate the light upon the screen to form a correct image.

The present invention, together with various objects and advantages thereof will best be understood from a description of a preferred form or example of a process and apparatus for television embodying the invention. For this purpose, I have hereinafter set forth one form of example of a method and apparatus for producing television in accordance with the present invention, and have illustrated said apparatus and method as it is adapted for television by wireless. It is to be understood, however, that the invention is capable of various and numerous modifications, changes, and substitutions, and is not necessarily limited to the transmission by wireless or radio.

The apparatus and method will best be understood from a description of the accompanying drawings, in which:

Figure 1 is a diagrammatic view of a complete television transmitter, including a circuit diagram therefor,

Figure 2 is a diagrammatic view of the television receiver,

Figure 3 is a circuit diagram of the electrical connections for the television receiver,

Figure 4 is an elevation of one of the oscillographs,

Figure 5 is a plan view of one of the oscillographs,

Figure 6 is a perspective view of the light diverting means,

Figure 7 is a sectional view of the photo-electric cell,

Figure 8 is a section of the line 8—8 of Figure 7,

Figure 9 is a section of the light rotator,

Figure 10 is an end view thereof,

Figure 11 is a representation of the form of electric current of the first oscillator employed in developing a potential for the photo-electric cell,

Figure 12 is a representation of the form of electric current produced in the second oscillator,

Figure 13 is a representation of the resulting straight lined potential,

Figure 14 is a view of the scanning path and also a view of the path of the light beam over the receiving screen,

Figure 15 is a perspective view of a bi-axial crystal showing the conical refraction of unpolarized light,

Figure 16 is a perspective view of a bi-axial crystal showing the refraction of polarized light, and,

Figure 17 is a diagrammatic illustration of the path of light through the gratings.

Referring to the drawings, 2 represents an object, an image of which is to be transmitted. Said object may be an actual scene or a photograph, a projection of a motion picture film, or any other object. The object 2 is preferably illuminated, for example, by means of an arc light 3 focused thereon by a lens 4. 5 indicates a lens for focusing an image of the object 2, upon the light sensitive plate 6 of a photo-electric cell 7.

The photo-electric cell is preferably constructed as follows:

The light sensitive plate 6 or cathode of the cell is preferably made flat and is formed of a fine mesh screen 8, and said screen 8 is covered or coated with a light sensitive material such as sodium, potassium, or rubidium. 10 is the anode of the photo-electric cell positioned at the other end of the cell. Between the sensitive plate 6 and anode 10 and closely adjacent to anode 10 is placed an electric shutter 11 formed by a metallic plate in which there is a small aperture 12. Between the shutter 11 and light sensitive plate 6, four plates 13, 14, 15, and 16 are placed at right angles to each other and outside the path of electrons from the plate 6 to the shutter 11. Each opposed pair of the plates are connected to a source of electrical potential of a different frequency. The photo-electric cell should be highly evacuated, such for example as to 10^{-7} cm. mercury to permit a high potential across the cell without ionization.

The necessity for employing a high potential across the cell arises from the fact that the photo electrons emitted from the cathode 6 have a small emission velocity which depends upon the color of the light causing their emission. This emission velocity is always small, of the order of that which an electron would acquire by falling through a volt or two, but it may have nearly any direction. This haphazard motion tends to distort the electric image and is only prevented from doing so by making the potential between the cathode 6 and the anode 10 high enough to insure that the time taken for an electron to traverse the distance between cathode 6 and anode 10 is so small that the small velocity transverse to this path produces no appreciable distortion. Hence the vacuum in the photo-electric cell 7 should be the highest obtainable.

The electrical potentials are provided by an oscillator 17, capable of developing two different high frequency electrical currents. Said oscillator 17 not only is required to provide a source of oscillating energy but is required to provide a form of oscillating energy, the wave form of which is composed of substantially straight lines, as will be hereinafter pointed out. Such a wave form is essential to accomplish a uniform lighting of all portions of the image which is to be produced.

The oscillator comprises a tri-electrode valve 20 connected in a circuit acting as an oscillator to produce an oscillating energy of low frequency, such for example as 10 cycles per second. It is understood that any customary or preferred form of circuit for this purpose may be employed, the particular circuit described being provided with a grid leak 21 connected with the grid 22 of the tube 20, and hence through a negative bias battery 23 to the filament 24. The filament 24 is indicated as heated by a battery 25. The plate 26 of the tube is connected through a battery 26' and the choke coil 25' to the filament 24. The plate 26 also connects through an inductance 27 and capacity 28' with the grid. The inductance 27 is shunted by a fixed capacity 28 and a variable capacity 29 in series, one end of the series being connected to the end of the inductance 27 and the other end having a variable connection with said inductance. Between these capacities 28 and 29, a lead 31 is connected which connects with filament 24 of the tube 20.

By this connection, the constants of the oscillating circuit may be any value of inductance and capacity to bring the oscillating circuit in resonance with the frequency of the desired circuit. Said oscillator in turn provides a source of potential for a second oscillating circuit of similar design, the second oscillator operating at a higher frequency such, for example, as 500 kilo-cycles. The second oscillator comprises the tube 32, the plate 33 of which is charged with the oscillatory energy of the first oscillator. The first oscillator is coupled thru the secondary coil 31' to plate 33, the inductance 34 being included in series therewith. The inductance 34 may be any suitable radio frequency choke to prevent the high frequencies in the second oscillating circuit from being imposed on the first oscillating circuit. The plate 33 is connected through the primary 40 of a radio frequency transformer and hence through the capacity 41 with the grid 42. Capacities 43 and 44 are shunted around all or part of the primary 40 and a lead is connected from their midpoint to the filament 39 of the tube 32. The grid 42 of the tube is connected through a suitable leak 45 and negative bias battery 46 with filament 39. It is understood that the second oscillating circuit thus described is only one example of a circuit adapted for this purpose and the various constants of the circuit may be of any value suitable for bringing the circuit into resonance with the frequency of the oscillations (500 kilo-cycles) desired to be produced therein.

The voltage of the first oscillator is adjusted to be well above the value required for maximum plate current of the second oscillator. Hence, since the second oscillator will generate oscillations only when the plate voltage is positive, the current generated by the second oscillator will be similar to that shown in Figure 12. The harmonic oscillating current developed by the first oscillator is represented in Figure 11. This current, when imposed upon the second oscillator, develops a current such as illustrated in Figure 12, in which it will be seen that each positive cycle of the first harmonic current produces a series of harmonic oscillations in the second oscillator of substantially equal intensity, while during the negative period of the first harmonic current, substantially no oscillations are developed in the second oscillator.

The output from the second oscillator is then imposed upon an audion circuit having a tube 48 with its grid 49 connected by a line through the grid leak and grid condenser 50 to an inductance 51 inductively coupled to the inductance 40. Said secondary 51 is connected to the filament 52 of the audion 48. Shunted across the secondary 52 is a condenser 53 of value suitable to produce resonance with the oscillations developed in the second oscillator. The plate 54 and the audion 48 is connected by the lead 55 with the plate 15 of the photo-electric cell, and the opposed plate 16 of the photo-electric cell is connected by a lead 56 through the battery 57 to the filament 52. The resistance 58 is shunted across the leads 55 and 56 to provide a potential for the plates 15 and 16.

The action of the audion circuit including the tube 48 is to produce an alternating current equal to the frequency developed in the first oscillator but the wave form of said frequency is of substantially straight lines such, for example, as indicated in Figure 13. In producing this wave form, the audion tube 48 operates due to the bias of the grid leak and condenser 50 to accumulate a charge during the passage of each wave train indicated in Figure 12, and such accumulated charge leaks off during the interval between successive trains, so that the output of the audion 40 into the plate circuit, indicated by the leads 55 and 56 (passing to the plates 15 and 16 of the photo-electric cells) assumes the straight line form of Figure 13.

[*A section of the patent has been removed for the sake of brevity.*]

While the process and apparatus for producing television herein described is well adapted for carrying out the objects of the present invention, it is understood that various modifications and changes may be made without departing from the invention, and the invention includes all such modifications and changes as come within the scope of the following appended claims.

I claim:

1. The method of television which includes forming an electrical image, and traversing each elementary area of the electrical image by an electric shutter at a velocity sufficient to cover the entire image within the optical period.
2. The process of television which comprises forming an electrical image, moving said electrical image in more than one direction by an analyzing potential, and varying the intensity of an electric current in accordance with the position of the electrical image.
3. The method of television which comprises focusing an image of an object upon the sensitive plate of a photo-electric cell, imposing a shutter in the path of the electrical discharge from said plate, and forming transverse to the electrical discharge two electrical potentials of different frequencies.
4. An apparatus for picture dissecting comprising a cell having a plate of photo sensitive material, an anode, a plurality of plates positioned between the photo sensitive plate and anode, and means for imposing upon said plates a plurality of electrical potentials of different frequencies.
5. An apparatus for dissecting an image comprising a cell having a photo sensitive plate, an anode, a shutter between the anode and plate, and electrical means for bending the electrical discharge from said plate.
6. The method of television which comprises forming an electrical discharge which corresponds in cross section in electrical intensity to the light intensity of an image to be transmitted, transmitting successive portions of said electric discharge, and modulating light thereby.

[*An additional section of the patent has been removed for the sake of brevity.*]

Signed at San Francisco, California, this 21st day of December, 1926.

PHILO T. FARNSWORTH

Television & Images and Cathode Ray Tube

Vladimir Kosma Zworykin (1903–1957)

3.4.2 Images of Objects (Zworykin, 1935)

UNITED STATES PATENT

November 26, 1935 2,021,907

Method of and Apparatus for Producing Images of Objects

VLADIMIR K. ZWORYKIN, HADDONFIELD, N.J., ASSIGNOR TO RADIO CORPORATION OF AMERICA, A CORPORATION OF DELAWARE
Application filed November 13, 1931, Serial No. 574,772. 4 Claims. (CL178-6)

My invention relates to improvements in methods of and apparatus for producing images of objects or phenomena which are invisible to the human eye.

In bacteriological work, for example, it is known that there are bacteria which cannot be seen with a microscope regardless of the degree of magnification. This is attributed to the fact that the wave length of the light used heretofore is of the same order or greater than the size of these bacteria. Recourse has, accordingly, been made to the use of ultra-violet rays of a wave length smaller than the size of the bacteria, and an ultra-violet light image of the bacteria projected on a photographic plate. This method is not entirely satisfactory, since it is quite difficult to photograph phenomena which cannot be observed. Furthermore, the bacteria or other objects can be observed only indirectly, that is, by studying the developed photographic plates.

With the foregoing in mind, it is one of the objects of my invention to provide an improved method whereby it is possible to obtain directly an image of objects so small that it has not been possible to observe them directly by the methods and apparatus proposed heretofore.

Another object of my invention is to provide improved apparatus from which a moving image can be developed and observed directly of an object whose size is of the order of the wave length of ultra-violet light.

Other objects and advantages will hereinafter appear.

In accordance with my invention, an ultra-violet light image of an object, such as bacteria whose size is of the order of the wave length of ultra-violet light, is developed and utilized to develop a corresponding electrical image, the electrical image is scanned to develop picture signals, and these signals are supplied to a transmitting system.

More particularly, in accordance with my invention, a moving image is produced of an object, whose size is of the order of the wave length of ultra-violet light, by projecting a magnified ultra-violet light image of the object on photo-sensitive structure, scanning this structure with a cathode ray to develop picture signals, and supplying these signals to a transmitting system.

My invention resides in the method, system and apparatus of the character hereinafter described and claimed.

For the purpose of illustrating my invention, an embodiment thereof is shown in the drawing, wherein Fig. 1 is a diagrammatic view of a system embodying my invention; and Fig. 2 is an enlarged fragmentary sectional view of a portion of the apparatus shown in Fig. 1

A practical embodiment of my invention, as shown in Fig. 1, comprises cathode ray apparatus, designated generally by the reference numeral 10, and having screen structure 11, upon which is projected an image of the object which is illuminated by ultra-violet light from a suitable source, such as a quartz mercury tube 12.

The screen structure 11 is of the same general type, and operates on the same principle as those disclosed in my copending applications, bearing Serial Nos. 448,834 and 468,610, and filed May 1, 1930 and July 17, 1930, respectively.

With reference to Fig. 2, this structure may comprise, specifically, a relatively thin sheet 13 of mica having deposited on one side thereof a thin film 14 of platinum. On the other side of the mica sheet are individual photosensitive elements 15, uniformly distributed over the surface. For the purpose of making these elements, it is proposed to place a suitable mask against this side of the mica sheet, and, with the assembly in a vacuum, evaporate cadmium to cause the individual elements 15 of this material to be deposited at the openings in the mask. As an alternative, a continuous film of cadmium may be applied to the mica sheet and scratched through in a ruling machine to form the individual elements 15, insulated from each other and from ground. Instead of cadmium; uranium, thorium, cerium or any other suitable photosensitive material, particularly sensitive to ultra-violet light, may be used. Some of the alkalis, such as lithium and sodium, are also suitable for the photosensitive material. The alkali metals can be in pure form, or in oxide form, such as is used in the preparation of the caesium oxide photocell.

When the light image is projected on the photosensitive surface comprised of the elements 15, the individual elements emit electrons and take on positive charges corresponding in value to the light intensities to which they are subjected. An electrical image of the object, in the form of electrostatic charges over the surface of the screen structure 11, is thereby developed.

For the purpose of utilizing the electrical image referred to in reproducing, at a receiving station, a visible image of the object, the electrostatic charges referred to are neutralized or discharged successively by causing a cathode ray 16 to scan the photosensitive surface, for which

Nov. 26, 1935. V. K. ZWORYKIN 2,021,907

METHOD OF AND APPARATUS FOR PRODUCING IMAGES OF OBJECTS

Filed Nov. 13, 1931

Fig.1.

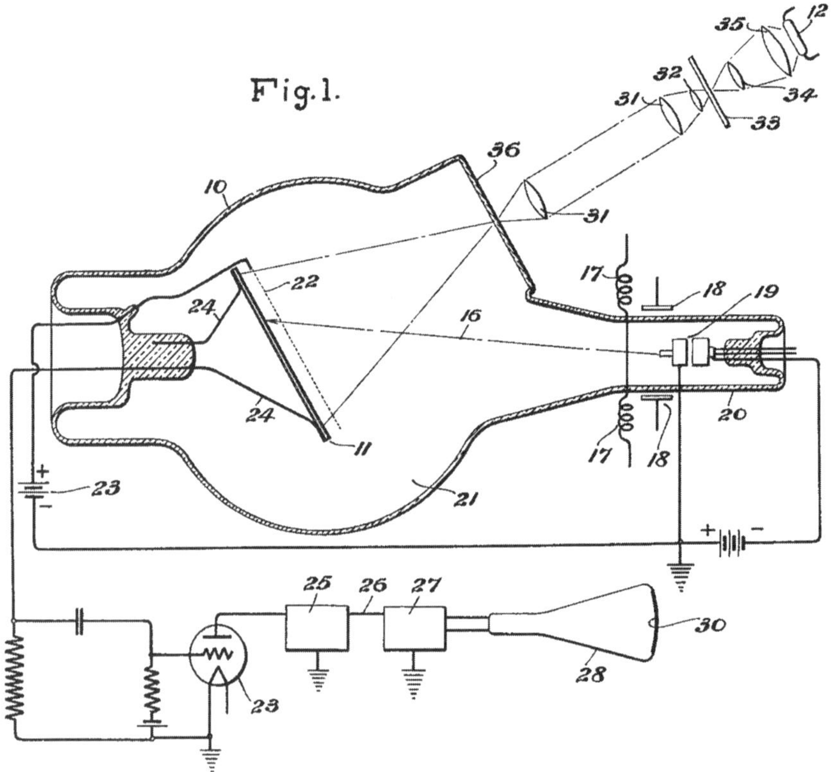

Fig.2.

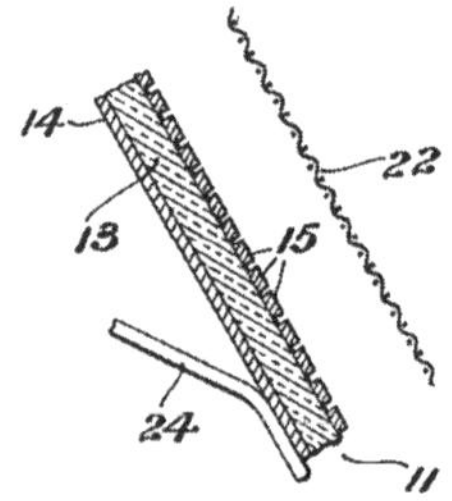

INVENTOR.
Vladimir K. Zworykin,

BY T. R. Goldsborough
HIS ATTORNEY.

purpose the ray is deflected horizontally in the usual manner by electromagnetic coils 17, and is deflected vertically by electrostatic plates 18. The ray 16 of electrons is developed by an electron gun 19 of usual construction supported in an elongated portion 20 extending from the bulb portion 21 of the tube 10.

For the purpose of removing the electrons of the ray 16 which are not required to neutralize the respective positive charges accumulated by the individual elements 15, as well as the electrons emitted by these elements in taking on these charges, a suitable collector 22, in the form of a screen, is supported in proximity to the photosensitive surface of the screen structure 11, and is maintained at a positive potential by a battery 23, leading to ground, as shown.

The successive discharge of the positive charges on the individual elements 15, as explained, causes corresponding current impulses or picture signals to be developed in the grid circuit of a suitable amplifier tube 23, for which purpose the platinum film 14 is connected to this circuit by way of one of the wire supports 24 carrying the screen structure 11.

The amplified picture signals are supplied to a suitable transmitting system 25 and over a connection 26 to a suitable receiving system 27, to which a cathode ray receiving tube 28 is connected.

The receiving system 27, which includes the tube 28, may be of the same general type described in detail in the copending application by Arthur W. Vance, bearing Serial No. 544,959, filed June 17, 1931 and assigned to the Radio Corporation of America. In operation, a moving image of the object is reproduced on the usual fluorescent screen 30 on the large end of the tube 28.

The light image of the object may be projected on the photosensitive surface 15 by any suitable form of microscope, shown diagrammatically as comprising quartz lenses 31 and 32, a quartz slide 33 containing the object, such as bacteria, and other quartz lenses 34 and 35 effective to focus the ultra-violet light from the source 12 upon the object. The tube 10 may be of glass, and provided with a suitable quartz or fused silica window 36 to admit the ultra-violet light, or the entire tube, or at least the entire bulb portion 29, may be made of quartz for this purpose.

From the foregoing it will be seen that by my improved method and system, it is possible to view moving images of objects which heretofore have been invisible on account of their size.

While my present improved system has been described more in connection with the observation of bacteria and other objects whose size is of the order of the wave length of ultra-violet light, the system is equally effective for the purpose of observing phenomena such as gas discharge, electric sparks, secondary emission under ultra-violet and x-ray radiation, and the effect of the so-called Lanard rays on different materials.

I claim as my invention:

1. The method of producing an image of an object whose size is of the order of the wave length of ultra-violet light, which comprises projecting a magnified image of the object illuminated by ultra-violet light on photosensitive structure, scanning said structure with a cathode ray to develop picture signals, and supplying said signals to a television receiving system.
2. In a system for producing an image of an object whose size is of the order of the wave length of ultra-violet light; cathode ray apparatus comprising a tube provided with a quartz window, screen structure disposed in said tube and having a photosensitive surface

facing said window, means for developing a ray of electrons and directing the same at said structure, and means for causing said ray to scan said structure; a source of ultra-violet light, an optical system of quartz interposed between said source and said window and operable to project a magnified ultra-violet light image of the object on the photosensitive surface of said structure, and a television receiving system having its input circuit connected to said apparatus.

3. The method of observing minute objects whose diameters are of the order of the wave length of ultra-violet light which comprises illuminating the objects with ultra-violet light, forming an optically invisible image of the objects, deriving an electrical image from the optically invisible image, utilizing the electrical image to produce a train of electrical impulses, translating the electrical impulses into an optically visible image, and adjusting the image-forming elements until the optically visible image is clear and distinct.

The method of observing the instantaneous condition of bacteria and the like whose diameters are of the order of the wave length of ultra-violet light, which comprises illuminating the objects with ultra-violet light, forming an optically invisible image of the objects, utilizing the optically invisible image to produce an electrical image, utilizing the electrical image to produce a train of electrical impulses, and translating the electrical impulses into an optically visible image, whereby living bacteria may be studied while irradiated with light that will ultimately kill them.

VLADIMIR K. ZWORYKIN

Cathode Ray Tube (Zworykin, 1938)

UNITED STATES PATENT

December 6, 1938 2,139,296

Cathode Ray Tube

VLADIMIR K. ZWORYKIN, PHILADELPHIA, PA., ASSIGNOR TO RADIO CORPORATION OF AMERICA, A CORPORATION OF DELAWARE

Application filed February 23, 1934, Serial No. 712,514, 4 Claims. (CL250-150)

My invention relates to cathode ray tubes and more particularly to tubes of the type now generally utilized in television transmitters.

This application forms an improvement upon the structure set forth by my copending application Serial No. 574,773, filed November 13, 1931, now listed as U.S. Patent #2,021,907 granted November 26, 1935 and assigned to Radio Corporation of America.

A cathode ray transmitting tube as disclosed in my copending application is constituted by an evacuated container having at least one transparent wall, a planar photo-sensitive cathode of the mosaic type, and an electron gun for directing a cathode ray toward and against the cathode.

When such a device is utilized in a television transmitter or in an ultra-microscope of the type disclosed in the aforesaid application, it is necessary to interpose an optical system, constituted by one or more lenses, between the object and the photo-sensitive electrode for the purpose of forming upon the said electrode an optical image of the object. An image formed by an uncorrected or non-aplanatic lens upon a plane surface is more or less distorted by reason of various phenomena, including that of chromatic aberration. To the end that such distortion may be compensated it has heretofore been necessary, in television transmitters, to use compound lens systems wherein different lenses or different portions of a single compound lens are made from glass having differing indices of reflection. Such lenses are expensive and it is, accordingly, the primary object of my invention to provide a cathode ray transmitting tube of such type that it may be utilized with plain or uncorrected lenses.

Another object of my invention is to provide a cathode ray tube of the type described whereby distortion, occasioned in the scanning operation shall be more or less compensated. This type of distortion is introduced by reason of the fact that the cathode ray, at the extreme ends of its vertical and horizontal paths upon the cathode, must traverse a greater distance between the gun and the cathode than at the midpoints of the several paths. This causes some defocusing of the ray at the extreme ends of its paths.

The foregoing objects and other objects ancillary thereto I accomplish, in a preferred embodiment of my invention, by giving to the cathode such configuration that when an image is focused thereon the phenomena of spherical and chromatic aberration are not made manifest. Specifically, I prefer to make the cathode concave toward the lens system, giving it either a spherical curvature or a curvature corresponding to a surface generated by rotating a parabola or a hyperbola around the optical axis of the lens system. In some instances the cathode may be trough shape, that is, it may have a surface equivalent to that produced by the movement of an arc of a circle, a parabola or a hyperbola in a direction at right angles to the optical axis of the system instead of a surface of rotation.

The novel features that I consider characteristic of my invention are set forth with particularity in the appended claims. The invention, however, both as to its organization and its method of operation, together with additional objects and advantages thereof will best be understood from the following description of a specific embodiment when read in connection with the accompanying drawing, in which – Fig. 1 is a highly conventionalized view, partly sectionalized, of my improved cathode ray tube exemplifying its utilization in an ultra-microscope, and Figs. 2 and 3 are views of modified cathodes.

Referring now to the drawing, my improved cathode ray tube is constituted by an evacuated envelope I, having a transparent end portion 3, in which envelope is mounted a target in the form of a photo-electric cathode 5 of the mosaic type and in which is also mounted a cathode ray source 7 commonly designated in the art as a "gun".

In order to simplify the drawing, I have purposely omitted some of the electrical connections through which energy is supplied to the tube and over which pulsations or impulses representing the optical image are led out from the photoelectric cathode. I have also omitted details of the mosaic surface of the cathode as well as the devices such as coils, electrostatic plated, or the like, by which the cathode ray is caused to rapidly scan each element of the photo-electric cathode horizontally and vertically.

In the drawing I have shown a sectional view of the cup-shaped cathode, it of course, being understood that the said electrode is preferably circular.

Dec. 6, 1938. V. K. ZWORYKIN **2,139,296**

CATHODE RAY TUBE

Filed Feb. 23, 1934

Fig. 1.

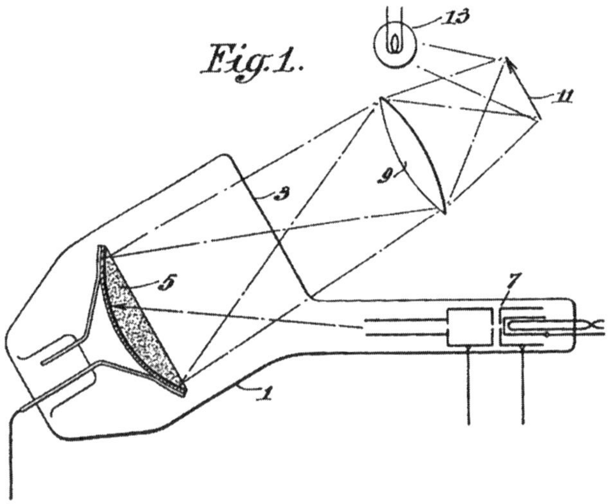

Fig. 2.

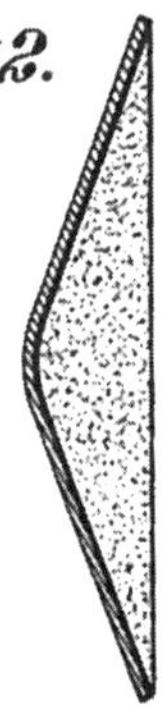

Fig. 3.

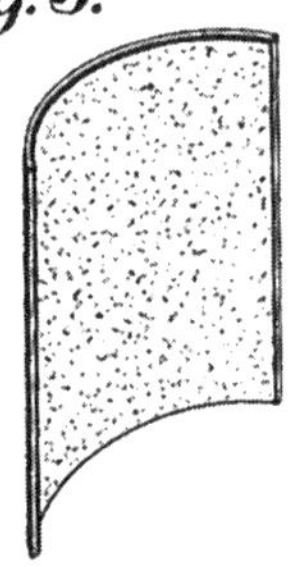

INVENTOR:
Vladimir K. Zworykin,
BY R. R. Goldsborough
ATTORNEY.

When my improved cathode ray tube is utilized, for example, in an ultra-microscope, I interpose a quartz lens or a quartz lens system, exemplified in the drawing by a single lens 9, between the photo-electric cathode and the object 11, which object is illuminated from a source 13 of monochromatic light, preferably ultra violet. By reason of the use of monochromatic light, chromatic aberration is not introduced into the optical image on the cathode by the lens system and the cathode, therefore, need only be curved to an extent sufficient to correct for other aberration.

It is, of course, apparent that the configuration of the photo-sensitive cathode depends upon the curvature of the lens or upon the curvature of the imaginary surface on which the lens system focuses sharply. It is not feasible to illustrate all possible cathode shapes since the curvature thereof is a matter that can best be determined through experiment with the particular lens system it is desired to use.

From a consideration of the foregoing, it will be apparent that I have provided an improved cathode ray tube through the use of which the optical system necessary to produce an optical image for television transmission, or to permit the tube to be utilized as an ultra-microscope, is materially cheapened.

An additional advantage accruing from the concave cathode is the minimization of de-focusing of the cathode ray during the scanning operation. Since the prevention of de-focusing of the cathode ray spot on the cathode is substantially equal in importance to the compensation for spherical or chromatic aberration of the optical image, I sometimes find it advisable to give to the cathode a curvature that is a compromise between the curvature which would give best de-focusing correction and best optical correction. As hereinbefore stated, under such conditions the cathode might be hyperbolic, parabolic, elliptical, and either trough or cup-shape as further exemplified by Figs. 2 and 3.

Although I have chosen a particular embodiment of my invention for purposes of explanation, it is to be understood that I am not restricted thereto since many modifications will be obvious to those skilled in optics and in electron-optics. My invention, therefore, is not to be limited except insofar as is necessitated by the prior art and by the spirit of the appended claims.

I claim as my invention:

In combination, a cathode ray image transmitting tube provided with an electrode in the form of a mosaic photosensitive target and with means for developing a ray of electrons and directing the ray at a surface of said target for scanning the same, a lens system for focusing an optical image of an object on the scanned portion of the surface of the target, such portion of said surface being concave toward the lens system with a curvature corresponding substantially to that of an imaginary spherical surface on which the lens system is able to focus the optical image sharply.

The invention set forth in claim 1 additionally characterized in that the lens system is permeable to monochromatic light.

The invention set forth in claim 1 additionally characterized in that the lens system is permeable to ultra-violet light.

A cathode ray image transmitting tube comprising a container provided with a photosensitive mosaic electrode in the form of a target and with means for developing a ray of electrons and directing the ray at a surface of said photosensitive mosaic target for scanning the same so as to produce image signals, the scanned portion of said surface being concave toward said ray-developing means to prevent distortion due to scanning, and additionally characterized in that the scanned portion of the surface of the photosensitive target is substantially a spherical surface of revolution.

VLADIMIR K. ZWORYKIN

3.5 COMPUTER SYSTEMS

Computing systems have a rich and varied history. Starting with the abacus about 1000–500 BC, the use of tables for multiplication was implemented by John Napier (1550–1617) during the early 1600s. His major inventions are still referred to as Napier's bones, Napier's rods, and table of logarithms (1614). Blaise Pascal initiated the use of mechanical devices and gadgets in 1640. His particular contribution was called the arithmetic machine (for addition/subtraction only), which was introduced in 1642. There is enough evidence (from his notebooks discovered in 1967) that Leonardo da Vinci had also conceived of a mechanical device for arithmetic calculations as early as 1500. In 1625, Wilhelm Schickard also supposedly wrote to a friend about his machine, which could add, subtract, multiply, and divide numbers. This claim, however, is not substantiated by any scientific documents. In 1670, Gottfried von Leinitz enhanced the range of functions of these mechanical devices to be able to perform addition, subtraction, multiplication, division, and computation of the square root of numbers. His device was called the step reckoner. In conjunction with Pascal's *older* devices, these *early* devices are the basis of the older mechanical desktop (mechanical) calculators [3.5.1].

These gadgets, which carried forth the concept that numbers could be generated by routine transversal and cyclic movements of the individual pieces, greatly benefited from the evolving technology. The synergy of computing concepts and the invention of devices,* such as the vacuum tube by Lee de Forest [3.5.2] in 1906, the transistor (see Section 3.6), integrated circuits, and VLSI and ULSI** [3.5.3], brought about the profound computer revolution that occurred during the last century. Many countries have evolved computing facilities over a period of time. Britain, the United States, Germany, and Japan have all made some significant contributions.

Going back to the 19th century in England, Charles Babbage started work on the difference engine [3.5.4] (in 1823) and the analytical engine (in 1834), but the work was never completed. These early mechanical types of gadgets were elaborate and intricate, but there was never any viable demand for the machines. They were used to generate trigonometric and other expansion tables using numerical addition as the only arithmetic operation. In his analytical engine, Babbage proposed to use one set of punched cards [3.5.5] to perform arithmetic operations (e.g., addition/subtraction, multiplication/division) and another set of cards for storing intermediate results and numbers within functions.

French (Blaise Pascal in the 1820s) and American (Dorr Eugene Felt in 1884) inventors devised mechanical devices for numerical calculation [3.5.6]. In the 1890s, Herman Hollerith [3.5.7] used a punch-card system for U.S. census tabulations. The punch-card concept was also used to add and tabulate other data. Hollerith founded the Tabulating Machine Company in 1896, which later became the Computer Tabulating Recording Company (CTRC). In 1924, CTRC changed its name to IBM.

In 1938, in Germany, Konrad Zuse [3.5.8] built the first mechanical computer (Z1) using binary arithmetic, and then in 1941 he built a decimal arithmetic-based general-purpose computer (Z3) system based on electromechanical devices. In the United States,

*Edison's electric lamp and Ambrose Fleming's valve. The Fleming valve had a plate in the tube, and de Forest added a grid to come up with the audion, as it was called then. This change in the basic structure of the tube permitted the device to be used as an amplifies, as a signal detector, and as an oscillator for radio-frequency (RF) transmitters.

**The tracking of prior ideas and inventions to develop new inventions is also evident in the semiconductor era, from the transistor to ULSI (see Section 3.6).

Howard Hathaway Aiken [3.5.9] of Harvard University also used electromechanical devices with programmable sequence-controlled calculations for his Harvard Mark I computer. Built by IBM it, too, used punch-cards and punched paper tape. Being mostly mechanical, it became operational only in 1944, although Aiken first proposed the machine in 1937.

During the early 1940s, the U.S. Army funded the Electronic Numerical Integrator and Calculator (ENIAC) [3.5.10] project under the direction of John Mauchly and J. Presper Eckert at the University of Pennsylvania. The electronic devices replaced most of the mechanical gadgets. The project lasted from 1943 to 1946 and resulted in a massive machine with some 18,000 vacuum tubes. It performed the typical 10 digit, 2 number multiplication in about 3 milliseconds, which was 1,000 times faster than the Harvard Mark I [3.5.9] machine.

This flurry of activity that erupted during the late 1940s resulted in the invention and reinvention of the basic concepts and algorithms for computing. Data processing and logical flow of control were emphasized. Many gadgets were conceived and improvised, built and rebuilt based on the technology of that time. The frenzy lasted for about two decades. The range of possibilities and the promise of new machines were so dramatic that large established institutions, corporations, and universities jumped on the programmable machine bandwagon. Since it is not our intent to suggest that some inventions were better than others, only the basic milestones in the realm of early computing are highlighted.

3.5.1 Stored Program Control (Von Neumann, 1946–1948)

Unlike other inventors, John von Neumann [3.5.11] did not write much nor document his work well [see 3.5.12] in the area of computers, even though his contributions in other areas (e.g., game theory and economic behavior, quantum mechanics, and integrals and functional operators) are well documented. In many cases, his work comes down secondhand with opinions and impressions rather than from his own manuscripts. In addition, his work is shrouded by the fact that a number of other scientists were investigating similar or very closely related topics. In a sense, the credit for the actual mainstream ideas is shared by a number of researchers from academia and from industry.

John Louis von Neumann (1903–1957)

Stored Program Control for Computers

Von Neumann made a series of contributions during the 1940s. The early contributions are related to the Electronic Numerical Integrator and Calculator (ENIAC) [3.5.10]. As a consultant, he was instrumental in proposing the stored program control (SPC) concept for computers, which was developed and used for the ENIAC computer.

In perspective, von Neumann proposed two embodiments for early computers: the Electronic Discrete Variable Computer EDVAC [3.5.13] and the Institute of Advanced Study at Princeton (IAS) machine. EDVAC deployed the SPC concept for controlling the functions within the machine and used binary arithmetic logic circuits within the CPU. It also used a two-level memory system for the computer: a 1,024-word faster memory block and a second block about 20 times this capacity (but slower) as the secondary memory. Some of these concepts are still in use today.

Von Neumann's concept for computers evolved amidst a flurry of innovations in vacuum tube engineering and the inception of semiconductors. A rigorous systems approach to overall design of the IAS computer system was launched rather than assembling different gadgets. Some of the basic gating functions of logic were already implemented. The quantum jump that von Neumann proposed was in the design of the entire system.

The IAS machine earned an even far greater reputation for von Neumann. The IAS computer [3.5.14] has all the features of any recent computer. The most distinguishing and innovative features of this machine are its memory size, word length, operation code (OPC) length, and addressable memory space in binary bits. The machine had 4,096 addressable locations in its memory, and the word length at each location was 40 bits, holding two instructions each with 1 byte for the OPC and a 12-bit address field. The machine also used fixed-bit, binary format (with a 40-bit representation), with the first bit denoting the sign of the number in the next 39 bits. Thus, two instructions or one data word would occupy one memory location, and the computer could execute one instruction after another and keep computing like any other modern machine.

To detail the IAS machine further, the memory size proposed had 4,096 40-bit words and an address space of 12 bits with $2^{12} = 4{,}096$. The way to store a program instruction was in 20-bit half-words, with 8 bits for the operation code and 12 bits for addressing. The data format was proposed to have 40 bits, and it could be moved and shuffled around in the same memory. The size of the address bus, and thus the number of bits (in the address register), were both proposed to have 12 bits. Likewise, the size of the instruction bus, and thus the number of bits (in the instruction register), were both proposed to have 8 bits. The memory word size, and thus the size of the data register and the data bus, each had 40 bits. As a matter of the detailed design, the IAS machine was deemed to have an instruction buffer register, which is seen as a necessity since the fetch cycle of the IAS machine would bring two instruction words (20 bits each) or data bit (40 bits long). A great many implications of this design can be derived based on the hardware parameters, but they are peripheral to the basic fabric of concepts systematically embedded in the design of the IAS machine. This extent of the detailed design provided some boundary conditions that helped users solve realistic problems during the 1950s and 1960s.

The CPU architecture was equally well designed with two sections: the program control unit (PCU) and the arithmetic logic unit (ALU). The PCU monitors and facilitates the flow of the program and the execution of the OPC. The ALU facilitates the flow of data through its data register and the execution of the arithmetic and logical instructions.

The choice of instruction sets (i.e., the collection and combination of the operation codes) for the IAS machine is optimally well selected. Five groups of instructions are used for the transfer of data, unconditional branching, conditional branching, and arithmetic and address modifications. Respectively, there are seven, two, two, eight, and two actual instructions in the five groups of instructions.

Von Neumann's most noteworthy contribution to the evolution of computing was the use of common memory for the storage of programs and data and the use of memory addresses and locations, while the earlier machines were using separate memories to hold programs and data. In addition, the spark of genius is tightly coupled with the rigor of the invention. Both are manifest in the extremely detailed design of the IAS computer [3.5.14].

The IAS computer was not the only possible implementation. Earlier electronic analog computing was attempted in 1938. In 1946, the era of digital computing started with the ENIAC [3.5.10] machine. In 1950 [3.5.13], the EDVAC was developed, followed by IBM's 701 calculator, based on electronic tubes, magnetic disk/drum, and tape [3.5.15]. This was followed by Bell Laboratories transitorized digital computer (TRADIC) in 1954 [3.5.16], using 800 transistors with printed circuits and magnetic core memories [3.5.16]. Software developments during this era were equally impressive with assemblers, FORTRAN compilers in 1956 [3.5.17], and ALGOL compilers [3.5.18] in 1960. Digital computing was firmly established at this time.

REFERENCES (C-3.5)

[3.5.1] See http://www.maxmon.com/1670ad.htm for a brief review of the contributions of Napier, Pascal, Leibnitz, etc., to the era of mechanical devices. (Accessed on March 28, 2004.)

[3.5.2] Levine, Israel E., *Electronics Pioneer: Lee de Forest.* Julian Messner, New York, 1964.

[3.5.3] Elsea, Art, The Impact of Lasers on Semiconductor Processing: A Strategic Report Analyzing the VLSI/ULSI Impact on Semiconductor Processing Equipment. Electronic Trend Publications, Cupertino, CA, 1985.

[3.5.4] Lindgren, Michael, Glory and Failure: The Difference Engines of Johann Muller, Charles Babbage and Georg and Edvard Sheutz, 2d ed. MIT Press, Cambridge, MA, 1990.

[3.5.5] Hyman, Anthony, *Charles Babbage.* Princeton University Press, New Haven, CT, 1982. Also see Swade, Doron, and Babbage, Charles, *The Difference Engine: Charles Babbage and the Quest to Build the First Computer, 1st* American ed. Viking Press, New York, 2001.

[3.5.6] Meehan, J. R., *How to Use the Calculator and Comptometer,* 5th ed. McGraw Hill Text, New York, 1964. Also see "Biography of a Machine," in *Comptometer,* on the Internet at http://www2.cruzio.com/~vagabond/CompHome.html. (Accessed on March 28, 2004.)

[3.5.7] Austrian, Geoffery, *Herman Hollerith: Forgotten Giant of Information Processing.* Columbia University Press, New York, 1982.

[3.5.8] Zuse, Konrard, *The Computer, My Life.* Springer Verlag, New York, 1993.

[3.5.9] Cohen, I. B., Welch, G. W., and Campbell R. V. D. (Editors.), *Makin' Numbers: Howard Aiken and the Computer.* MIT Press, Cambridge, MA, 1999. Also see "Biography of a Machine," in *Comptometer,* on the Internet at http://www2.cruzio.com/~vagabond/ComptHome.html. (Accessed on March 28, 2004.)

[3.5.10] McCartney, Scott, ENIAC: The Triumphs and Tragedies of the World's First Computer. Walker and Co., New York, 1999.

[3.5.11] MacRae, Norman, John von Neumann: The Scientific Genius Who Pioneered the Modern Computer, Game Theory, Nuclear Deterrence and Much More, 2d ed. American Mathematical Society, Providence, RI, 2000.

[3.5.12] von Neumann, John, *The Computer and the Brain.* Yale University Press, New Haven, CT, 2000.

[3.5.13] Stern, Nancy B., From ENIAC to UNIVAC: An Appraisal of the Eckert-Mauchly Computers. Digital Press, Bedford, MA, 1981.

[3.5.14] Burks, A. W., Goldstine, H. H., and von Neumann, J., "Preliminary Discussion of the Logical Design of a Electronic Computing Instrument." *Report to the U.S. Army Ordnance Department,* 1946, on the Internet at http://www.informatik.uni-stuttgart.de /ipvr/ise/Skripte/Prof_B_Skripte/CSO_Skript/CSO_01.pd. (Accessed on March 28, 2004.) Also see Estrin, Gerald, "The Electronic Computer at the Institute for Advanced Study." *Mathematical Tables and Other Aids to Computation (MTAC)*, Vol. 7, No. 42, April 1953, pp. 108–114.

[3.5.15] Hayes, J. P., *Computer Architecture and Organization, 2d ed.* McGraw-Hill, New York, 1988. Also see Stone, H. S., et al. (Editor), *Introduction to Computer Architecture, 2d ed.* Science Research Associates, Chicago, 1980.

[3.5.16] Koffman, Elliot, and Friedman, Frank L., *Fortran*, 5th ed. Addison-Wesley, Reading, MA, 1997. Also see IBM Fortran Manuals from the 1950s.

[3.5.17] Bell Laboratories, *TRADIC: The First Transistor Calculator*, Bell Laboratories, Internal Communications, 1955. Also see *1955: TRADIC—The First Fully Transistorized Computer*, on the Internet at http://www. ced-magic.com/history/tradic-transistorized.html. (Accessed on April 3, 2004.)

[3.5.18] Lindsey, C. H., and van der Meulen, S. G., *Informal Introduction to ALGOL 68*, revised ed. North-Holland Pub. Co., London, 1977.

3.6 SEMICONDUCTOR ERA

During the mid-1940s, deliberate experiments were initiated at Bell Laboratories [3.6.1] to duplicate the migration of electrons and holes in solid-state materials. The migration of electrons was observed and documented in the gaseous plasma of electron tubes [Lee de Forest (1873–1961)] [3.6.2] and is in fact responsible for the rectification and amplification of electrical signals. The devices were already being used and deployed. The spark of genius in starting the revolution lies in the belief that such duplication is feasible within semiconductors at room temperature. These effects are well documented and have been proven to be consider valid, scientific facts. These seminal patents are presented in this section.

The inventors behind these patents are reasonably exhaustive about the use and the application of their inventions in a realistic and practical sense. However, the coverage and the thoroughness are even more evident when compared with the earlier patents presented in Sections 3.1, 3.2, and 3.3. These differences are consistent with our observations listed in Section 4.2. Especially commendable are the observations of Shockley (Section 3.6.1) about the possible use of these devices for solid-state power conversion and power-frequency conversion, in addition to their use for amplification of signals and modulation of information-bearing signals. Their use in performing logic and gating functions is an aspect that has seriously influenced the control systems and computer industries.

In the scientific community, Bardeen's spectacular contributions [3.6.3] are recognized by his two Nobel Prizes (1956 and 1972). In addition, the detailed drawings in Bardeen's and Brattain's patent (see Section 3.6.2) present the mathematical basis for their reasoning and a very sound analytical approach to inventing in the 1940s. This approach for being precise, mathematical, and convincing is consistent with our observation (in Section 4.2) of the newer patents. In essence, the collaborative role of the three scientists (Shockley, Bardeen, and Brattain) has led to the revolutionary breakthrough that resulted from the transistor. Shockley's* insight into the device's physics appears to be as dominant as the symbolism and the derived mathematics of Bardeen (compare Shockley's patent with Bardeen's and Brattain's patent).

We have retained as much of these two patents as possible in order to see the inventions as the inventors saw them and thus to gain insight into their thinking. The three inventors worked together closely and have documented their contributions as patents.

*These two patents provide a window into the creative minds of the three scientists, thus alluding to the fact that synergy of thought (see Sections 3.6 and 3.7) leads to an escalation of the mental processes. In the verses of Omar Khayyam, the question still remains, "Who is the Potter and who is the Pot?" or who created the transistor and who enhanced it?

REFERENCES (C-3.6)

[3.6.1] "Transistorized!" on the Internet at http://www.pbs.org/transistor/album1/index.html. This website details the events that triggered the deliberate search for new patents for the final push into the semiconductor era. (Accessed on January 3, 2003, and April 3, 2004.)

[3.6.2] Levine, Israel E., *Electronics Pioneer: Lee de Forest.* Julian Messner, New York, 1964. Also see *The Complete Lee de Forest*, on the Internet at http://www.leedeforest.org/inventor.html. (Accessed on April 6, 2004.)

[3.6.3] Hoddeson, Lillian, and Vicki, Daitch, *True Genius: The Life and Science of John Bardeen—*The Only winner of Two Nobel Prizes in Physics. Joseph Henry Press, Washington, DC, 2002.

Semiconductor Era

William Bradford Shockley (1910–1989)

Transistor: Semiconductor Amplifier (Shockley, 1950)

UNITED STATES PATENT

April 4, 1950 2,502,488

Semiconductor Amplifier

WILLIAM SHOCKLEY, MADISON, N.J., ASSIGNOR TO BELL TELEPHONE LABORATORIES, INCORPORATED, NEW YORK, N.Y., A CORPORATION OF NEW YORK
Application filed September 24, 1948, Serial No. 50,895. 9 Claims. (CL179-171)

This invention relates to means for and methods of translating or controlling electrical signals and more particularly to circuit elements utilizing semiconductors and to systems including such elements.

One general object of this invention is to provide new and improved means for and methods of translating and controlling, for example, amplifying, generating, modulating, etc., electric signals.

Another general object of this invention is to enable the efficient, expeditious and economic translation or control of electrical energy.

In accordance with one broad feature of this invention, translation and control of electric signals is effected by alteration or regulation of the conduction characteristics of a semiconductive body. More specifically, in accordance with one broad feature of this invention, such translation and control is effected by control of the characteristics, for example the impedance, of a layer or barrier intermediate the two portions of a semiconductive body in such a manner as to alter advantageously the flow of current between the two portions.

One feature of this invention relates to the control of current flow through a semiconductive body by means of carriers of charge of opposite sign to the carriers which convey the current through the body.

April 4, 1950 W. SHOCKLEY **2,502,488**

SEMICONDUCTOR AMPLIFIER

Filed Sept. 24, 1948

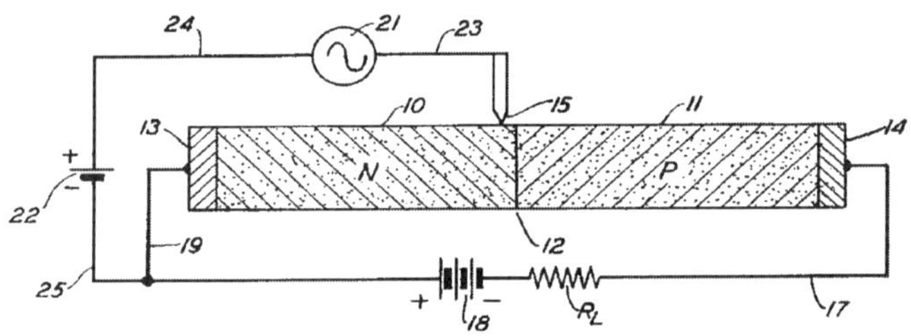

INVENTOR
W. SHOCKLEY
BY
ATTORNEY

Another feature of this invention relates to a body of semiconductive material, means for making electrical connection respectively to two portions of said body, means for making a third electrical connection to another portion of the body intermediate the first two portions, and circuit means including power sources whereby the influence of the third connection may be made to control the flow of current between the other connections.

A further feature of this invention resides in a body of semiconductive material comprising two zones of material of opposite conductivity type separated by a barrier, means for making external electrical connections respectively to each zone and means for making a third connection to one of the zones adjacent the barrier for controlling the flow of current between the other two connections.

Another feature of this invention involves a semiconductive body which may be used for voltage and power amplification when associated with means for introducing mobile carriers of charge to the body at relatively low voltage and extracting like carriers at a relatively high voltage.

Other objects and features of this invention will appear more fully and clearly from the following description of illustrative embodiments thereof taken in connection with the appended drawings in which:

The single figure shows in section a device suitable for practicing the invention, in connection with an appropriate circuit.

The device of this invention is in some respects like devices illustrated and described in the application of W. Shockley Serial No. 35,423 filed June 26, 1948. Reference is made to said application for a general description and discussion of pertinent principles, suitable materials and certain terms which may be employed in the description and claiming of this invention.

The terms N-type and P-type are applied to semiconductive materials which tend to pass current easily when the material is respectively negative or positive with respect to a conductive contact thereto and with difficulty when the reverse is true, and which also have consistent Hall and thermoelectric effects.

The conductivity type (either N- or P-type) of a semiconductor may be determined in one way by minute quantities of significant impurities as discussed in the previously noted application Serial No. 35,423. Energy relations within the semiconductor may also determine the conductivity type as more fully discussed in the application of J. Bardeen and W. H. Brattain Serial No. 33,466 filed June 17, 1948.

The term "barrier" or "electrical barrier" used in the description and discussion of the device of this invention is applied to a high resistance interfacial condition between contacting semiconductors of respectively opposite conductivity type or between a semiconductor and a metallic conductor whereby current passes with relative ease in one direction and with relative difficulty in the other.

The device shown in the figure comprises a body or block of semiconductive material, for example germanium, containing significant impurities. The block comprises two zones 10 and 11 respectively of N- and P-type materials separated by the barrier 12. The opposite ends of the block are provided with relatively large area connections 13 and 14 respectively, which may be metallic coatings such as solder, vapor deposited metal coatings, electroplated metal coatings or the like.

Means for making connection to one of the zones closely adjacent to the barrier 12 may comprise a point contact 15 bearing on the surface of the block close to the barrier. The contact point may be of tungsten, Phosphor bronze or the like.

Conductor 17 leads from connection 14 to a load R_L and thence through a power source, such as a battery 18, and back through conductor 19 to the body at connection 13. A source 21 of signal voltage and a biasing source 22 are connected from the contact 15 adjacent the barrier to connection 13 by conductors 23, 24, 25 and 19. With N and P zones as shown, the negative pole of source 18 is connected to the P zone and the positive pole to the N zone. The positive pole of biasing source 22 is connected to the contact 15 in order that current may be easily introduced to the N-type material at this point.

With the device and circuit as shown the connections 13, 14 and 15 may be called the base, collector and emitter respectively in accordance with nomenclature which has been applied to devices of this type. The circuit between the base 13 and the emitter 15 then is the input circuit and that between the base 13 and the collector 14 is the output circuit. The battery 18 in the output circuit may have a voltage of the order of 10 to 100 volts and the bias 22 in the input circuit may be of the order of 0:1 to 1:0 volt.

The current introduced to the N zone at the point contact 15 comprises "holes" as more fully discussed in application Serial No. 35,423 to which reference has been made heretofore. The "holes" injected from the point 15 will diffuse in the N-type material and if they approach the N-P junction they will be drawn across into the P region by the field existing there. Under the conditions of operating bias, there is a steady small current across the junction and this enhances the natural field so that the "holes" entering the N region will tend to be pulled toward the junction by the biasing field produced by the battery. For this reason if the "holes" are injected within a few mils of the junction, they will substantially all be drawn across the barrier into the P region. Consequently, substantially all of the injected current which consists of "holes" will flow into the P region.

Since the point 15 is operated in the forward direction or direction of easy current flow, it has relatively low input impedance. On the other hand, the N-P junction has relatively high impedance so that there is power gain across the device in the usual way as discussed in the previous application Serial No. 35,423.

If a contact such as 15 is made so that it will inject electrons, it should be placed on the P side of the barrier and the device operated with the input and output circuits interchanged and with all of the polarities reversed. Other semiconductive materials than germanium may be used so long as high impedance N-P junctions are obtained and so long as the point injects carriers of the sign not normally present in the region to which it makes contact.

It is to be understood that the specific embodiments of the invention shown and described are but illustrative and that various modifications may be made therein without departing from the scope and spirit of this invention.

What is claimed is: A translating device comprising a body of semiconductive material having zones of opposite conductivity type separated by a barrier, an ohmic connection to each zone remote from the barrier, and a rectifying contact on one zone closely adjacent to the barrier.

A translating device comprising a body of germanium material having zones of opposite conductivity type separated by a barrier, an ohmic connection to each zone remote from the barrier, and a metallic rectifying point on one of said zones closely adjacent to the barrier.

A translating device comprising a body of semiconductive material having zones of N- and P-type material separated by a barrier, an ohmic connection to each zone remote from the barrier, and a metallic rectifying point on the N-type zone closely adjacent to the barrier.

A translating device comprising a body of germanium material having zones of N- and P-type material separated by a barrier, an ohmic connection to each zone remote from the barrier, and a metallic rectifying point on the P-type zone closely adjacent to the barrier.

An amplifier comprising a body of semiconductive material having zones of opposite conductivity type separated by a barrier, a base connection to one zone remote from the barrier, means for introducing current to said one zone including a rectifying contact closely adjacent to the barrier and a source of relatively low voltage between the base connection and the rectifying contact, said voltage poled in the direction of easy current flow into said one zone, and means for extracting current from the other zone including an ohmic connection and a source of relatively high voltage between the base connection and the ohmic connection, said voltage poled in the direction of difficult current flow through said barrier.

An amplifier comprising a body of germanium material having zones of N- and P-type material separated by a barrier, an input circuit including a source of relatively low voltage, a rectifying point on the surface of the N-type zone closely adjacent the barrier, and an ohmic base connection to the same zone remote from the barrier, and an output circuit including a source of relatively high voltage, an ohmic connection to the P-type zone remote from the barrier and the base connection.

A translating device comprising a body of semiconductive material having zones of opposite conductivity type separated by a barrier, means for introducing current at low impedance into one of said zones including a metallic rectifier point on the surface of said one zone closely adjacent said barrier, and means for extracting current from the other zone comprising an ohmic connection to said other zone remote from said barrier.

A circuit element comprising a slab of semiconductive material having zones of opposite conductivity type separated by a barrier intermediate its end, a large area ohmic connection to each end of the slab, and a metallic point contact on the surface of one of said zones closely adjacent to said barrier.

A translating device comprising a body of semiconductive material having zones of opposite conductivity type separated by a barrier, means for introducing at low impedance to one of said zones current carriers of opposite sign to those normally found in said zone including a metallic rectifier point on the surface of said zone closely adjacent said barrier, and means for extracting current from said other zone comprising an ohmic connection to said zone remote from said barrier.

WILLIAM SHOCKLEY

John Bardeen (1908–1991)

Walter Hauser Brattain (1902–1987)

The Transistor

3.6.2 Transistor: Three-Electrode Circuit Element (Bardeen, Brattain, 1950)

UNITED STATES PATENT

October 3, 1950 2,524,035

Three-Electrode Circuit Element Utilizing Semiconductive Materials

JOHN BARDEEN, SUMMIT, AND WALTER H. BRATTAIN, MORRISTOWN, N.J., ASSIGNORS TO BELL TELEPHONE LABORATORIES, INCORPORATED, NEW YORK, N.Y., A CORPORATION OF NEW YORK
Application filed June 17, 1948, Serial No. 33,466, 40 Claims. (CL179-171)

This application is a continuation-in-part of application Serial No. 11,156, filed February 26, 1948, and thereafter abandoned.

This invention relates to a novel method of and means for translating electrical variations for such purposes as amplification, wave generation, and the like.

The principal object of the invention is to amplify or otherwise translate electric signals or variations by use of compact, simple, and rugged apparatus of novel type.

Another object is to provide a circuit element for use as an amplifier or the like which does not require a heated thermionic cathode for its operation, and which therefore is immediately operative when turned on. A related object is to provide such a circuit element which requires no evacuated or gas-filled envelope.

It has also been proposed to pass a current from end to end of a strip of homogenous isotropic semiconductive material and, by the application of a strong transverse electrostatic field, to control the resistance of the strip, and hence the current through it.

Attempts have been made in the past to convert solid rectifiers utilizing selenium, copper sulfide, or other semi-conductive materials into amplifiers by the direct expedient of embedding a grid-like electrode in a dielectric layer disposed between the cathode and the anode of the rectifier.

The grid is supposed, by exerting an electric force at the surface of the cathode, to modify its emission and so alter the cathode-anode current. As a practical matter it is impossible to embed a grid in a layer which is so thick as to insulate the grid from the other electrodes and yet so thin as to permit current to flow between them.

So far as is known, all of such past devices are beyond human skill to fabricate with the fineness necessary to produce amplification. In any event they do not appear to have been commercially successful.

It is well known that in semiconductors there are two types of carriers of electricity which differ in the signs of the effective mobile charges. The negative carriers are excess electrons which are free to move, and are denoted by the term conduction electrons or simply electrons. The positive carriers are missing or defect "electrons," and are denoted by the term "holes." The conductivity of a semiconductor is called excess or defect, or N or P type, depending on whether the mobile charges normally present in excess in the material under equilibrium conditions are electrons (Negative carriers) or holes (Positive carriers).

When a metal electrode is placed in contact with a semiconductor and a potential difference is applied across the junction, the magnitude of the current which flows often depends on the sign as well as on the magnitude of the potential. A junction of this sort is called a rectifying contact. If the contact is made to an N-type semiconductor, the direction of easy current flow is that in which the semiconductor is negative with respect to the electrode. With a P-type semiconductor, the direction of easy flow is that in which the semiconductor is positive. A similar rectifying contact exists at the boundary between two semiconductors of opposite conductivity types.

This boundary may separate two semiconductor materials of different constitutions, or it may separate zones or regions, within a body of semiconductor material which is chemically and stoichiometrically uniform, which exhibit different conductivity characteristics.

The present invention in one form utilizes a block of semiconductor material on which three electrodes are placed. One of these, termed the collector, makes rectifier contact with the body of the block. The other, termed the emitter, preferably makes rectifier

Oct. 3, 1950 | J. BARDEEN ET AL | 2,524,035

THREE-ELECTRODE CIRCUIT ELEMENT UTILIZING SEMICONDUCTIVE MATERIALS

Filed June 17, 1948 | 3 Sheets-Sheet 1

FIG. 1

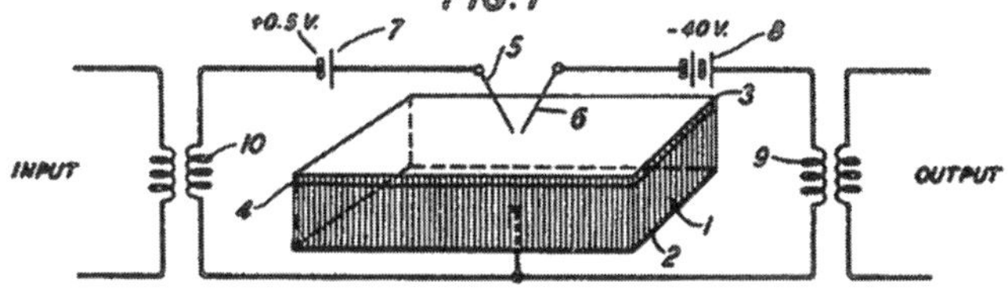

FIG. 1A

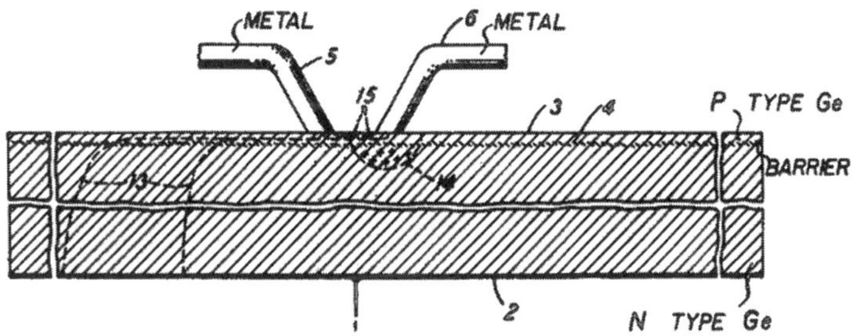

FIG. 2

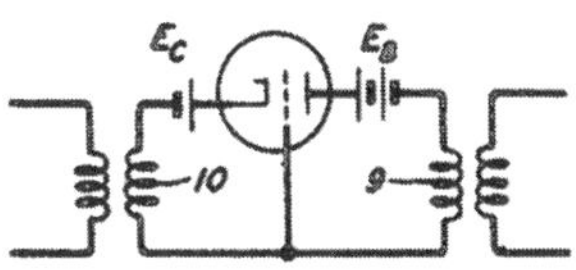

FIG. 10

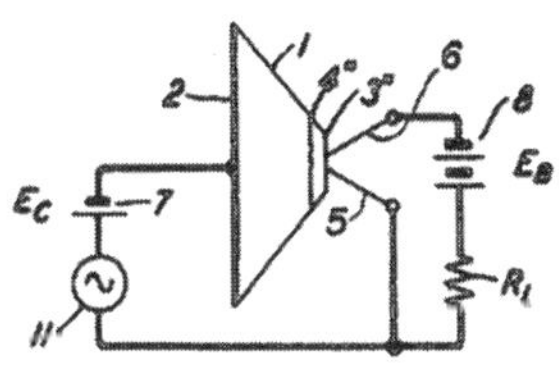

FIG. 11

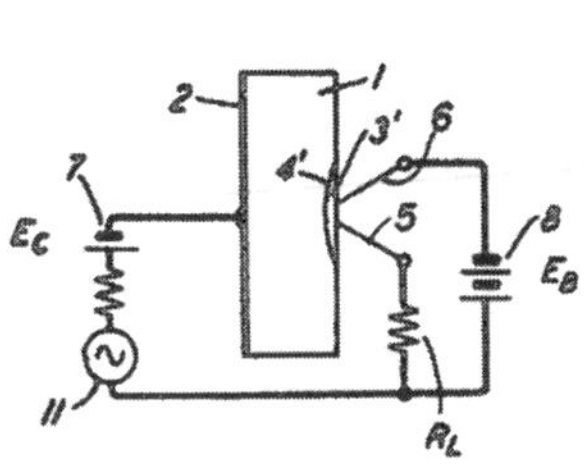

FIG. 12

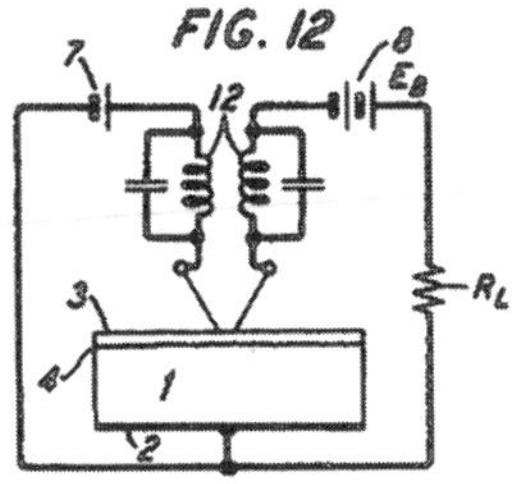

INVENTORS: J. BARDEEN
W. H. BRATTAIN
BY Harry C. Hart
ATTORNEY

Oct. 3, 1950 J. BARDEEN ET AL 2,524,035

THREE-ELECTRODE CIRCUIT ELEMENT UTILIZING SEMICONDUCTIVE MATERIALS

Filed June 17, 1948 3 Sheets-Sheet 2

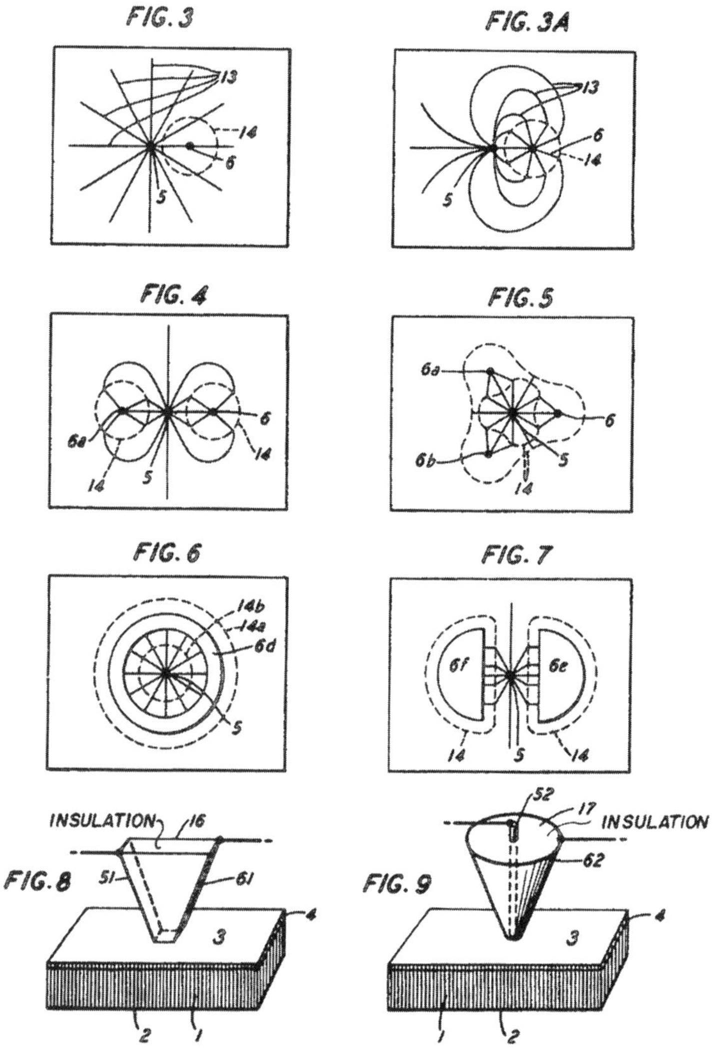

INVENTORS: J. BARDEEN
W. H. BRATTAIN
BY Harry C. Hart
ATTORNEY

Oct. 3, 1950 J. BARDEEN ET AL 2,524,035
THREE-ELECTRODE CIRCUIT ELEMENT UTILIZING
SEMICONDUCTIVE MATERIALS
Filed June 17, 1948 3 Sheets-Sheet 3

FIG. 13

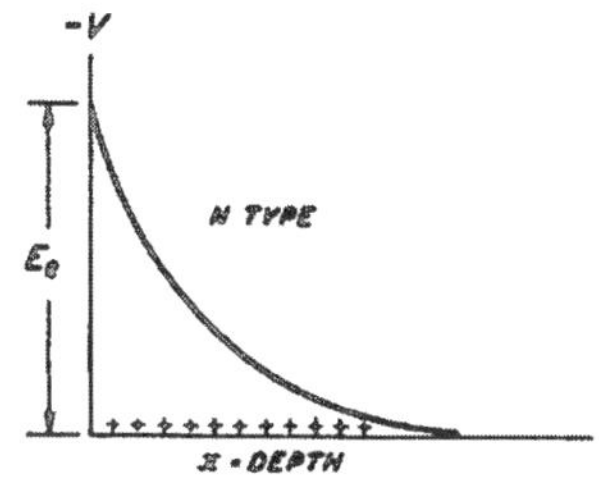

FIG. 14

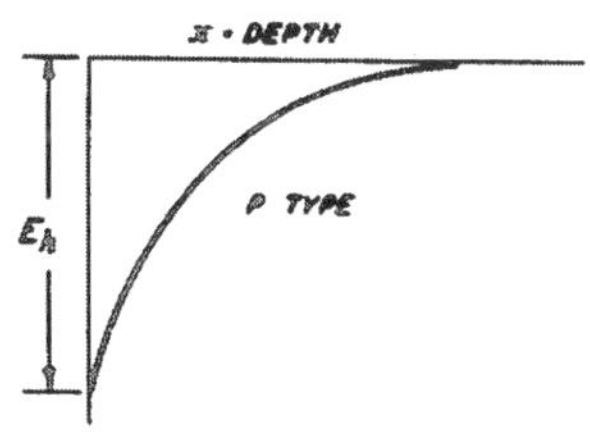

FIG. 15

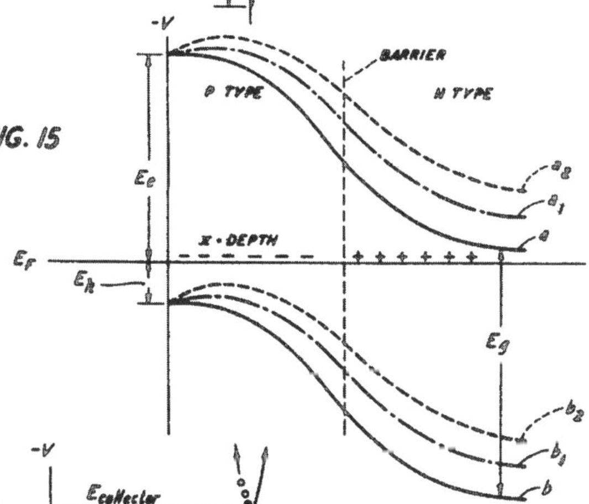

FIG. 16

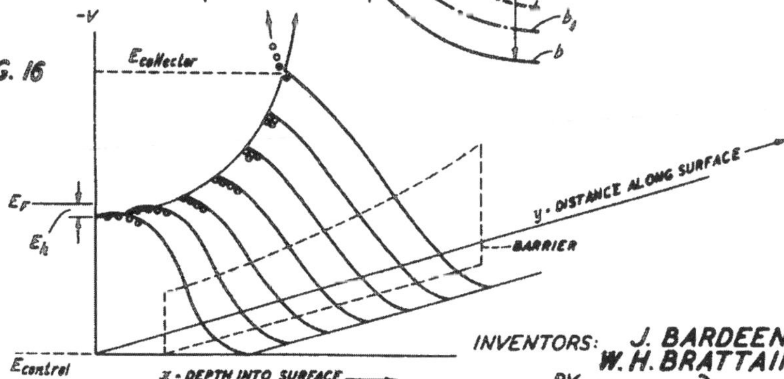

INVENTORS: J. BARDEEN
W. H. BRATTAIN
BY Harry C. Hart
ATTORNEY

contact with the body of the block also. The third electrode, which may be designated the base electrode, preferably makes a low resistance contact with the body of the block. When operated as an amplifier, the emitter is normally biased in the direction of easy current flow with respect to the body of the semiconductor block. The nature of the emitter electrode and of that portion of the semiconductor which is in the immediate neighborhood of the electrode contact is such that a substantial fraction of the current from this electrode is carried by charges whose signs are opposite to the signs of the mobile charges normally in excess in the body of the semiconductor. The collector is biased in the reverse, or high resistance direction relative to the body of the semiconductor. In the absence of the emitter, the current to the collector flows exclusively from the base electrode and is impeded by the high resistance of this collector contact. The sign of the collector bias potential is such as to attract the carriers of opposite sign which come from the emitter. The collector is so disposed in relation to the emitter that a large fraction of the emitter current enters the collector. The fraction depends in part on the geometrical disposition of the electrodes and in part on the bias potentials applied. As the emitter is biased in the direction of easy flow, the emitter current is sensitive to small changes in potential between the emitter and the body of the semiconductor, or between the emitter and the base electrode. Application of a small voltage variation between the base electrode and emitter causes a relatively large change in the current entering the semiconductor from the emitter, and a correspondingly large change in the current to the collector. One effect of the change in emitter current is to modify the total current flowing to the collector, so that the overall change in collector current may be greater than the change in the emitter current. The collector circuit may contain a load of high impedance matched to the internal impedance of the collector, which, because of the high resistance rectifier contact of the collector, is high. As a result, voltage amplification, current amplification, and power amplification of the input signal are obtained.

In one form, the device utilizes a block of semiconductor material of which the main body is of one conductivity type while a very thin surface layer or film is of opposite conductivity type. The surface layer is separated from the body by a high resistance rectifying barrier. The emitter and collector electrodes make contact with this surface layer sufficiently close together for mutual influence in the manner described above. The base electrode makes a low resistance contact with the body of the semiconductor. When suitable bias potentials are applied to the various electrodes, a current flows from the emitter into the thin layer. Owing to the conductivity of the layer and to the nature of the barrier, this current tends to flow laterally in the thin layer, rather than following the most direct path across the barrier to the base electrode. This current is composed of carriers whose signs are opposite to the signs of the mobile charges normally in excess in the body of the semiconductor. In other words, when there is a thin layer of opposite conductivity type immediately under the emitter electrode, the current flowing into the block in the direction of easy flow consists largely of carriers of opposite sign to those of the mobile charges normally present in excess in the body of the block; and the presence of these carriers increases the conductivity of the block. The bias voltage on the collector which, as stated above, is biased in the reverse or high resistance direction relative to the block, produces a strong electrostatic field in a region surrounding the collector so that the current from the emitter which enters this region is drawn in to the collector. Thus, the collector current, and hence the conductance of the unit as a whole, are increased. The size of the region in which this strong field exists is comparatively insensitive to variations in the collector potential so that the impedance of the collector circuit is high. On the other hand, the current from the emitter to the layer is extremely sensitive to variations of the emitter potential, so that the impedance of the emitter circuit is low.

It is a feature of the invention that the input and output impedances of the device are controlled by choice and treatment of the semiconductor material body and of its surface, as well as by choice of the bias potentials of the electrodes.

From the standpoint of its external behavior and uses, the device of the invention resembles a vacuum tube triode; and while the electrodes are designated emitter, collector and base electrode, respectively, they may be externally interconnected in the various ways which have become recognized as appropriate for triodes, such as the conventional, the "grounded grid," the "grounded plate" or "cathode follower", and the like. Indeed, the discovery on which the invention is based was first made with circuit connections which are extremely similar to the so-called "grounded grid" vacuum tube connections. However, the analogies among the circuits is, of course, no better than the analogy between emitter and cathode, base electrode and grid, collector and anode.

By feeding back a portion of the output voltage in proper phase to the input terminals, the device may be caused to oscillate at a frequency determined by its external circuit elements, and, among other tests, power amplification was confirmed by a feedback connection which caused it to oscillate.

It has been found that the performance of the device is expressed, to a good approximation, by the following functional relations:

$$I_e = f(V_e + R_F I_c) \tag{1}$$

$$I_c = I_c^0(V_c) + \alpha I_e \tag{1a}$$

where

I_e = emitter current

I_c = collector current

$I_c^0(V_c)$ = collector current with emitter disconnected

V_e = voltage of emitter electrode measured with respect to the base electrode

V_c = voltage of collector electrode measured with respect to the base electrode

R_F = an equivalent resistance independent of bias

α = a numerical factor which depends on the bias voltages $f(V_e)$gives the relation between emitter current and emitter voltage with the collector circuit open.

The interpretation of the foregoing Equation 1 is that the collector current lowers the potential of the surface of the block in the vicinity of the emitter relative to the base electrode by an amount $R_F I_c$, and thus increases the effective bias voltage on the emitter by the same amount. The term $R_F I_c$ thus represents positive feedback.

The invention will be fully apprehended from the following detailed description of one embodiment thereof, taken in connection with the appended drawings, in which:

Fig. 1 is a schematic diagram, partly in perspective, showing a preferred embodiment of the invention;

Fig. 1a is a cross-section of a part of Fig. 1 to a greatly enlarged scale;

Fig. 2 is the equivalent vacuum tube schematic circuit of Fig. 1;

Fig. 3 is a plan view of the block of Fig. 1, showing the disposition of the electrodes;

Fig. 3a is like Fig. 3 but shows the influence of the collector in modifying the emitter current;

Figs. 4, 5, 6 and 7 show electrode dispositions alternative to those of Fig. 1;

Figs. 8 and 9 show electrode structures alternative to those of Fig. 1.

The materials with which the invention deals are those semiconductors whose electrical characteristics are largely dependent on the inclusion therein of very small amounts of significant impurities. The expression "significant impurities" is here used to denote those impurities which affect the electrical characteristics of the material such as its resistivity, photosensitivity, rectification, and the like, as distinguished from other impurities which have no apparent effect on these characteristics. The term "impurities" is intended to include intentionally added constituents as well as any which may be included in the basic material as found in nature or as commercially available. Germanium is such a material which, along with some representative impurities, will furnish an illustrative example for explanation of the present invention. Silicon is another such material. In the case of semiconductors which are chemical compounds such as cuprous oxide (Cu_2O) or silicon carbide (SiC), deviations from stoichiometric composition may constitute significant impurities.

Small amounts, i.e., up to 0.1 per cent of impurities, generally of higher valency than the basic semiconductor material, e.g., phosphorus in silicon, antimony and arsenic in germanium, are termed "donor" impurities because they contribute to the conductivity of the basic material by donating electrons to an unfilled "conduction energy band" in the basic material. In such case the donated negative electrons constitute the carriers of current and the material and its conductivity are said to be of the N-type. Similar small amounts of impurities, generally of lower valency than the basic material, e.g., boron in silicon or aluminum in germanium, are termed "acceptor" impurities because they contribute to the conductivity by "accepting" electrons from the atoms of the basic material in the filled band. Such an acceptance leaves a gap or "hole" in the filled band. By interchange of the borrowed electrons from atom to atom, these positive "holes" effectively move about and constitute the carriers of current, and the material and its conductivity are said to be of the P-type.

Fig. 1 shows a block 1 of germanium which has been prepared in the foregoing manner, and Fig. 1a shows the central part of the block 1 in section and to an enlarged scale. Referring to Figs. 1 and 1a together, the lower part of the block 1, whose surface is plated with a metal film 2 serving as the base electrode, is known to be of N-type and Fig. 1a shows the central part of the block 1 in section and to an enlarged scale. Referring to Figs. 1 and 1a together, the lower part of the block I, whose surface is plated with a metal film 2 serving as the base electrode, is known to be of N-type. The thin layer 3 at the upper surface manifests P-type conductivity in which case, as is well known, the boundary 4 separating this P-type layer from the N-type material of the main body of the block behaves like a high resistance rectifying barrier. A first electrode 5, denoted the emitter, makes contact with the upper face of the block, i.e., with the P-type layer 3, preferably somewhere near its center, or at lest several point diameters removed from the nearest edge. This contact is preferably of the rectifier type with respect to the body of the block 1. It may comprise a bent wire of springy material, from 0.5 to 5 mils in diameter, preferably pointed at the contact end electrolytically or by grinding. Processes for forming the points on such wires are described in United States Patent 2,430,028 to W.G. Pfann, J.H. Scaff and A.H. White. The point of the wire is brought into contact with the upper surface 3 of the block with a force of 1 to 10 grams, whereupon a cold flow of the metal of the point takes place, causing it to conform to any minute irregularities of the block surface. To this end the wire of the point should be ductile as compared with the material of the block. Tungsten, copper and phosphor bronze are examples of suitable materials.

A second electrode 6, denoted the collector, makes contact with the upper face 3 of the block very close to the emitter 5. Best results have been obtained when the separation, measured along the surface of the block, between the collector and the emitter, is from 1 to 10 mils. This electrode 6 should make rectifier contact with the block and may be a pointed spring wire, formed and placed as above described in connection with the emitter 5. On the other hand, it may comprise a small spot of metal, for example, gold, which has been evaporated onto the upper surface of the block in the course of the final drying operation, and through which a central hole has been pierced (see Fig. 6) or across which a diametral slot has been cut (see Fig. 7). Evaporation of such a spot or film of metal onto the upper face after completion of the anodic oxidation process described above results in a non-ohmic rectifier junction or connection.

A third connection, termed the base electrode, is made, by soldering or other-wise, to the metal film 2 which has been plated onto the lower surface of the block I.

While the unit is now ready for use, its operation can generally be improved by an electrical forming process, in which a potential in excess of the peak back voltage is applied to either one or both of the point electrodes 5, 6 i.e., between it and the base electrode 2. The unit is protected from injury by heavy currents by inclusion of a resistor in series. The effect of this treatment is believed to lie in a concentrated application of electric field and heat to the material in the immediate neighborhood of the point, and so in an improvement of the electrical characteristics of the contact.

Bias voltages are now applied to the electrodes, a small bias, usually positive, on the emitter of the order of a fraction of a volt and a larger negative bias on the collector, usually in the range from −5 to −50 volts, measured, in each case, from the body of the block to the point electrode. These bias potentials may be obtained from batteries 7, 8 connected as shown or otherwise, as desired.

As shown in Fig. 1, it is preferred that the area of contact of each of the two point electrodes with the surface of the block be large as compared with the layer thickness. This reduces the actual contact resistance as compared with the resistance encountered by the current flowing laterally in the surface layer itself; i.e., the spreading resistance of the layer.

When the collector electrode 6 is a single pointed wire or an evaporated metal spot, a fraction of the emitter current, after spreading out laterally in the P-type layer 3, eventually finds its way across the barrier 4 to the plated electrode 2 on the lower face of the block, i.e., to the base electrode. This situation is depicted in Fig. 3 which is a plan view of the block showing current stream lines 13 diverging in all directions from the emitter. The current stream lines 13 are straight in the absence of the collector field. When the collector field 14 is present the current field is distorted as in fig. 3A which shows that even with a single collector electrode 6 more than half of the emitter current can be collected. In fact, the fraction of the emitter current which reaches the collector may in favorable cases be as high as 90 per cent.

To increase this ratio, especially in the case of units in which this ratio is less favorable, requires a modified electrode arrangement. Obviously, if the strong field 14 surrounding the collector 6 were to overlap or include the emitter 5, substantially all of the emitter current would be collected. This, however, would involve a loss of control. A solution is to provide two collectors 6, 6a, 6b, as in Fig. 4, symmetrically disposed about the emitter 5. Evidently with such an arrangement a considerably greater fraction of the emitter current is collected. In each case the boundaries of the collector field are indicated by broken lines 14. The several collectors may be connected together and as many may be employed as may seem desirable. Pursuing this solution still further leads to the ring collector 6d of Fig. 6, in which case the collector field 14 bears the shape of a semitorus. Its trace on the plane of the block surface is shown by the broken lines 14a, 14b. The two semicircular spots 6e, 6f, of Fig. 7 are the substantial equivalent of the circle of Fig. 6.

Further understanding of the considerations which govern the thickness of the P-type surface layer may be had from the following considerations, which apply specifically to a chemical layer. Fig. 13 is a plot of the electrostatic potential within the body of an N-type semiconductor in contact with a metal. As above stated, the N-type material of the semiconductor contains fixed or bound positive charges. They are believed to be distributed with fair uniformity in depth to a certain distance, beyond which the material is electrically neutral, because the bound positive charges are balanced by equal negative (electron) charges. In accordance with Poisson's equation:

$$\frac{d^2V}{dx^2} = -\frac{4\pi\rho}{\varepsilon} \tag{2}$$

where

V is the potential
x is the distance, measured from the metal into the semiconductor
ρ is the charge density, and
ε is the dielectric constant of the material.

Assuming the charge density ρ to be uniform, two integrations give the potential as a function of depth. When plotted, it is a parabola. In the figure, negative potential has been plotted upward. The vertical rise Ee from the Fermi level to the terminus of the curve, i.e., to its intercept with the potential axis, represents the energy which must be imparted to an electron to cause it to move from the metal to the semiconductor. These matters are fully explained in the literature, for example, in "Schottky's Theories of Dry Solid Rectifiers." by J. Joffe, published in "Electrical Communication," vol. 22 (1944–1945) at page 217.

[*A section of the patent has been removed for the sake of brevity.*]

From the foregoing description it will be clear that if it is desired to employ a semiconductor block of which the main body is of P-type so that the conductivity of the thin surface layer, whether due to impurities or to space charge effects, is of N-type, the polarities of all the bias sources of Figs. 1, 10, 11 and 12 are to be reversed. It is also to be understood that the magnitudes of the biases for best operation will depend on the semiconductor material employed and on its heat treatment and processing. Furthermore, it is possible to use a P-type layer of one semiconductor material on an N-type body of some other semiconductor material or vice versa. All such variations are contemplated as being within the spirit of the invention.

The invention is not to be construed as limited to the particular forms disclosed herein, since these are to be regarded as illustrative rather than restrictive.

What is claimed is:

1. A circuit element which comprises a block of semiconductive material of which the body is of one conductivity type and a thin surface layer is of the opposite conductivity type, an emitter electrode making contact with said layer, a collector electrode making contact with said layer disposed to collect current spreading from said emitter electrode, and a base electrode making contact with the body of the block.
2. Apparatus as defined in claim 1 wherein the surface layer is of the same chemical material as the block.
3. Apparatus as defined in claim 1 wherein the block is of germanium.
4. Apparatus as defined in claim 1 wherein the block is of N-type germanium and the surface layer is of P-type germanium.
5. Apparatus as defined in claim 1 wherein the block is of high back voltage germanium and to at least a part of one surface of which an anodic oxidation treatment has been applied.

JOHN BARDEEN.

WALTER H. BRATTAIN.

REFERENCES CITED

The following references are of record in the file of this patent:

UNITED STATES PATENTS

Number	Name	Date
1,745,175	Lilienfeld	Jan. 28, 1930
1,900,018	Lilienfeld	Mar. 7, 1933
1,949,383	Weber	Feb. 17, 1934
2,173.904	Hoist	Sept. 26,1939
2,402,662	Ohl	June 25, 1946
2,438,893	Bieling	Apr. 6, 1948
2,441,603	Storks et al.	May 18, 1948
2,447,829	Whaley	Aug. 24, 1948
2,464.807	Hansen	Mar. 22, 1949

FOREIGN PATENTS

Number	Country	Date
439,457	Great Britain	Dec. 6, 1935

Certificate of Correction

Patent No. 2,524,035 October 3, 1950

JOHN BARDEEN ET AL.

It is hereby certified that error appears in the printed specification of the above numbered patent requiring correction as follows:

Column 6, line 54, for "ar" read are; column 8, line 73, for "and" read end; column 17, line 51, for the word "side" read sign;

And that the said Letters Patent should be read as corrected above, so that the same may conform to the record of the case in the Patent Office.

Signed and sealed this 2nd day of January, A.D. 1951.

[SEAL]

THOMAS F. MURPHY,
Assistant Commissioner of Patents.

3.7 MASERS

Highly stable sources of very-high frequency (VHF) power were becoming increasingly desirable as communication scientists [3.7.1] were pushing the limits of transmission [3.7.2]. In addition, modulation techniques had to be refined to pass more information over communication channels. The amplification of these signals was also equally essential. The extraneous factors limiting the stability of generation and amplification of the power signals posed challenges to the early inventors. They attempted to study and overcome the limitations of earlier devices [3.7.3]. The maser subsystems presented in this section are an example of the devices that pushed the limits of communication systems [3.7.4] in the 1960s. This desire to meet social needs is discussed in Section 2.4 and in Chapter 4.

Unlike earlier patents, patents awarded after 1950 tend to have a significant amount of mathematical content. In fact, the tendency to use well-proven yet very abstract concepts and constants is very apparent Schawlow's and Townes's patent. This patent relates to the generation and amplification of a very wide spectrum of waves (from infrared to ultraviolet) by using devices with media and cavities in which the simulated emission of radiation can occur. The invention also offers methods and devices that are efficient and more practical and that offers high-quality (low-noise) radiation or amplification that is more coherent over the electromagnetic spectrum from infrared to ultraviolet. Such a breakthrough could have resulted from a very well-planned combination of mathematics behind the concept and the very fine control of the devices that actually deliver the quality. A parallel development occurred when the concept of jet propulsion was implemented by multistage ignition of fuel in the very early jet engines. Such engines combine the concept of partial combustion of jet fuel and fluid mechanics very elegantly.

Schawlow's contributions were recognized by the Nobel Committee in 1981 for his outstanding work. The Nobel Prize (with Townes) was awarded for the development of laser spectroscopy, even though the first patent (with Townes) was granted in 1960. Townes's (with Nikolai Basov and Aleksander Prokhorov) contributions were recognized by the Nobel Committee as early as 1964 for "fundamental work in the field of quantum electronics which has led to the construction of oscillators and amplifiers based on the laser-maser principle." Townes' contributions were recognized again in 1981 (with Schawlow), and the Nobel Prize was awarded in 1964 and again in 1981.

The complexity of these structures and the accuracy in building these devices needed the technology and the scientific understanding that the 1960s could provide. The inventors needed to take advantage of both of these aspects to build their devices, which is supported by our general observations regarding the differences between earlier inventions and more recent inventions. A detailed discussion of the differences is presented in Chapter 4.

REFERENCES (C-3.7)

[3.7.1] Townes, Charles H., *How the Laser Happened: Adventures of a Scientist.* Oxford University Press, New York, 1999.

[3.7.2] Vittoria, Carmine, *Elements of Microwave Networks: Basics of Microwave Engineering.* World Scientific Publishing Company, River Edge, NJ, 1998.

[3.7.3] Gilmour, A.S., *Principles of Traveling Wave Tubes.* Artech House Publishers, Boston, 1994.

[3.7.4] Cripps, Steve C., *RF Power Amplifiers for Wireless Communications.* Artech House, Boston, 1999. Also see [3.7.1].

Aurthur L Schawlow (1921–1999)
Masers

Charles Hard Townes (1915–)

Maser Communications System (Schawlow, Townes, 1960)

UNITED STATES PATENT

March 22, 1960 2,929,922

Masers and Maser Communications System

ARTHUR L. SCHAWLOW, MADISON, N.J., AND CHARLES H. TOWNES, NEW YORK, N.Y., ASSIGNORS TO BELL TELEPHONE LABORATORIES, INCORPORATED, NEW YORK, N.Y., A CORPORATION OF NEW YORK.

Application filed July 30, 1958, Serial No. 752,137, 11 Claims. (Cl. 250-7)

This invention relates to the generation and amplification of infrared, visible, and ultraviolet waves, and more particularly to the generation and amplification of such waves by means of devices including media in which the stimulated emission of radiation occurs; devices of this type are now generally termed "masers."

It is characteristic of a maser that it employs a medium in which there is established at least intermittently a nonequilibrium population distribution in a pair of spaced energy levels of its energy level system. In particular, the population of the higher of the selected pair of energy levels may be made larger than that of the lower. It is now usual to describe a medium which is in such a state of nonequilibrium as exhibiting a negative temperature. It is known that a competing process known as relaxation tends to return the system to equilibrium.

It is characteristic that if there be applied to a medium which is in a negative temperature state a signal of a frequency which satisfies Planck's law with respect to the two energy levels which are in nonequilibrium

$$\left(v = \frac{E_2 - E_1}{h} \text{ where } h \text{ is Planck's constant} \right)$$

then the applied signal will stimulate the emission of radiation at the signal frequency from the medium and the signal will be amplified.

Among the more promising forms of masers known is one which employs as the negative temperature medium a material whose energy level system is characterized by at least three energy levels, with the separations of these three energy levels falling within desired operating frequency ranges. To this material, there is supplied pumping power which effects transitions from the lowest to the highest of the selected three energy levels. By power saturation of the highest energy level, whereby the populations of the highest and lowest energy levels tend to be equalized, there is established in one of these two energy levels a nonequilibrium population distribution with respect to the intermediate energy level of the selected three, whereby a negative temperature results in the material. Thereafter a signal of appropriate frequency can be amplified by being applied thereto in a manner such that the emission of radiation is stimulated therefrom.

It is to be noted that the process of relaxation from randomly overpopulated states may give rise to spontaneous emission, that is, emission caused by radiative transitions in a mode other than the desired or stimulated one.

Generators and amplifiers employing atomic and molecular processes, as do the various known varieties of masers, may in principle be extended in operation far beyond the range of frequencies which have been generated and amplified by electronic processes. As, however, the maser concept is applied to the translation of wavelengths in the infrared, visible, and ultraviolet regions of the electromagnetic wave spectrum, it is found that conventional or microwave maser techniques and structures are suitable neither for the generation of monochromatic radiation nor to provide coherent amplification.

Accordingly, an object of the present invention is a system, including a maser, for translating infrared, visible, and ultraviolet energy.

A maser designed for operation in the microwave range of the spectrum might, for example, comprise a cavity having therein an ensemble of atomic or molecular systems, the cavity being characterized by being able to support only one mode near the frequency which corresponds to the desired radiative transitions of the systems. Alternatively, such an ensemble might be located in a waveguide, which similarly would be characterized by one, or a very few, preferred modes of propagation in the frequency range of interest.

Thus, the energy emitted by a maser operating in the microwave range is typically monochromatic, due to the energy produced by stimulated emission being very much larger than the background of radiation caused by spontaneous emission. In other words, such devices are inherently monochromatic because stimulated emission produces completely coherent amplification, and spontaneous emission, which is not so coherent, is characteristically small by comparison with the stimulated emission.

On the other hand, the maintenance of a single isolated mode is a master cavity operating at frequencies above those in the microwave range requires an impractically small cavity structure (of the order of one wavelength) and/or a high and not easily realizable density of pumping power. Hence, one is led to consider, in these higher frequency ranges, cavities which are large compared to a wavelength, and which are accordingly capable of supporting a large number of modes within the frequency ranges of interest. A disadvantage of this approach, however, is that masers including such cavities must be operated at relatively high power levels in order that the emission stimulated therefrom be at least as large as that spontaneously emitted therefrom.

Accordingly, another object of this invention is a practically realizable, efficient, low-noise maser structure which is capable of the generation of monochromatic radiation, or coherent amplification, in the infrared, visible, and ultraviolet portions of the electromagnetic spectrum.

The above and other objects of the present invention are realized in an illustrative embodiment thereof wherein a negative temperature medium is disposed between two spaced parallel reflecting plates in a configuration which is of practical size and which may be pumped by readily available power sources, and wherein a single mode corresponding to the stimulated emissions can be effectively isolated.

More particularly, one specific illustrative embodiment of the present invention comprises a maser including a chamber having reflective end parallel plates and side walls. Positioned within the chamber is a negative temperature medium, which is pumped by an energy source disposed about the chamber. The side walls are transparent to the pumping energy and either transparent to or absorptive of other energy radiated thereat. Further, an optical configuration is arranged adjacent to one of the end plates of the chamber for isolating the one mode of those supported within the chamber which it is desired to selectively utilize.

The principles of the present invention may illustratively be embodied in a communications system which comprises a maser device capable of generating monochromatic radiation, a second maser device capable of modulating and coherently amplifying the output of the maser generator, and a device for detecting the output of the second maser. Alternatively, such a system may include a maser generator whose output is modulated by a nonamplifying device, or a system in which the maser generator itself is modulated.

Thus, a feature of the present invention is a system for communicating information by means of energy having wavelengths in the infrared, visible, or ultraviolet portions of the electromagnetic spectrum, comprising a monochromatic generator, a modulatable coherent amplifier, and a detector.

Another feature of this invention is a maser generator including a chamber comprising reflective parallel end members and side members, a negative temperature medium within said

March 22, 1960 A. L. SCHAWLOW ET AL 2,929,922

MASERS AND MASER COMMUNICATIONS SYSTEM

Filed July 30, 1958

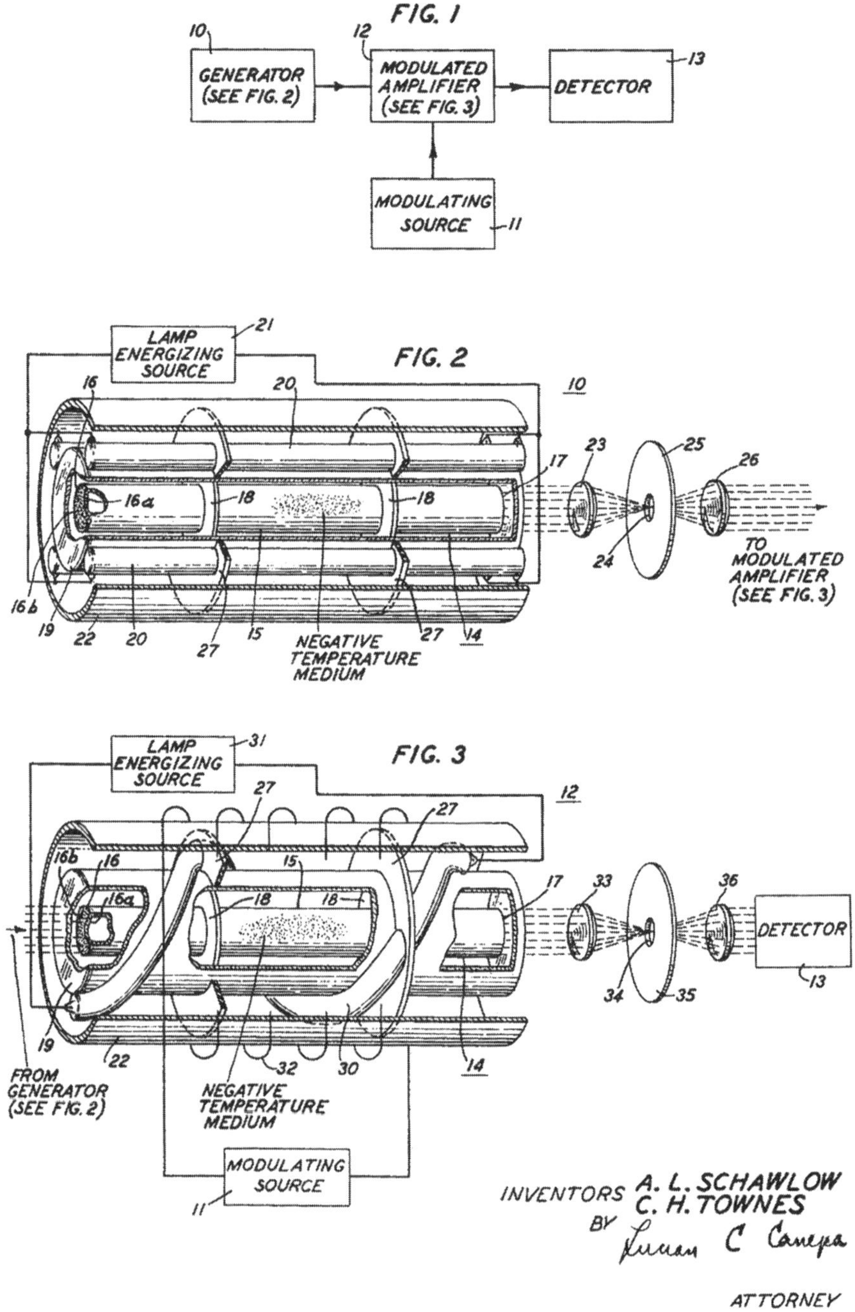

chamber, a pumping power source disposed about the side members, the side members being transparent to the pumping energy and either transparent to or absorptive of other energy radiated thereat, and a configuration arranged adjacent to one of the end members for abstracting from the chamber a selected one of the modes supported therein, whereby there is provided efficient, low-noise, monochromatic generation of infrared, visible, or ultraviolet waves.

A further feature of the present invention is a maser amplified, including a chamber comprising reflective parallel, end members and side members, a negative temperature medium within said chamber, a pumping power source disposed about the said members, the side members being transparent to the pumping energy and non-reflective of other energy radiated thereat, and a configuration arranged adjacent to one of the end members for abstracting from the chamber an amplified replica of a wave fed through the other end member thereof, whereby there is provided an efficient, low-noise, coherent amplifier of infrared, visible, or ultraviolet waves.

A still further feature of this invention is an arrangement for modulating the signal output of a maser of the type herein-described comprising a structure for establishing a magnetic field parallel to the longitudinal axis of the chamber thereof, and an information source capable of varying the magnetic field in correspondence with the output of the source.

The principles of the present invention will be better understood from the following more detailed discussion taken in conjunction with the accompanying drawing, in which:

Fig. 1 is a block diagram of a communications system illustratively embodying aspects of the principles of the present invention;

Fig. 2 is a perspective view of a generator made in accordance with the principles of this invention; and

Fig. 3 is a perspective view of an amplifier embodying the principles of the present invention. Also, Fig. 3 depicts a modulating source and a detector, arranged in typical relationship to the amplifier.

Referring now to Fig. 1, there is shown a communications system in which the principles of the present invention are illustratively embodied. The system includes a generator or oscillator 10, a modulating source 11, a modulated amplifier 12, and a detector 13.

The generator 10, which is shown in detail in Fig. 2, includes a chamber 14, which typically may be about one centimeter in diameter and ten centimeters in length. The chamber 14 comprises a hollow cylinder 15 having its ends capped by two flat parallel assemblies 16 and 17. Disposed within the chamber 14 is a negative temperature material whose radiative energy level separations correspond to frequencies in the ranges of interest, namely, the infrared, visible, and ultraviolet ranges.

Various materials are suitable for use as the active or negative temperature medium of maser devices of the general type described herein. For example, vapors of the alkali metals; namely, lithium, sodium, potassium, rubidium and cesium, and some solid rare earth salts, for example, anhydrous chlorides of europium and samarium, may be so used.

In particular, potassium maintained at a temperature of about 435 degrees Kelvin, at which temperature it exhibits a vapor pressure of about 0.001 millimeter, may advantageously be included in a specific illustrative embodiment of the principles of the present invention as the active medium thereof.

Each of the flat parallel assemblies 16 and 17 of the device shown in Fig. 2 advantageously includes as a component part thereof a material which reflects most of the energy incident thereupon. Thus, for example, an assembly comprising sapphire 16a, which material is characterized by good chemical inertness and excellent transmission properties, particularly for infrared wavelengths, and having a coating of gold 16b, typically about 500 angstrom units thick, on the outer surface of the sapphire member, may be included in specific embodiments of this invention. Such an assembly exhibits 97 percent reflectivity, 2 percent absorptivity, and 1 percent transmitivity to wavelengths in the infrared range.

The inner and outer parallel faces of each sapphire plate reflect a small portion of the radiation directed thereat. Therefore, the thickness of the sapphire plates should advantageously be chosen such that the reflections from the two faces of each plate add in phase.

It is noted that the phase angle between the reflections from the two faces or surfaces of each sapphire member depends on the thickness and refractive index thereof. Since sapphire is crystalline and the refractive index is different for ordinary and extraordinary rays, the thickness may be chosen to give constructive interference for one polarization and destructive interference for the polarization, perpendicular thereto, in that manner discriminating between modes traveling in the same direction but having different polarizations.

The cylinder 15 of the chamber 14 is advantageously of a material which is transparent to the pumping energy and either transparent to or absorptive of other radiation impinging thereupon, thereby both to allow the negative temperature medium within the cylinder 15 to be pumped and to eliminate from the chamber radiation occurring in all modes except those corresponding to waves which travel back and forth between the reflective assemblies 16 and 17. These reflected modes are coupled much more strongly to the excited atomic systems of the negative temperature medium than any other modes and hence would be strongly favored for maser oscillations.

In these specific embodiments of the present invention in which the negative temperature medium within the chamber 14 is at a pressure other than atmospheric, as in the case of the potassium vapor, for example, it is advantageous to support the chamber 14, by means of spacer elements 18, within a protective shell 19, typically of glass, within which shell a pressure approximately equal to that within the chamber 14 is maintained. In this manner, the resultant forces acting on the opposing faces of the end assemblies 16 and 17 are made so small as not to be capable of distorting the assemblies and thereby disturbing their parallelism.

Arranged around the protective shell 19 are a plurality of pumping sources 20 which, in a maser generator including potassium vapor as the active medium thereof, may advantageously comprise an assembly of potassium lamps, which lamps 20 are energized by a source 21.

The maser generator shown in Fig. 2 further includes a housing 22 in which the protective shell 19 is supported by spacer elements 27. The inner surface of the housing 22 is of a material which is capable of reflecting a major part of the energy radiated thereupon from the pumping power-sources 20, thereby to aid in directing a substantial portion of the energy emitted by the sources 20 toward the chamber 14 and into the negative temperature medium therein.

The process of oscillation within a maser generator made in accordance with the principles of the present invention depends on the selective regeneration with the chamber 14 of a component of the energy spontaneously emitted by the negative temperature medium therein.

The process of oscillation within a maser generator made in accordance with the principles of the present invention depends on the selective regeneration with the chamber 14 of a component of the energy spontaneously emitted by the negative temperature medium therein.

Mode selection in the maser generator shown in Fig. 2 is based on the phenomenon that, when energy is radiated from a chamber of the type herein-described through an end plate member which is large compared to the wavelength of the radiation, each mode radiates in a characteristic direction. Thus, if the emitted radiation is focused by a lens, each point in the focal plane thereof will correspond to a mode of a particular direction, affording thereby a separation of modes. And, if radiation falling on a very limited area in the focal plane is detected, that radiation will represent spontaneous and stimulated emission from a selected and limited number of modes, the large background of spontaneous emission produced in other modes being thereby effectively isolated.

The principles of this phenomenon are utilized in the maser generator shown in Fig. 2. Radiation in the desired mode is transmitted through the end assembly 17 and focused by a double-convex lens 23 arranged such that the desired energy is directed through an aperture 24 in an absorptive sheet 25 which lies in the focal plane of the lens 23. A second double-convex lens 26 is employed to reconvert the selected energy to the form of a plane wave, in which form the desired energy radiates to the modulated amplified 13.

The maser amplifier shown in Fig. 3 is similar in structure to the generator described above. The amplifier includes a chamber 14 comprising a hollow cylinder 15 supported within a protective shall 19 by supporting members 18 and within which cylinder 15 there is disposed a suitable negative temperature medium. The shell 19 is supported within a reflective housing 22 by spacer or supporting members 27.

Arranged about the protective shell 19 of the amplifier 12 shown in Fig. 3 is a pumping power assembly 30 which may advantageously comprise, in a specific illustrative embodiment of the present invention wherein the negative temperature medium is potassium vapor, a potassium lamp formed in the shape of a spiral, which spiral lamp is energized by a source 31.

Energy which is directed from the generator 10 through the left-hand end of the cylinder 15 of the amplifying device of Fig. 3 may be modulated by an assembly including a coil 32 for establishing a magnetic field parallel to the longitudinal axis of the cylinder 15 and a source 11 for varying the strength of the longitudinal magnetic field, whereby broadening or splitting of the spectral lines emitted by the device 12 in correspondence with the variation of the magnetic field results, which phenomenon is generally termed the Zeeman effect.

The device 12 shown in Fig. 3 radiates through the right-hand end of the cylinder 15 an amplified counterpart of the energy directed at the device 12 by the generator 10. The radiated energy is directed by two lenses 33 and 36 through an aperture 34 is an absorptive member 35 and to a detector 13. The detector 13 may, for example, include a photomultiplier tube.

It is noted that the admission of a signal into the region between the two end parallel plates of the amplifying device 12 is similar to the process involved in a microwave cavity. More particularly, the partially reflecting surfaces of the end plates re analogous to coupling holes; and, if a monochromatic plane wave strikes the outside of one of the partially reflecting surfaces, energy will build up in the region between the plates, and the relations between input wave, energy in the "cavity", and output wave correspond to those for a microwave impinging on an appropriate cavity with input and output coupling holes.

Thus, it is seen that the principles of the present invention may illustratively be embodied in monochromatic maser generators of infrared, visible or ultraviolet wavelengths. It is feasible to tune such generators by varying the pressure or temperature of the negative temperature media thereof. Alternatively, tuning of such devices may be based on the Stark effect (i.e., observed changes in the spectrum of a system when the system is subjected to an electric field) or on the Zeeman effect.

Further, it has been shown that embodiments of this invention include coherent maser amplifiers of infrared, visible, or ultraviolet wavelengths. It is to be noted that these devices are capable of amplifying energy of these wavelengths with no significant change in the wavefront or phase thereof.

It is to be noted that maser devices embodying the principles of the present invention may advantageously be utilized in various spectroscopy and measurement applications, as well as in the communications field.

It is to be understood that the various specific embodiments disclosed are merely illustrative of the general principles of the invention. Thus, although the amplifying chamber 14 has been shown and described as including a hollow cylinder, it is of course clear that any other transparent structure (more specifically, transparent to the pumping energy and transparent to or absorptive of other radiation) suitable for retaining the negative temperature medium and including reflective end assemblies may be easily substituted thereof.

What is claimed is:

1. A communications system for operation in the infrared, visible, or ultraviolet regions of the electromagnetic wave spectrum comprising a monochromatic maser generator, a coherent modulated maser amplifier, a modulating source, and a detector; said generator comprising a chamber having end reflective parallel members and transparent side members, a negative temperature medium disposed within said chamber, and means arranged about said chamber for pumping said medium; said amplifier comprising a chamber having end reflective parallel members and transparent side members, a negative temperature medium disposed within said chamber, means arranged about said chamber for pumping said medium, and coupling means for abstracting from one end of said chamber an amplified counterpart of the energy transmitted into the other end thereof and for directing said amplified counterpart at said detector.
2. A communications system for operation in the infrared, visible or ultraviolet regions of the electromagnetic wave spectrum comprising a monochromatic maser generator, a coherent maser amplifier, said generator and amplifier including means for modulating the output of said generator in accordance with signal information, and a detector; said generator comprising a chamber having a length which is substantially greater than its transverse dimension and having partially reflective parallel end members and non-reflective side members, a negative temperature medium disposed within said chamber and characterized by at least three distinct energy levels, two of which have a separation in the frequency range of interest, means for pumping said medium so that a population inversion is produced therein between said two separated energy levels, and means for abstracting from said chamber and directing at the amplifier input the energy of a particular mode of electromagnetic vibration; said amplifier comprising a chamber having a length which is substantially greater than its transverse dimension and having partially reflective parallel end members and non-reflective side members, a negative temperature medium disposed within said chamber and characterized by at least three distinct energy levels, two of which have a separation in the frequency range of interest, means for pumping said medium so that population inversion is produced therein between said two separated energy levels, means for abstracting from said chamber the energy of a particular mode of electromagnetic vibration representing an amplified and modulated replica of the generator output, and means for direction said replica at said detector.

3. A communications system for operation in the infrared, visible, or ultraviolet regions of the electromagnetic wave spectrum comprising a monochromatic maser generator, a coherent modulated maser amplifier, a modulating source, and a detector; said amplifier including means defining an amplifying chamber, and means for establishing a magnetic field parallel to the longitudinal axis of said chamber, said modulating source being coupled to said magnetic means, the radiative output of said generator being directed at said amplifier, and the radiative output of said amplifier, constituting an amplified and modulated counterpart of the energy radiated thereinto, being directed at said detector.
4. A maser generator comprising a chamber having end reflective parallel members and side members, a negative temperature medium disposed within said chamber, and means arranged about said chamber for pumping said medium, said side members being transparent to the pumping energy and transparent to or absorptive of other energy radiated thereat.
5. A maser generator for operation in the infrared, visible or ultraviolet regions of the electromagnetic wave spectrum comprising a chamber having a length which is substantially greater than its transverse dimension and having partially reflective parallel end members and non-reflective side members, a negative temperature medium disposed within said chamber and characterized by at least three distinct energy levels, two of which have a separation in the frequency range of interest, means for pumping said medium so that a population inversion is produced therein between said two separated energy levels, and means for abstracting from said chamber and directing at an amplifier input the energy of a particular mode of electromagnetic vibration.
6. A maser generator as in claim 5 wherein said mode selecting means includes an absorptive member having an opening therethrough, said absorptive member being positioned adjacent to one end of said chamber, and means for directing a selected portion of the energy radiated by said generator through said opening.
7. A maser generator as in claim 5 wherein said negative temperature medium comprises potassium, and said pumping means comprises an assembly of potassium lamps.
8. A maser amplifier comprising a chamber having end reflective parallel members and said members, a negative temperature medium disposed within said chamber, means arranged about said chamber for pumping said medium, said side members being transparent to the pumping energy and non-reflective of other energy radiated thereat, and coupling means for abstracting from one end of said chamber an amplified counterpart of the energy directed into the other end thereof.
9. A maser amplifier for operation in the infrared, visible or ultraviolet regions of the electromagnetic wave spectrum comprising a chamber having a length which is substantially greater than its transverse dimension and having partially reflective parallel end members and non-reflective side members, a negative temperature medium disposed within said chamber and characterized by at least three distinct energy levels, two of which have a separation in the frequency range of interest, means for pumping said medium so that a population inversion is produced therein between said two separated energy levels, and means for abstracting from said chamber and directing at a detector the energy of a particular mode of electromagnetic vibration.
10. A maser amplifier as in claim 9 wherein said negative temperature medium comprises potassium, and said pumping means comprises an assembly of potassium lamps.

11. A modulated maser amplifier comprising a chamber having end reflective parallel members, a negative temperature medium disposed within said chamber, means arranged about said chamber for pumping said medium, means coupled to and under the control of a modulating source for establishing a magnetic field parallel to the longitudinal axis of said chamber, and means for abstracting from one end of said chamber an amplified counterpart of the energy directed into the other end thereof, which counterpart is modulatable in accordance with the output of said source.

References Cited in the file of this patent

UNITED STATES PATENTS

2,836,722 Dicke et al. May 27, 1958

3.8 CANCER DETECTION

The detection of cancerous tissue in biological systems was made feasible after the magnetic resonance imaging (MRI) techniques developed in the 1970s. It is now a refined and valuable diagnostic procedure in the medical field. Tumors and other tissue mass that develop without inflammation (generally associated with cancer) usually have abnormal structures [3.8.1] when compared with the structure of normal tissue. The rate of growth is also different from that of normal tissue. From a historical perspective, the use of X-rays to penetrate the body and photograph the internal organs was the general practice to detect abnormalities. In many cases, some tumors are not always detected by X-rays.

However, Raymond Damadian's invention [3.8.2] in 1974 was much more elegant and powerful than the conventional X-ray technique [3.8.3]. It was based on comparing the different structures of healthy and malignant cells. The logic behind cancer detection by nuclear magnetic resonance techniques is derived from the fundamental differences (i.e., the organization and the structure) between healthy cells and cancerous or malignant tissue. The technique penetrates nuclei within the tissue to evaluate quantitatively the difference between the structure, composition, and organization of the cells.

The concept is unique to the extent that two magnetic fields are induced by external induction devices. One of the fields is essentially static, and the other is oscillating in an orthogonal direction to the fixed field. The energy from the oscillating field is absorbed to a higher energy state. At switch-off, of the oscillating field, the tissues return to their normal state, and the two relaxation time measurements (spin-lattice [3.8.4] and spin-spin [3.8.5]) are made as they return to the normal state.

The relaxation times are measured, and they provide quantitative data about the extent of disorganization of the selected nuclei [3.8.6]. By comparison with similar measurements of the healthy cells, the extent of malignancy [3.8.7] may be calculated. Both the location and extent of malignant cells are determined by a three-dimensional embodiment [3.8.5] of the Damadian invention. In this invention, all of the features necessary to make it a classic contribution (and a breakthrough) can be detected. The social need, its timeliness, the use of the most current technology of the times, and then the exploitation of the most fundamental properties of the biological structures to give an accurate and exact portrayal of the cancerous tissue make this invention a contribution and a breakthrough. The use of structured creative thinking can be identified in this patent as compared with some of the early patents.

We would also like to acknowledge the pioneering work in MRI techniques by Paul Lauterbur, who received the 2003 Nobel Prize in Medicine with Peter Mansfield.

REFERENCES (C-3.8)

[3.8.1] Brooks Susan A., and Schumacher, Udo (Editors), *Metastasis Research Protocols*. Volume 1: Analysis of Cells & Tissues. Humana Press, Totowa, NJ, 2001.

[3.8.2] Kaplan, Phoebe A., Dussault, Robert, Helms, Clyde A., and Anderson, Mark W. *Musculoskeletal MRI*. W B Saunders Co, 2001.

[3.8.3] Koningsberger, D. C., and Prins R. (Editors), *X-Ray Absorption: Principles, Applications, Techniques of EXAFS, SEXAFS and XANES*. Wiley-Interscience, New York, 1988.

[3.8.4] Kubo, R., and Hanamura , E. (Editors), "Relaxation of Elementary Excitations." *Proceedings of the Taniguchi International Symposium, Susono-shi, Japan, October 12–16, 1979*. Springer-Verlag, New York, 1980.

[3.8.5] Poole Charles P., and Farach, Horacio A. (Editors), *Handbook of Electron Spin Resonance: Data Sources, Computer Technology, Relaxation, and ENDOR*. Springer Verlag, New York, 1999.

[3.8.6] Parsonnet Julie, and Horning, Sandra J. (Editors), *Microbes and Malignancy: Infection as a Cause of Human Cancers*. Oxford University Press, New York, 1999.

[3.8.7] Al-Sam, Salam Z., Lakhani, Sunil R., and Davis, Jack (Editors), *Practical Atlas of Pseudomalignancy: Benign Lesions Mimicking Malignancy*. Oxford University Press, Oxford, UK, 1998.

Raymond D. Damadian (1936–)

MRI and Cancer Detection

3.8.1 Cancer Detection (Damadian, 1974)

UNITED STATES PATENT

February 5, 1974 3,789,832

Apparatus and Method for Detecting Cancer in Tissue

RAYMOND V. DAMADIAN, 64 SHORT HILL RD., FOREST HILL, N.Y. 11375
Application filed March 17, 1972, Application No. 235,624

U.S. Cl. 128/2 R, 128/2 A, 324/.5 R
Int. Cl. A61B 5/05
Field of Search 128/2 R, 2 A, 1.3; 324/.5 A, 324/.5 B
References Cited
UNITED STATES PATENTS
3,691,455,9/1972 Moisio et al. 324/.5 R
3,557,777,1/1971 Cohen 128/2R
3,530,371,9/1970 Nelson, et al 324/.5 AC

OTHER PUBLICATIONS
Singer, J. R., Journ. of Applied Physics, Vol. 31, No. 1, Jan., 1960, pp. 125–127
Primary Examiner – Kyle L. Howell
Attorney, Agent, or Firm – Brumbaugh, Graves, Donohue & Raymond
Sixteen claims, three drawing figures, CDF

Abstract An apparatus and method in which a tissue sample is positioned in a nuclear induction apparatus whereby selected nuclei are energized from their equilibrium states to higher energy states through nuclear magnetic resonance. By measuring the spin-lattice relaxation time and the spin-spin relaxation time as the energized nuclei return to their equilibrium states, and then comparing these relaxation times with their respective values for know normal and malignant tissue, an indication of the presence and degree of malignancy of cancerous tissue can be obtained.

Background of the Invention This invention relates to an apparatus and method of detecting cancer in tissue and, more particularly, to the use of nuclear magnetic resonance techniques to detect alterations in the organization and structure of selected nuclei in the tissue which alterations are believed to be caused by cancer.

At present, early detection of internal cancerous tumors is hampered by the relatively high permeability of the many tumors to X-rays. Detection in these cases depends mainly on the indirect evidence provided by the displacement of normal radiopaque structures pushed aside by the tumor. Therefore, the present methods are directed to a qualitative analysis only and lack a quantitized aspect such that a firm basis for decision does not exist as to whether cancer is present and if so how malignant it is.

Summary of the Invention It is an object of this invention to provide apparatus and a method for the quantitative detection of cancer in tissue and in particular to provide apparatus for use by clinical and diagnostic laboratories and in the operating room for rapid screening of cancer in small samples of surgically removed tissue.

Furthermore, it is an object of this invention to provide criteria for determining the degree of malignancy of the cancerous tissue detected.

It is a further object of this invention to provide apparatus and method for the detection of cancer in humans which does not require tissue to be surgically removed and can be done with the probes entirely external to the body of the human who is being examined.

PATENTED FEB 5 1974 3,789,832

SHEET 1 OF 2

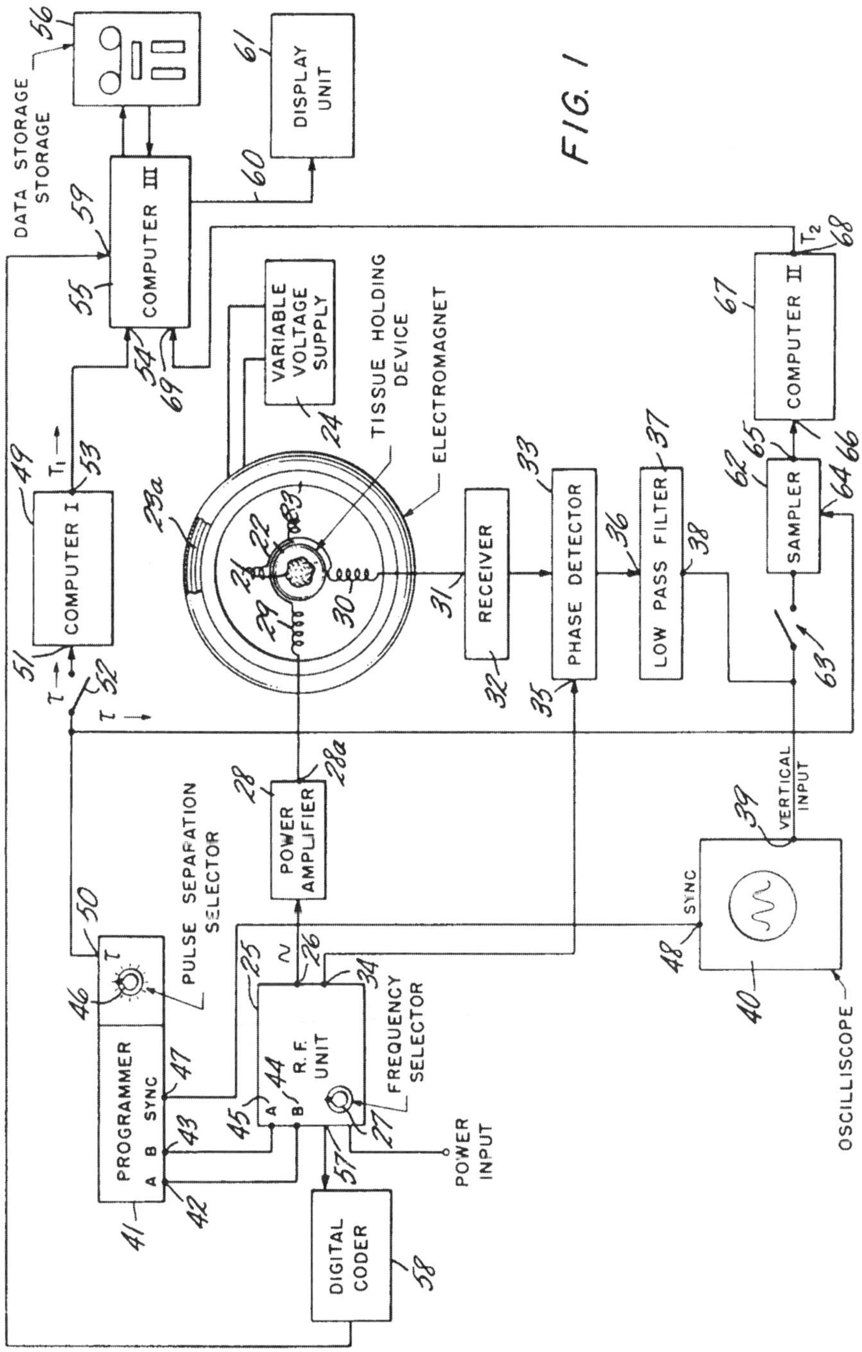

FIG. 1

PATENTED FEB 5 1974 3,789,832

SHEET 2 OF 2

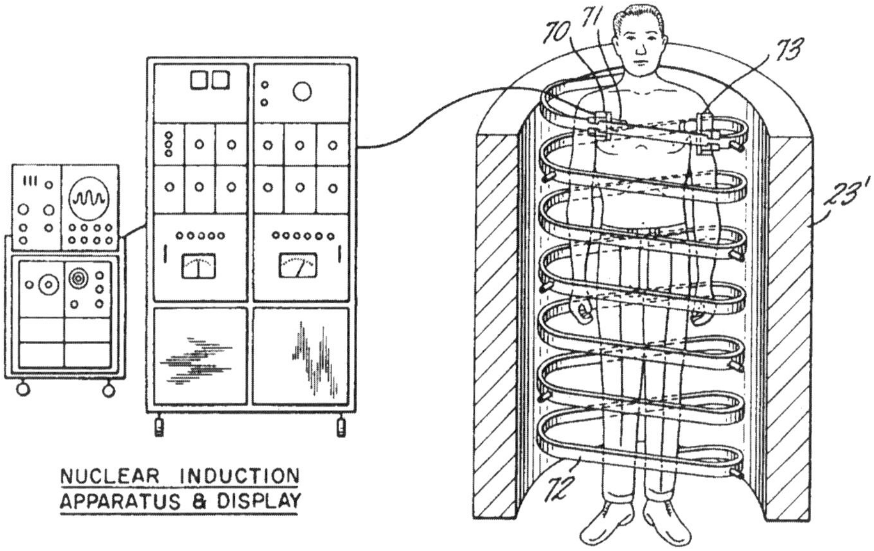

FIG. 2

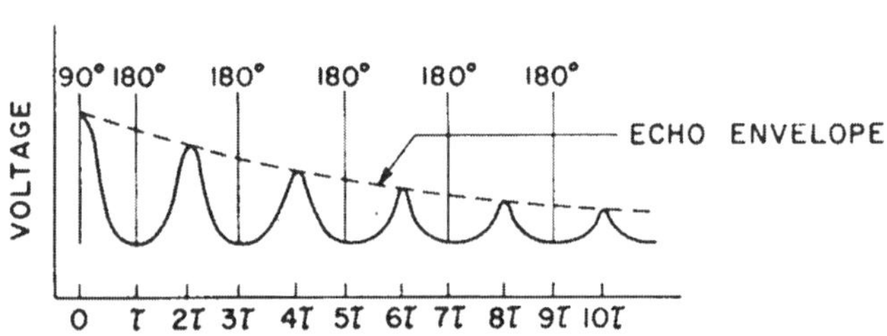

FIG. 3

These and other objects are attained, in accordance with the invention, by measuring the degree of organization of selected nuclei in the tissue being analyzed.

It has now been found that certain molecules in cancerous cells are more disorganized and arranged with less structure than the same molecules in normal cells. It is known, for example, that the water molecules present in normal cells exist in multiple polarized layers absorbed onto cell protein chains. In cancerous tissue, however, the structure is broken down and the water molecules become disassociated from the protein chains, the result being that water molecules in cancerous tissue are not as compacted as they are in normal tissue.

It has now been found that, by measuring the degree of organization of these selected molecules in cells being studied and comparing this with the degree of organization in a known cancerous cell, cancer cells can be detected. Furthermore, it has now been found that the less the organization the greater the malignancy, therefore, a scale can be made to provide a standard for basing a decision on the degree of malignancy.

The present invention utilizes nuclear magnetic resonance (NMR) techniques to determine this degree of organization. Essentially, the tissue to be analyzed is placed in a tissue holder located in a nuclear induction apparatus and positioned to be exposed to two magnetic fields provided by two sources located in the induction apparatus. One source is an electromagnet in the preferred embodiment and provides a static magnetic field of known strength. The other source provides an oscillating magnetic field and is positioned so that the direction of its radiated field is orthogonal to the field provided by the electromagnet. The frequency of the oscillating magnetic field is adjusted to cause selected nuclei in the cells to be energized from their equilibrium states to higher energy states through NMR absorption. After a predetermined length of time, the oscillating source is turned off and measurements are made of the spin-lattice relaxation time and spin-spin relaxation time for the energized nuclei returning to their equilibrium states.

It is known that for any given nuclei, the relaxation times increase in proportion to the degree of motional freedom between the nuclei. For example, the relaxation times for hydrogen nuclei in distilled water is several orders of magnitude greater than the corresponding relaxation times for the hydrogen nuclei in ice.

In the present invention, the measurement of relaxation times for selected nuclei in the tissue cells being examined provide a way of measuring the degree of disorganization of the selected nuclei in these cells. By comparing the relaxation times for this tissue with relaxation times obtained for known normal tissue and known cancerous tissue, it can be determined whether cancer is present and if so the degree of malignancy.

In a second embodiment, apparatus is described which can be used to detect cancer in mammallian tissue and in particular humans without having to remove surgically a sample of the tissue to be analyzed. In this embodiment, the entire mammal or human being is placed in a magnetic field provided by an electromagnet sufficiently large to surround the mammal or human. The second magnetic field is directed to a unit which focuses the oscillating magnetic field radiation into a beam. Further apparatus is provided for scanning throughout the entire body during which time the relaxation times are measured for selected nuclei and compared with standards. In this way a determination can be made of the existence of cancer together with the location and degree of malignancy of the cancerous cells present.

For a better understanding of the invention, reference may be made to the following detailed description of the representative embodiments, taken in conjunction with the figures of the accompanying drawings in which:

Fig. 1 is a schematic diagram of suitable apparatus for detecting cancerous tissue in accordance with the invention;

Fig.2 is a diagram of a particular application of this invention to the detection of cancer in humans; and

Fig. 3 is a diagram showing the signal output of the low pass filter used in the determination of the spin-spin relaxation time.

Description of a Preferred Embodiment In the embodiment illustrated in FIG. 1, apparatus is shown for measuring the spin-lattice relaxation time and spin-spin relaxation time of a tissue sample 21 held by tissue holding means 22.

An electromagnet 23, having a wire coil 23a energized by a variable voltage supply 24, provides the necessary magnetic field for causing selected nuclei, generally protons, in the tissue sample 21 to separate into discrete energy levels. This separation is commonly known as nuclear Zeeman splitting.

Now if a second field is applied to the tissue sample in a direction orthogonal to the first with a frequency determined by inherent characteristics of the selected nuclei and the strength of the field provided by the electromagnet 23, the selected nuclei can be energized from their equilibrium states to higher energy states through nuclear magnetic resonance (NMR) absorption.

In the present invention this second magnetic field is generated by a radio-frequency oscillator 25 which provides a signal at its output terminal 26 having a frequency which can be adjusted manually by frequency selector 27.

The signal on output terminal 26 is directed to a power amplifier 28 whereby the amplitude of the signal generated by the oscillator 25 is increased. The amplified signal appearing at the output terminal 28a is then directed to a coil transmitter probe 29 which radiates magnetic energy into the magnetic field space formed by the electromagnet 23. The orientation of the transmitter probe 29, in the preferred embodiment, is adjusted to be orthogonal to the direction of the field formed by electromagnet 23.

In order to determine the spin-lattice relaxation time and the spin-spin relaxation time of the selected nuclei returning to their equilibrium states from the higher energy states, a coil receiver probe 30 is provided for detecting signals generated by the decaying nuclei. This probe is positioned essentially at right angles with the transmitter probe 29 and also the direction of the field provided by the electromagnet 23.

The signals detected by the receiver probe 30 are directed to the input terminal 31 of the receiver amplifier 32. The output of the receiver amplifier 32 is directed to a phase detector 33 which is commonly used in NMR applications to determine when the frequency of the radio frequency source 25 has been properly adjusted to supply the conditions necessary for NMR absorption. A signal having the frequency of the radio frequency source 25 is supplied to the output terminal 34 of unit 25 and then supplied to the input terminal 25 of the phase detector 33.

The signal output of the phase detector 33 is in turn directed to the input terminal 36 of a low pass filter 37. This filter filters out the resonance frequencies and allows the envelope responsive to the amplitude of the signal detected by the receiver probe 30 to appear at output terminal 38.

The signal appearing at 38 is directed to the vertical input 39 of oscilloscope 40 so that the signal can be displayed visually.

In the experiments described below, a programmer 41 (See Fig.1) was used to provide a sequence of pulses necessary in NMR applications. By way of example, a Nuclear Magnetic Resonance Specialties, Model P-102A Pulse Programmer, can be used. This Programmer is part of the Nuclear Magnetic Resonance Specialties, PS-60A NMR Spectrometer.

In programmer 41, an A pulse output terminal 42 and a B pulse output terminal 43 are provided and they are connected with their respective input terminals 44 and 45 of radio frequency generator 25. Under operator control the programmer 41 through circuitry not shown signals the radio frequency generator 25 to produce pulse sequences necessary in measuring spin-lattice and spin-spin relaxation times. These pulse sequences corresponding to the Carr-Purcell sequences will be described in more detail below. The separation time between pulses is controlled manually by the pulse separation selector 46. The programmer 41 also provides a sync signal appearing on output terminal 47 which is directed to input terminal 48 of the oscilloscope 40 for synchronizing the oscilloscope with the pulse sequences.

Example 1

Several experiments were performed using Sprague-Dawley rats. Five of the rats were normal, five had previously been infected with Walker sarcoma, three had been infected with Novikoff hepatoma and two of the rats had Fibroadenoma (benign tumors). The rats ranged in weight from 150 to 500 grams and were selected at random to exclude variations in the age of the rat as a material consideration.

The animals were killed by cervical fracture. The infected rats were killed when tumor size reached approximately 1.5 ml in volume. This was 4 to 5 days after inoculation in the case of animals with Novikoff hepatoma and 10 days after inoculation in the case of animals with Walker sarcoma. The samples were excised and packaged in cellulose nitrate tubes. In all cases, NMR measurements were obtained within 5 minutes after the death of the animal.

NMR measurements were taken with an apparatus comprising a Varian electromagnet, 12 inches in diameter operating at approximately 5,610 gauss, a pulse spectrometer model PS-60AW, made by Nuclear Magnetic Resonance Specialties Corporation, together with a probe of cross-coil design operation at 24 Mhz.

Two types of NMR measurements were taken for each tissue sample. The first measurement was the so-called spin-lattice relaxation time (T_1) and this was measured by the method of Carr and Purcell, *Phys. Rev.* 94, 630 (1954). In accordance with this method, a sequence of two pulses is used with pulse widths for the two pulses set to produce a 180° nutation followed by a 90° nutation.

With this sequence, one observes a free induction decay after the second pulse whose amplitude is given by

$$M(\tau) = M_0(1 - 2_e^{-\tau/T_1})$$

where M_0 is the equilibrium value of the pulse amplitude and τ is the interval between pulses. This equation implies that, if T_1 is multiplied by the natural logarithm of 2, the product is equal to the pulse interval that produces no free induction decay after the 90° pulse. In actual practice, once the two pulses were phased and pulse widths set for the proper nutation angle, an oscilloscope 40, Fairchild 766 H/F, 25 and 50 Mhz, was synchronized to trigger on the second pulse and the pulse interval was adjusted until the null free induction decay was obtained. The interval between the two pulses was obtained from a frequency counter, Computer Measurement Company 200 CN, interfaced with the output of the PS-60 spectrometer programmer 41.

The second NMR measurement taken was the so-called spin-spin relaxation time. This was accomplished by using a 90° to 180° pulse sequence, making use of the Carr-Purcell modification identified in the reference noted above to obtain the echo decay envelope.

This method for measuring T_2 is free of diffusion effects and field inhomogeneities. Since the envelope height E is given by

$$E(2n\tau) \approx e^{-2n\tau/T_2}$$

where n represents integral multiples of the pulse separation τ, T_2 was estimated from the oscilloscope trace as the time required for the envelope height to decay to 1/e.

Five normal rats were sacrificed, and various types of normal tissue were subject to NMR measurements. The results of these tests are summarized in Table 1 below.

In addition, tumor tissue from a number of infected rats was subjected to NMR measurements and the results of these tests are given in Table 2. As a standard comparison the NMR T_1 values for distilled water are also given.

The contrast between the relaxation rates of malignant Novikoff hepatoma and normal liver illustrates the degree of perturbation of endosolvent structure that can accompany malignant transformation. The considerable increase in relaxation times for the hepatoma (T_1, 0.826 second; T_2, 0.118 second) relative to normal liver (T_1, 0.293 second; T_2, 0.050 second) suggests a significant decrease in the degree of ordering of intra-cellular water in malignant tissue. In addition, it is apparent from the prolonged relaxation times of the two malignant tumors reported in Table 2 that NMR values make it possible to one to detect the presence of metastatic infiltrates of the liver from either Walker sarcoma or Novikoff hepatoma.

The differences between the relaxation rates of malignant tumors and normal liver may be used to distinguish the two malignancies from all of the normal tissues studied [P values less than 0.01 (Table 2)]. The values of T_1 in Walker sarcoma (0.736 second) and Novikoff hepatoma (0.826 second) were significantly greater than the values of T_1 in any of the normal tissues (0.293 to 0.595 second). The values of T_2 in the malignant tumors (0.100 and 0.118 second) were about twice the values of T_2 in rectus muscle (0.055 second) and liver (0.052 second). Furthermore, replicate measurements of T_1 in the malignant tissues were found to be highly reproducible (standard error of the mean, <0.02) and the normal tissues had a standard error of the mean of 0.03 or less despite deliberate scrambling of the ages and weights of the animals in the experimental colony.

The measurements were also unaffected by any change in the elevation of the sample position in the probe, packing and repacking of the specimen, or the stepwise rotation of the sample tube in the probe through 360°. In fact T_1 proved to be even relatively unchanged after the specimens stood overnight at room temperature (Tables 1 and 2, parenthetical values for rats 4 and 8).

TABLE 1 Spin-lattice (T_1) and spin-spin (T_2) relaxation times (in seconds) of normal tissue

		Tissue							
		. Rectus Muscle .		. Liver .		. Stomach .	.Small Intestine.	. Kidney .	. Brain .
Rat No.	Weight (g.)	T_1	T_2	T_1	T_2	T_1	T_1	T_1	T_1
1.....	156	0.493	0.050	0.286	0.050	0.272	0.280	0.444	0.573
2.....	150	0.548	0.050	0.322	0.060	0.214	0.225	0.503	0.573
3.....	495	0.541	0.050	0.241	0.050	0.260	0.316	0.423	0.596
4.....	233	0.576(*0.600)	0.070	0.306(*0.287)	0.048	0.247(*0.159)	0.316(*0.280)	0.541(*0.530)	
5.....	255	0.531	.....	0.300	.....	0.360	0.150	0.489	0.612
Mean and standard error									
		0.538 ± 0.015	0.055 ± 0.005	0.203 ± 0.010	0.052 ± 0.003	0.270 ± 0.016	0.257 ± 0.030	0.480 ± 0.026	0.595 ± 0.007

* Spin-lattice relaxation time after the specimen stood overnight at room temperature.

TABLE 2 Spin-lattice (T_1) and spin-spin (T_2) relaxation times (in seconds) in tumors

Rat No.	Weight (g)	T_1	T_2
		Walker sarcoma	
6	156	0.700	0.100
7	150	0.750	0.100
8	495	0.794(0.794)*	0.100
9	233	0.688	
10	255	0.750	
	Mean and S.E.	0.736 ± 0.022	0.100
	P	<.01+	
		Novikoff hepatoma	
11	155	0.798	0.120
12	160	0.852	0.120
13	231	0.827	0.115
	Mean and S.E.	0.826 ± 0.013	0.118 ± 0.002
	P	<.01+	
		Fibroadenoma (benign)	
14		0.448	
15		0.537	
	Mean	0.492	
		Distilled water	
		2.691	
		2.690	
		2.640	
Mean and S.E.		2.677 ± 0.021	

*Spin-lattice relaxation time after the specimen stood overnight at room temperature.

+The P values are the probability estimates of the significance of the difference in the means of T_1 for the malignant tumor and for brain.

The above experiments were conducted without the aid of equipment which could automatically compute T_1 and T_2 and automatically compare these values with their respective standards for the detection of cancerous tissues. However, a computer 49 could be provided to calculate $T_1 = \tau$ null / ln2 where τ is the time interval between the 180° pulse and the 90° pulse in the Carr-Purcell method described above that produces no free induction decay after the 90° pulse. This value of τ null is the value τ selected manually on the pulse separation selector 46 to produce the above condition. An electrical signal responsive to the setting of τ on selector 46 is provided by the programmer 41 and appears at output terminal 50. The signal is directed to an input terminal 51 of computer 49 through a switch 52 which is closed when it is desired to compute and compare values of T_1 with a standard.

The value of T_1 appearing as a signal at an output 53 of computer 49 is directed to an input terminal 54 of a computer 55. This computer compares the value of T_1 at input terminal 54 with a table of standard values for non-cancerous and cancerous tissues stored on a data storage unit 56. This comparison provides an indication whether cancer is present and if so its degree of malignancy. The comparison is made for the particular nuclei selected, each type of nuclei having a separate table of standards. For a given field strength provided by electromagnet 23 the frequency of the radio frequency generator 25 determines which nuclei are selected. Thus, the selected nuclei are known from the setting of the frequency selector 27. An electrical signal responsive to the setting of the frequency selector 27 appears at output terminal 57 of the radio frequency generator 25. This signal is converted into a digital code by a digital coder 58 and supplied to an input terminal 59 of computer 55.

After comparing the values of T_1 for the sample 21 with the standards on a data storage unit 56, the computer 55 supplies a signal via line 60 to a display unit 61. This signal is converted by display unit 61 into a visual signal indicating whether cancer is present in tissue sample 21 and if so the degree of malignancy.

For automatically calculating values of T_2 for comparing with a standard, a sampler 62 is provided for sampling the signal appearing at terminal 38 of the low pass filter when the switch 63 is closed. The sampler 62 samples the signal at terminal 38 at times 0, 2τ, 4τ, 6τ, etc., where τ is again the time interval between the 180° pulse and the 90° pulse generated by the programmer 41 for determining T_2. A signal responsive to the τ selected on selector 46 appears at output terminal 50 of programmer 41 and is supplied to input terminal 64 of sampler 62. To better understand the technique used to determine T_2, reference is made to Fig. 3 wherein a plot of the signal appearing at terminal 38 is shown. This plot shows the signal when the programmer 41 has been set to supply a 90° pulse and a train of 180° pulses pursuant to the Carr-Purcell method of measuring T_2. The 900 pulse is applied at time 0, and a 180° pulse is applied at times τ, 3τ, 5τ, 7τ, etc. The sampler 62 samples the signal at times 0, 2τ, 4τ, 6τ, etc., thus obtaining values of the signal which lie on the echo envelope.

The sampled values detected by sampler 62 appear on an output terminal 65 of sampler 62 and are directed to an input terminal 66 of computer 67. The sampled values appearing on input terminal 66 are stored on a real time basis in the computer 67 as is well known in the art.

[*A section of the patent has been removed for the sake of brevity.*]

I claim:

1. A method for detecting cancer comprising:
   - **a.** measuring and establishing standard NMR spin-lattice relaxation times and spin-spin relaxation times for both normal and cancerous tissue of the type under analysis using as an indicator nuclei at least one nuclei which exhibits deviant behavior in cancerous tissue;
   - **b.** measuring the NMR spin-lattice relaxation times and spin-spin relaxation times for the suspected tissue to determine the extent of deviant behavior of the indicator nuclei; and
   - **c.** comparing the values obtained in (b) against the standards obtained in (a).
2. The method of claim 1, wherein the indicator nuclei are cell water protons.
3. A method for distinguishing cancerous tissue from normal, healthy tissue which comprises:
   - **a.** positioning a tissue sample in a nuclear induction apparatus having two sources of magnetic energy;
   - **b.** actuating the first source to expose the sample to a magnetic field;

   c. actuating the second source to expose the sample to oscillating magnetic radiation having a frequency selected to permit the absorption by selected nuclei of energy from the oscillating field through nuclear magnetic resonance absorption, thereby inducing transistors between equilibrium states and higher energy states;
   d. terminating the radiation from the second source after a predetermined duration;
   e. measuring the spin-lattice relaxation time and spin-spin relaxation time as the energized nuclei return to their equilibrium states; and
   f. comparing the relaxation times with respective predetermined standards for normal and cancerous tissue, said comparisons indicating the presence and degree of malignancy of cancerous tissue.

4. The method according to claim 3, wherein the first source of magnetic energy is actuated to provide a static magnetic field orthogonal to the second magnetic field.
5. A method according to claim 4, wherein the selected nuclei comprise water protons.
6. A method according to claim 2, wherein the selected nuclei are phosphorus.
7. A method of detecting cancerous mammalian tissue comprising:
   a. providing a magnetic field space having a locus therein of equal field strength;
   b. positioning a mammal within said space;
   c. directing a beam of magnetic radiation having a predetermined frequency to impinge on said locus in a direction orthogonal to the direction of the magnetic field, the frequency being selected to permit absorption by selected nuclei in the region of impingement of energy from the beam through nuclear magnetic resonance absorption, thereby inducing transitions in the selected nuclei between equilibrium states and higher energy states;
   d. terminating the beam after a predetermined duration;
   e. measuring the spin-lattice relaxation time and the spin-spin relaxation time values as the energized nuclei return to their equilibrium states; and
   f. comparing the values obtained in (e) with predetermined like values for normal and for cancerous tissue, said comparison indicating the existence and degree of malignancy.

A section of the patent has been removed for the sake of brevity.

RAYMOND V. DAMADIAN

3.9 GLASS AND FIBER-OPTICS TECHNOLOGY

The practice of using light to convey rudimentary information is many centuries old [3.9.1]. Light, visible to the naked eye, had been encoded as information-bearing signals and relayed between distant hills. Telegraph hills, scattered around the European countryside, date back to the 1790s. The concept was documented early; however, the development and fabrication of fibers and associated devices was long delayed.

All the same, glass manufacturers had been actively engaged in altering the properties of glass as far back as 1915. U.S. Patent No. 1,151,911 [3.9.2], assigned to Corning Glass Works, attempts to adjust the composition of glass that adheres to metal and its sustained durability. The basic questions that were being addressed revolved around the strength and bondage of glass and its durability.

The pursuit for fluidity in molten glass, for the "formation of glass articles and especially glass fibers," came much later, in the 1950s. Details of glass composition are listed in [3.9.3] in U.S. Patent No. 2,664,359, issued in 1953 and assigned to Owens-Corning Fiberglas Corporation. The patent addresses the issues of drawing very fine fibers with desired fluidity, durability, and melting rate. Questions on the effects of absorption of moisture by the fibers are also addressed, however, in view of the dominant hydra-oxyl peaks [3.9.4] that influence the wavelength of light and the attenuation constants for the different types of fibers.

During the early years, the composition of glass and control of the contaminants were undertaken based on chance rather than on a scientific approach [3.9.5]. However, later the precise effect of each of the impurities and their effect on the refractive index and the attenuation coefficient were necessary to draw very practical fibers strands as a compound glass [3.9.6] with both core and cladding together. The practice was in its infancy during the early 1970s and reached near perfection during the 1980s. Other types of fibers (e.g., "dark fibers") are still being perfected for extremely low attenuation.

The possibility of using optical signals for communication was proposed as far back as 1880 by Alexander Graham Bell. The device, "Apparatus for Signaling and Communicating Called 'Photophone,' is described in U.S. Patent No. 235,199, issued on December 7, 1880. Novel as it was, it was not practical enough to be implemented. This invention came about 26 years after John Tyndell [3.9.7] performed light-transmission experiments over a dielectric conductor before the Royal Institution in England in 1854.

The transmission properties of glass fiber were seriously investigated during the early sixties for building a viable fiber-optic system. The invention of lasers at Bell Laboratories in 1960 triggered this study of glass. As we stressed earlier, the acceptance and success of any concept depends on the supporting developments, such as the social demand for the new device/technology, the capital outlay and commitment to make the system a reality, and the win-win combination that makes two or more powerful entities bring the system to a level of commercial acceptability.

First, there a was a fusion of laser [3.9.8] technology, a powerful optical source (from Bell Laboratories), and fiberglass refinement and fabrication technology, which reduces the attenuation and the consistency of the characteristics of the fiber (from Corning Glass). During the 1970s, the practicality of optical systems was firmly established. (Warren Wolf's patent on fiber optics is presented here.)

The second part is still being played out by the fabrication of synthetic materials for stable quantum-well devices [3.9.9] and the fabrication of wideband, wavelength, glass material for DWDM applications on fiber strands. The concepts of optical amplification are also being explored at the quantum level to enhance the very highest quality of optical-signal restoration. Very successful experimental results and pragmatic tests are the reason for its deployment in several projects at Bell Labs. (See Chapter 17 of [3.9.10].) This network spans over 33,000 km and is seen as a major factor in connecting all of the countries in the information revolution already underway in the west.

Historically, from the perspective of glass technology for fibers, the multilayered glass structure [3.9.11] was suggested during the mid-sixties, and a rough specimen of fiber with about 20 dB/km loss was produced by Corning Glass in 1970. Optical systems have been functioning with ever-increasing capacity and sophistication. Large capacity, immunity from stray electromagnetic interference, low weight, and low cost are the chief attractions of the optical-fiber waveguide (simply stated, optical fiber) medium. The highest quality single-mode fiber drawn during the late eighties offers 0.16 dB/km loss at 1.55 μm wavelength and less than 0.35 dB/km at 1.30 μm wavelength. These lower loss limits are due to intrinsic scattering within the silica glass rather than manufacturing tolerances in extruding the fiber glass. The chromatic dispersion is also equally attractive at approximately 15 ps/km-nm. The explanation of the units of measurements for chromatic dispersion in the low-loss window of 1.5 μm (with zero dispersion at 1.3 μm) are explained in [3.9.10]. Specially designed fibers further reduce the overall dispersion to about 17 ps/km-nm at 1.5 μm band, and they prove satisfactory for laser sources without chirp or drifting wavelength of the photonic energy emitted. These two effects, light loss (or attenuation) and chromatic smearing (or delay distortion effects), eventually limit the product of bit rate and distance through the fiber carrying information. Since there is only one mode in the single-mode fibers, the light through these fibers does not suffer modal-dispersion effects, but instead is susceptible to chromatic-dispersion effects.

Fibers are dielectric waveguides. A subset of the classic Maxwell's equations* leads to the fields that coexist in the fiber. Since the material is (largely) linear, isotropic without free circulating currents, and devoid of free charges, the basic equations are written as

$$\nabla \times \mathbf{E} = \frac{\partial B}{\partial t}, \qquad \nabla \times \mathbf{H} = \frac{\partial D}{\partial t},$$

$$\nabla \cdot \mathbf{B} = 0, \qquad \text{and} \qquad \nabla \cdot \mathbf{D} = 0,$$

with the traditional relations between **D** and **E** defining ε **B** and **H** defining μ as follows:

$$\mathbf{D} = \varepsilon \mathbf{E} \qquad \text{and} \qquad \mathbf{B} = \mu \mathbf{H}.$$

The solution of the preceding two equations leads to the two scalar wave equations

$$\nabla^2 \mathbf{E} = \mu\varepsilon \frac{\partial^2 \mathrm{E}}{\partial t^2} \qquad \text{and} \qquad \nabla^2 \mathrm{H} = \mu\varepsilon \frac{\partial^2 \mathrm{H}}{\partial t^2}.$$

From the point of view of applications, the structure is cylindrical. The cylindrical structures are the core and the cladding. A fiber with distinctly different optical densities makes up the core and cladding. At the interface, there are two effects: refraction and reflection. Both affect the propagation of light. When the optical density is graded as a function of its diameter, a combination of refraction and reflection along the radius affects the recapture and propagation of light through the fiber core.

Three crucial layers of the fiber are the core or the innermost layer, the cladding, and the outer protective jacket [3.9.10]. The core and cladding together provide the optical environment for the light wave(s) through the fiber. The cladding is tailored to be an integral part of the fiber geometry that contains and propagates the light. The core has a slightly higher optical density (index of refraction) than that of the cladding. The optical density of the core material or silica glass is higher (1.46 to 1.5) than that of the air where the light is incident, which affects the propagation in two ways:

*For an excellent discussion of Maxwell's equations and their generalizations, including relativistic corrections, see Kraus, J. D., and Carver, K. R., *Electromagnetics*. McGraw-Hill, New York, 1973.

First, the velocity in the denser medium is reduced, and second, the direction is changed by the conventional rules of refraction. The net effect of the two is a realignment of the rays of light towards the axis of reference or the axis of the fiber in a single (H_{11}) mode. Maurer, Keck, and Schultz present a technique to realize this efficient mode by altering the geometry of the optical fiber and the molecular chemistry of the core and cladding glass. Their patents are presented in Sections 3.9.1 and 3.9.2. The concept of total internal reflection is captured in these two patents to reduce the loss of optical energy as it traverses axially through the fiber in the most efficient H_{11} mode.

The contributions of three inventors, R.D. Maurer, P.C. Schultz, and D. B. Keck, are (in our opinion) as significant to fiber-optic silica-glass technology as the contributions of the W. B. Shockley, J. Bardeen, and W. H. Brattain to semiconductor technology. An excellent example of synergy of goal-oriented thought (see Sections 2.5.1 and 2.5.2) within a creative process of blending silica glass as a multilayered optical medium such as to only permit the HE_{11} mode of propagation (and to also minimize the formation of the "light scattering centers" that disperse the optical energy) is evident in these two patents (see Sections 3.9.1 and 3.9.2). In the former case, Shockley, Bardeen, and Brattain made a conscious and concerted effort to duplicate the amplification process of an electronic vacuum tube in semiconductor materials and at room temperatures. Another example of this confluence of the two very creative minds working in close cooperation is documented by the contributions of A. L. Schawlow and C. H. Townes (see Section 3.7.1). We reiterate this pattern of convergence of coherent creative thinking in a small community of researchers working toward a well-directed industrial/corporate goal once again in Sections 4.1.3 and 4.1.4.

REFERENCES (C-3.9)

[3.9.1] Tubbs, Stewart L., and Moss, Sylvia, *Human Communication*, 8th ed. McGraw-Hill Humanities/Social Sciences/Languages, New York, 1999.

[3.9.2] Sullivan Eugene C., and Taylor, William C., "Glass." Letter Patent, Patent Number 1,151,911, August 31, 1915.

[3.9.3] Dingley, David P.," "Glass Composition." Patent Number 2,664,359, June 1, 1951.

[3.9.4] Barnoski, Michael K. (Editor), *Fundamentals of Optical Fiber Communications*, 2d ed. Academic Press, New York, 1981.

[3.9.5] Othonos, Andreas, and Kalli, Kyriacos, *Fiber Bragg Gratings: Fundamentals and Applications in Telecommunications and Sensing*. Artech House Optoelectronics Library, Artech House, Boston, 1999.

[3.9.6] Marcuse, D., *Light Transmission Optics*, Van Nostrand Reinhold Company, New York, 1972.

[3.9.7] Tyndell, J., "On Some Phenomenon Connected with the Motion of Liquids." *Light Transmission Optics*, Vol. 1, 1854, p. 446.

[3.9.8] Abraham, Neal B., Arecchi, F. T., Morradian, Aram, and So, Alberto (Editors), *Physics of New Laser Sources*. NATO ASI Series; Series B, Physics, Vol. 132, Plenum Press, New York, 1986.

[3.9.9] Nag, B. R., *Physics of Quantum Well Devices*. Vol. 7, Kluwer Academic Publishers, Boston, 2001.

[3.9.10] Ahamed Syed V., and Lawrence, Victor B., *Design and Engineering of Intelligent Communication Systems*. Kluwer Academic Press, Boston, 1997.

[3.9.11] "Undersea Communications Technology." *AT&T Technical Journal*, Vol. 74, No. 1, Jan–Feb. 1995, pp. 75–82.

R.D. Maurer (1924–)

P.C. Schultz (1942–)

3.9.1 Fiber Optics (R.D. Maurer and P.C. Schultz, 1972)

UNITED STATES PATENT

May 2, 1972 3,659,915

Fused Silica Optical Waveguide*

Inventors: Robert D. Maurer; Peter C. Schultz, both of Painted Post, N.Y.

Assignee: Corning Glass Works, Corning, N.Y.
Filed: May 11, 1970, Application No.: 36,109

U.S. CI 350/96 WG, 65/30, 65/DIG. 7, 65/DIG. 8
Int. CI GO2b 5/14, HOlp 3/00
Field of Search 350/96 WG, 175 GN
U.S. Cl. 65/30 10 Claims

References Cited

UNITED STATES PATENTS

3,542,536	11/1970	Flam et al 350/96 WG X
3,445,785	5/1969	Koester ct al 350/96 WG UX
3,533,013	10/1970	Seitz 350/175 GN X

Primary Examiner—John K. Corbin
Attorney—Clarence R. Patty, Jr. and Walter S. Zebrowski

*The entire patent (with 10 claims and 2 figures) is presented here.

Abstract An optical waveguide having a high purity fused silica cladding layer, and a core of high plurality fused silica doped with a sufficient amount of a multivalent metal oxide so as to increase the index of refraction of the core above that of the cladding layer.

Background of the Invention

1. Field of the Invention
   The increase in the amount of traffic that communication systems are required to handle has forced the development of high capacity systems. The presently used systems which operate between 10^4 hz and 10^9 hz, although relatively new, have already become saturated in some frequency bands due to excessive traffic. To allow for needed increases in traffic which the present systems cannot accommodate, commercial communication groups are presently installing high capacity systems that operate between 10^9 hz and 10^{12} hz. Even with the increased capacity that these new systems will offer, traffic growth is so rapid that saturation of such systems is anticipated in the very near future. Therefore, to accommodate even more traffic higher capacity communication systems operating around $l0^{15}$ hz are needed. These systems will hereinafter be referred to as "optical communication systems" since 10^{15} hz is within the frequency spectrum of light. Waveguides are desirable for effective transmission of information by systems operating at frequencies above 10^9 hz.. Systems operating at frequencies between 10^9 hz and 10^{12} hz normally use electrical conductive waveguides a transmitting medium. However, conventional electrical conductive waveguides are not satisfactory for transmitting information at frequencies around 10^{15} hz.
2. Description of the Prior Art
   Producing a satisfactory transmitting media for frequencies around 10^{15} hz has been one of the more difficult problems in the development of an effective optical communication system. Such transmitting media are hereinafter referred to as "optical waveguides." To be an effective transmitting media for an optical communication system an optical waveguide should transmit light without excessive attenuation and should not cause dispersion or scattering of the transmitted light. In addition, an optical waveguide should allow only preselected modes of light to propagate along the fiber.

   A very thorough and complete discussion concerning the operational theories of optical waveguides is contained in U.S. Pat. No. 3,157,726 issued to Hicks et al. and in the publication "Cylindrical Dielectric Waveguide Mode" by E. Snitzer, *Journal of the Optical Society of America*, Vol. 51, No. 5, pages 491-498, May 1961. Another excellent source of information concerning optical waveguides is *Fiber Optics - Principles and Applications* by N. S. Kapany, Academic Press, 1967.

   Optical waveguides are a unique type of optical fiber in that many of the physical characteristics and parameters must be carefully coordinated. In the conventional optical fiber substantially all of the transmitted light is retained within the core, and the light transmission qualities of the cladding layer are therefore of no consequence. Some fibers, in fact, are purposely produced with a core possessing good transmission qualities and a cladding layer that is highly absorptive. In order to prevent transmitted light from escaping the fiber, one of the more basic considerations in producing conventional optical fibers is that the index of refraction of the core be larger than the index of refraction of the cladding layer. In fact, if the difference between the two indices of refraction is increased, the amount of light

Patented May 2, 1972 3,659,915

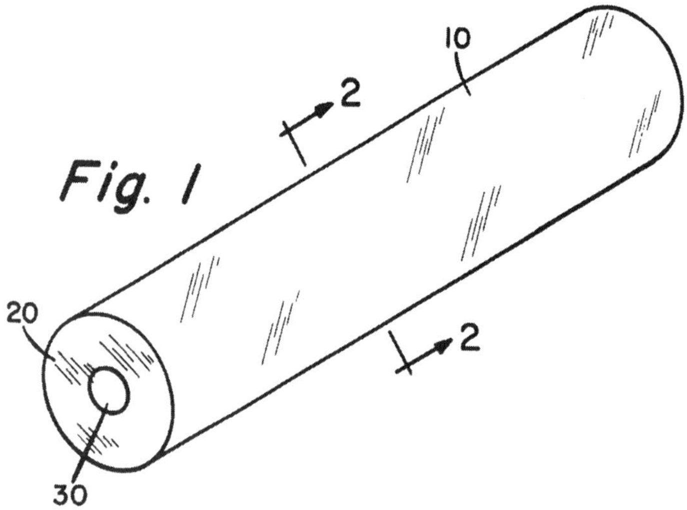

Fig. 1

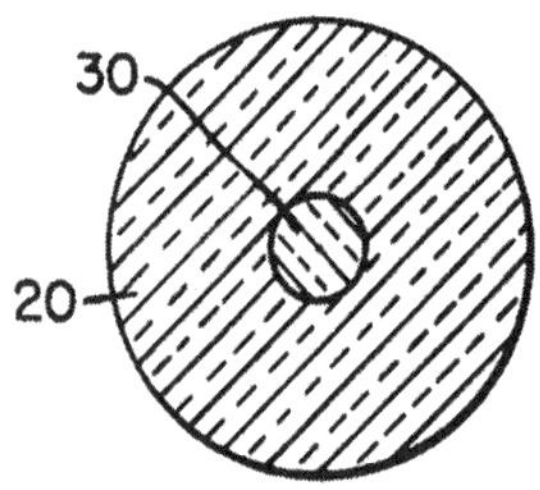

Fig. 2

INVENTORS
ROBERT D. MAURER,
PETER C. SCHULTZ

BY *Walter S. Zebrowski*
ATTORNEY

escaping from the fiber will decrease. Therefore, it is common practice in the production of conventional fibers to make the difference between the two indices of refraction as large as possible. However, as is explained by N. S. Kapany if an optical fiber is to function as an optical waveguide, that is, limiting the transmitted light to preselected modes, the diameter of the core, the index of refraction of the core and the index of refraction of the cladding layer must be carefully coordinated. An abbreviated and simplified discussion of optical waveguide theory follows so as to assist understanding of this invention.

Explanations of the physics of electrical and magnetic microwave transmission are often based on the concept that such waves are made up of an infinite number of modes. Each of these modes has its own propagation and distribution characteristics. The propagation of light waves is governed by the same laws of physics that govern microwave propagation and therefore can also be studied in terms of modes.

Since each mode of light traveling along a glass fiber structure propagates at its own characteristic velocity it can be shown that if the same information is initially supplied to all modes there will be a dispersion of this information after a given length of fiber due to the different propagation velocities. It then follows that if light propagation along the optical fiber could be restricted to preselected modes, more effective information transmission would result.

The conditions under which propagation of a particular mode will no longer be localized within the core of an optical fiber can be expressed in terms of a cutoff value U. An exceptionally complex equation, and an explanation thereof, from which the value of U for a particular mode can be determined may be found on page 55 of *Fiber Optics Principles and Applications* by N. S. Kapany. On the same page of this book Kapany expresses a "fiber characteristic term" — "R" in terms of the optical fiber variables by the equation

$$R = (2\pi a/\lambda)\sqrt{n_1{}^2 - n_2{}^2} \qquad 1.$$

where

a = core radius of the waveguide
λ = wavelength of light to be transmitted
n_1 = core index of refraction
n_2 = cladding index of refraction

which can be rewritten as

$$R = (2\pi a/\lambda)\sqrt{(n_1 + n_2)(n_1 - n_2)} \qquad 2.$$

Then as is explained in Kapany, for a particular mode to propagate within an optical fiber having a particular "fiber characteristic term" – R. R must be greater than or equal to the cutoff value U for said mode.

The mode HE_{11}, the definition and physical characteristics of which can be found in the aforementioned cited sources, is the only mode of light that will propagate along a fiber which has an R value of less than 2.405. Therefore, fi R is set equal to 2.405, and equation (2) is evaluated it can be seen that a method of limiting light propagation of a desired wavelength to one mode is to coordinate the parameters a, n_1 and n_2 of the waveguide. That is, if the difference between the two indices of refraction $(n_1 - n_2)$ increases, the core radius a must decrease and if $(n_1 - n_2)$ decreases the core radius a must increase. Producing waveguides having core and cladding indices of refraction within limits necessary to maintain single mode propagation is difficult even for waveguides with very small cores. The difficulty is markedly increased in producing waveguides with larger cores. As an example, if the optical waveguide is to have a small core, that is a core diameter of approximately one micron, the required difference in the two indices of refraction will be on the order of 10^{-2}; and if the optical waveguide is to have a large core, that is a core diameter of approximately 1 millimeter, the required difference in the two indices of refraction would be even smaller; that is, on the order of 10^{-4}. The difficulty of very accurately controlling the diameter of the core, the index of refraction of the core and the index of refraction of the cladding layer has in the past made the production of optical waveguides very slow and expensive.

Summary of the Invention It is an object of this invention to provide an economical and more easily-formed optical waveguide that overcomes the heretofore noted disadvantages.

Another object of this invention is to provide an optical waveguide that will not absorb an excessive amount of transmitted light.

Briefly, according to this invention, an optical waveguide is produced comprising a cladding layer formed substantially from pure fused silica, and a core formed from fused silica doped with one or more doping materials so as to selectively increase the index of refraction above that of the cladding.

Brief Description of the Drawings FIG. 1 shows an oblique view of an optical waveguide in accordance with the invention.

FIG. 2 shows a sectional view taken at liner 2-2 of FIG. 1.

Detailed Description of the Invention According to this invention a completely new and novel approach has been taken as to the type of material to be used in the production of optical waveguides. Contrary to the soft and easily worked materials normally used in the production of optical waveguides, I have discovered that substantially pure fused silica which is extremely hard and difficult to work can be economically and readily used as material from which a superior optical waveguide can be produced. The term "pure fused silica" when used herein is hereby defined as fused silica containing no impurities in an amount greater than 0.1 percent by weight except for hydrogen-oxygen groups which may be present in amounts up to 5 percent by weight. The following discussion is set out to aid understanding as to why fused silica, apparently so unsuitable, is in reality an excellent material for this purpose. Many of the difficulties normally encountered in the formation of waveguides can be substantially eliminated if both the core and the cladding layer possess similar physical characteristics such as, viscosity, softening point and coefficient of expansion. Further, if a very small yet precise difference can be maintained between the index of refraction of the core and the index of refraction of the cladding layer the diameter of the core may be made proportionally larger and therefore more easily controlled, and the waveguide will still maintain the ability to limit light propagation to preselected modes. Fused silica is readily attainable with exceptionally high purity, and it has been found that fused silica in such pure form has a very predictable index of refraction. Therefore, adding a precise percentage by weight of doping materials to fused silica will produce "doped fused silica" with an index of refraction predictably higher than that of pure fused silica, and because of its high purity level only a minimal amount of doping material will be necessary to cause an appreciable change in the index of refraction. Referring to

FIGS. 1 and 2, there is shown an optical waveguide 10 having a cladding layer 20 and a core 30. According to this invention pure fused silica or slightly doped fused silica is used as the cladding layer 20, and fused silica doped with the necessary amount of dopant to increase the index of refraction to a desired level above that of the cladding layer is the material used for the core 30. As a result of the minimum amount of dopant material necessary to increase the index of refraction of fused silica, at any given temperature, the physical characteristics of the doped fused silica, such as viscosity, softening point and coefficient of expansion, used for the core 30, are almost identical to the more nearly pure fused silica used for the cladding layer 20, thereby substantially eliminating many of the difficulties in drawing the waveguide and also reducing the possibility of devitrification. To make certain that doped fused silica possesses optical and physical characteristics almost identical to those of pure fused silica, doping materials should not exceed 15 percent by weight.

In addition, fused silica has excellent light transmission qualities in that absorption of light energy and intrinsic scattering of light by the material is exceptionally low. Scattering of light that does occur within fused silica is primarily caused by the presence of impurities rather than the intrinsic nature of the material itself. Furthermore, fused silica is such a hard material that an optical waveguide once formed possesses the quality of being highly resistant to damage from high temperatures, corrosive atmosphere and other severe environments.

A waveguide possessing a pure fused silica cladding, and a doped fused silica core may be produced by any suitable method including but not limited to: (a) inserting a rod of fused silica, doped as required to increase the index of refraction to the desired level above that of pure fused silica, into a tube of pure fused silica, raising the temperature of the rod and tube combination until said combination has a low enough viscosity for drawing, and then drawing said rod and tube until the tube collapses around and fuses to the rod and the cross-sectional area of the combination is decreased to the desired dimension; or (b) the method set out in co-pending application "Method of Producing Optical Waveguide Fibers" by Donald B. Keck and Peter C. Schultz, Ser. No. 36,267, filed on May 1 1, 1970, and assigned to a common assignee, which method comprises first forming a film of doped fused silica on the inside wall of a tube of pure fused silica, and then drawing this composite structure to reduce the cross-sectional area and to collapse the film of doped fused silica to form a fiber having a solid cross-section of the desired diameter.

There are many dopant materials that can be added to fused silica in minute quantities to increase its index of refraction to a predetermined level. However, the diffusion properties of the dopant material must be considered. For most methods of producing optical waveguides suitable dopants are those having minimum diffusion properties so that the dopant will not diffuse out of the core and into the cladding layer. Such diffusion would effectively increase the diameter of the core and thereby alter the mode selection abilities of the waveguide. Suitable dopants having minimum diffusion properties include, for example, such multivalent metal oxides as titanium oxide, tantalum oxide, tin oxide, niobium oxide, zirconium oxide, ytterbium oxide, lanthanum oxide, and aluminum oxide. However, in certain methods of producing optical waveguides the core is drawn with a small diameter, and subsequent diffusion increases the core diameter to the desired size. Doping materials containing alkali ions will readily diffuse into the fused silica cladding of an optical waveguide and increase the effective core diameter. Cesium and rubidium are examples of especially suitable materials for such use.

In one embodiment of the present invention, pure fused silica is used as the cladding material, and fused silica doped with a desired doping material or combination of doping materials in the amount effective to increase the index of refraction to the desired level above that of the cladding layer is used as the core material.

In a second embodiment, fused silica slightly doped with a desired doping material or combination of doping materials is used as the cladding material, and fused silica more heavily doped with a desired doping material or a combination of doping materials, either the same or different than those used in the cladding, to increase the index of refraction to the desired level above that of the cladding layer is used as the core material.

It has been found that light absorption properties may be decreased and light transmission qualities improved in titanium oxide doped fused silica formed into optical wave-guides, if the waveguides are drawn in an oxygen atmosphere and then "heat treated" in an oxygen atmosphere. The "heat treatment" referred to consists of heating the waveguide in an oxygen atmosphere to between 500° and 1,000°C for not less than thirty minutes; the length of treatment being related to the treatment temperature. Lower temperatures require longer treatment periods, while treatment at the higher temperature allows shorter time periods.

A specific example of a waveguide produced by the practice of this invention follows. A 1 ½ to 2 micron film of fused silica doped with titanium oxide was bonded to the inside wall of a ¾ inch outside diameter, ¼ inch inside diameter, substantially pure fused silica tube by the method described in said co-pending Keck-Schultz application. The deposited doped fused silica essentially consisted of 94.75 percent fused silica and 5.25 percent titanium oxide. The composite structure was then heated in substantially an oxygen atmosphere until it reached a temperature at which the materials had low enough viscosity for drawing (approximately 1,900°C). The composite structure was then drawn to reduce the diameter thereof until the film of titanium oxide doped fused silica was collapsed, that is, it sealed the longitudinal hole to form a solid core surrounded by pure fused silica. The resulting composite rod was then further drawn to reduce the diameter thereof to a final diameter of approximately 100 microns. The core of the optical waveguide was measured at approximately 3 microns in diameter. The core index of refraction was approximately 1.466 while the cladding had an index of refraction of approximately 1.4584. After the fiber was drawn, it was heat treated at 800' C in an oxygen atmosphere for approximately three hours.

Although the present invention has been described with respect to specific limitations, it is not intended that these limitations affect the scope of the invention except insofar as set forth in the following claims.

Claims:

1. An optical waveguide comprising a cladding layer formed of a material selected from the group consisting of pure fused silica and fused silica to which a dopant material on at least an elemental basis has been added, and a core formed of fused silica to which a dopant material on at least an elemental basis has been added to a degree in excess of that of the cladding layer so that the index of refraction thereof is of a value greater than the index of refraction of said cladding layer, said core being formed of at least 85 percent by weight of fused silica and an effective amount up to 15 percent by weight of said dopant material.
2. The waveguide of claim 1 wherein the cladding layer is substantially pure fused silica.
3. The waveguide of claim 9 wherein the core is formed of at least 94.75 percent by weight of fused silica and an effective amount up to 5.25 percent titanium oxide.
4. An optical waveguide comprising a cladding layer formed of a material selected from the group consisting of pure fused silica and doped fused silica, and a core formed of fused

silica doped to a degree in excess of that of the cladding layer so that the index of refraction thereof is of a value greater than the index of refraction of said cladding layer, said core being doped with at least one multivalent metal oxide selected from the group consisting of titanium oxide, tantalum oxide, tin oxide, niobium oxide, zirconium oxide, ytterbium oxide, lanthanum oxide, and aluminum oxide.

5. An optical waveguide comprising a cladding layer formed of a material selected from the group consisting of pure fused silica and doped fused silica, and a core formed of fused silica doped to a degree in excess of that of the cladding layer so that the index of refraction thereof is of a value greater than the index of refraction of said cladding layer, said core being doped with at least one material selected from the group consisting of cesium and rubidium,

6. An optical waveguide comprising a cladding layer formed of fused silica doped with at learnt one multivalent metal oxide selected from the group consisting of titanium oxide, tantalum oxide, tin oxide, niobium oxide, zirconium oxide, ytterbium oxide, lanthanum oxide, and aluminum oxide, and a core of fused silica doped to a degree in excess of that of the cladding layer so that the index of refraction thereof is of a value greater than the index of refraction of said cladding layer.

7. An optical waveguide comprising a cladding layer formed of fused silica doped with at least one material selected from the group consisting of cesium and rubidium, and a core formed of a fused silica doped to a degree in excess of that of the cladding layer so that the index of refraction thereof is of a value greater than the index of refraction of said cladding layer,

8. An optical waveguide comprising a cladding layer formed of fused silica doped with titanium oxide, and a core formed of fused silica doped to n degree in excess of that of the cladding layer so that the index of refraction thereof is of a value greater than the index of refraction of said cladding layer, said core being doped with titanium oxide.

9. An optical waveguide comprising a cladding layer formed of substantially pure fused silica, and a core formed of fused silica doped to a degree so that the index of refraction thereof is of a value greater than the index of refraction of said cladding layer, said core being doped with titanium oxide.

10. An optical waveguide comprising a cladding layer formed of substantially pure fused silica, and a core formed of fused silica doped to a degree so that the index of refraction thereof is of a value greater than the index of refraction of said cladding layer, said core being doped with at least one material selected from the group consisting of cesium and rubidium.

PO-10t0 **UNITED STATES PATENT OFFICE**

(5/69) **CERTIFICATE OF CORRECTION**

Patent No. 3,659,915 Dated . May 2,1972.

Inventor(s) Robert D. Maurer and Peter C. Schultz

It is certified that error appears in the above-identified patent and that said Letters Patent are hereby corrected as shown below:

In the Abstract, line 2, delete "plurality" and instead insert — purity —.

Claim 3, line l, delete "9" and instead insert — 1 —.

Signed and sealed this 9th day of January 1973.

(SEAL)
Attest:

EDWARD H.FLETCHER, JR.
Attesting Officer

ROBERT GOTTSCHALK
Commissioner of Patents

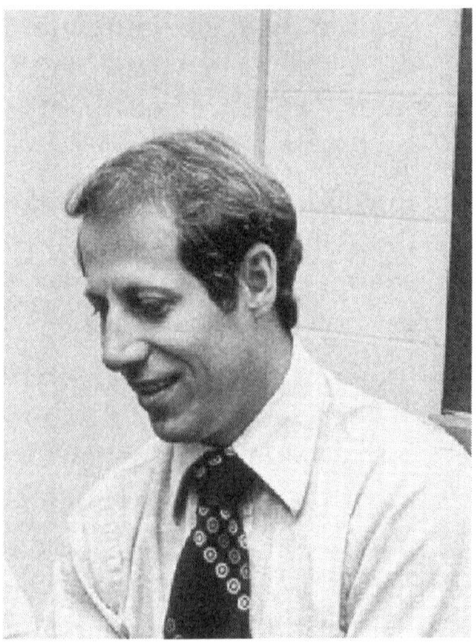

D.B. Keck (1941–)

3.9.2 Fiber Optics (Donald B. Keck and Peter C. Schultz, 1973)

UNITED STATES PATENT

January 16, 1973 3,711,262

Method of Producing Optical Waveguide Fibers*

Donald B. Keck, Corning, and Peter C. Schultz, Painted Post, N.Y., assignors to Corning Glass Works, Corning, N.Y.

Filed May 11, 1970, Ser. No. 36,267
Int. Cl. C03c *25/02;* C03b *21/00*
U.S. Cl. 65—3 28 Claims

The entire patent (with all 28 claims and 4 figures) is presented here.

Abstract of the Disclosure A method of producing an optical waveguide by first forming a film of glass with a preselected index of refraction on the inside wall of a glass tube having a different preselected index of refraction. This glass tube and glass film combination is then drawn to reduce the cross-sectional area and to collapse the film of glass to form a fiber having a solid cross-sectional area; the core being formed from the glass film, and the cladding being formed from the glass tube.

Background of the Invention ***(I) Field of the invention*** The increase in the amount of traffic that communication systems are required to handle has forced the development of high capacity systems. The presently used systems which operate between 10^6 Hz. and 10^9 Hz., although relatively new, have already become saturated in some frequency bands by excessive traffic. To allow for needed increases in traffic which the present systems cannot accommodate, commercial communication groups are presently installing high capacity systems that operate between 10^9 Hz. and 10^{12} Hz. Even with the increased capacity that these new systems will offer, traffic growth is so rapid that saturation of these systems is anticipated in the very near future. Therefore, to accommodate even more traffic, higher capacity communication systems operating around 10^{15} Hz. are needed. These future systems will hereinafter be referred to as "optical communication systems" since 10^{15} Hz is within the frequency spectrum of light. Wave-guides are desirable for effective transmission of information by systems operating at frequencies above 10^9 Hz. Systems operating at frequencies between 10^9 Hz. and 10^{12} Hz. normally use electrically conductive waveguides as their transmitting medium. Waveguides used by optical communication systems operating around 10^{15} Hz. are herein referred to as "optical waveguides" and are normally constructed from a transparent dielectric material such as glass or plastic. Producing a satisfactory optical waveguide has been one of the more difficult problems in the development of an effective optical communication system.

It is well known to one skilled in the art that light can be forced to propagate along a transparent fiber structure which has a higher refractive index than its surroundings. The ordinary use of such optical fibers is to transmit light or an image from one point to another. Optical fibers produced for these purposes must avoid excessive attenuation of the transmitted light to be effective. However, to be an effective transmitting media for an optical communication system an optical waveguide should not only transmit light without excessive attenuation, but also should not be subject to "cross-talk" from adjacent wave-guides, nor cause dispersion of the transmitted light. In addition, an optical waveguide should allow only pre-selected modes of light to propagate along the fiber.

A very thorough and complete discussion concerning the operational theories of optical waveguides is contained in U.S. Pat. No. 3,157,726 issued to Hicks et al. and in a publication "Cylindrical Dielectric Waveguide Mode" by E. Snitzer, *Journal of the Optical Society of America*, vol. 51, No. 5, pages 491–498, May 1961. Another excellent source of information concerning optical waveguides is *Fiber Optics—Principles and Applications* by N. S. Kapany, Academic Press (1967). An abbreviated and simplified discussion of some of these theories follows so as to assist understanding of this invention.

Explanations of the physics of electrical and magnetic microwave transmission are often based on the concept that such waves are made up of a finite number of modes. Each of these modes has its own propagation and distribution characteristics. The propagation of light waves is governed by the same laws of physics that govern microwave propagation and therefore can also be studied in terms of modes.

Jan. 16, 3,711,262

METHOD OF PRODUCING OPTICAL WAVEGUIDE FIBERS

Filed May 11, 1970

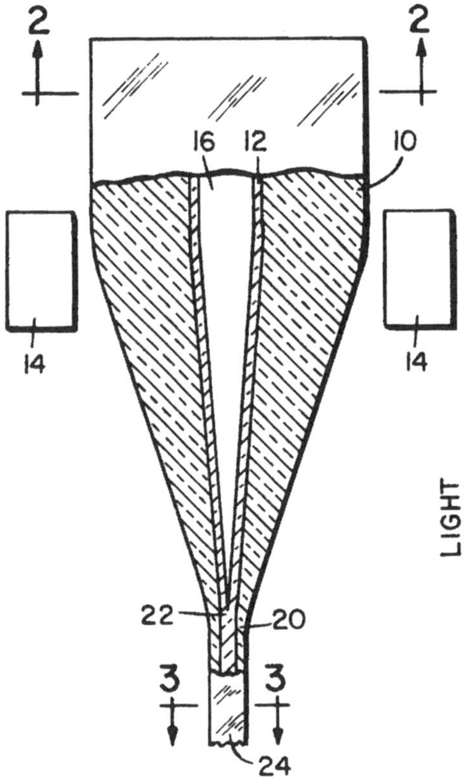

Fig. 1

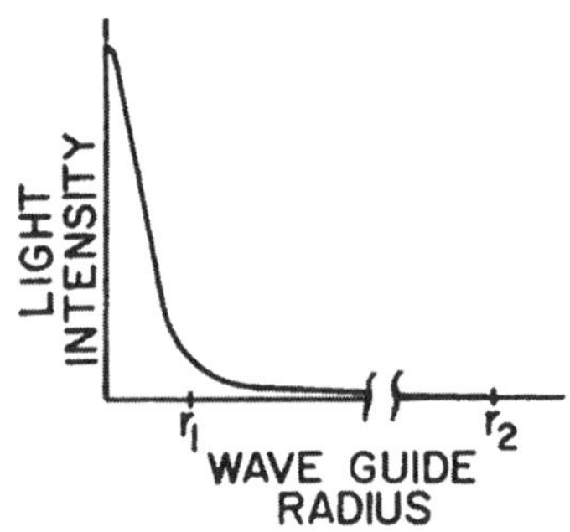

Fig. 4

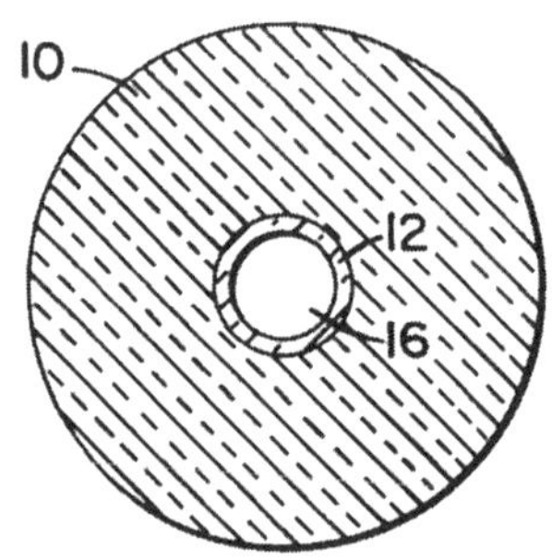

Fig. 2

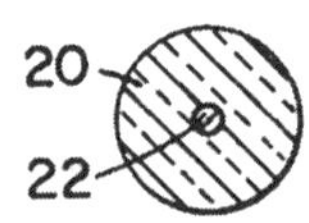

Fig. 3

INVENTORS.
Donald B. Keck
Peter C. Schultz
BY
Walter S. Zebrowski
ATTORNEY

Since each mode of light traveling along a glass fiber structure propagates at its own characteristic velocity, it can be shown that if the same information is initially supplied to all modes there will be a dispersion of this information after a given length of fiber due to the different propagation velocities. It then follows that if light propagation along the optical fiber could be restricted to preselected modes, more effective information transmission would result.

The conditions under which propagation of a particular mode will no longer be localized within and around the core of an optical fiber can be expressed in terms of a cutoff value U. An exceptionally complex equation, and an explanation thereof, from which the value of U for a particular mode can be determined may be found on page 55 of *Fiber Optics—Principles and Applications* by N. S. Kapany. On the same page of this book Kapany expresses a "fiber characteristic term"—"R" in terms of the optical fiber variables by the equation

$$R = \frac{2\pi a}{\lambda}\sqrt{n_1^2 - n_2^2} \qquad 1.$$

where

a = core radius of the waveguide
λ = wavelength of light to be transmitted
n_1 = core index of refraction
n_2 = cladding index of refraction

which can be rewritten as

$$R = \frac{2\pi a}{\lambda}\sqrt{(n_1 + n_2)(n_1 - n_2)} \qquad 2.$$

Then as is explained in Kapany, for a particular mode to propagate within an optical fiber having a particular "fiber characteristic term" - R. R must be greater than or equal to the cutoff value U for said mode.

The mode HE_{11}, the definition and physical characteristics of which can be found in the aforementioned cited sources, is the only mode of light that will propagate along a fiber which has an R value of less than 2.405. Therefore, fi R is set equal to 2.405, and equation (2) is evaluated it can be seen that a method of limiting light propagation of a desired wavelength to one mode is to coordinate the parameters a, n_1 and n_2 of the waveguide. That is, if the difference between the two indices of refraction $(n_1 - n_2)$ increases, the core radius a must decrease and if $(n_1 - n_2)$ decreases the core radius a must increase. Producing waveguides having core and cladding indices of refraction within limits necessary to maintain single mode propagation is difficult even for waveguides with very small cores. The difficulty is markedly increased in producing waveguides with larger cores. As an example, if the optical waveguide is to have a small core, that is a core diameter of approximately one micron, the required difference in the two indices of refraction will be on the order of 10^{-2}; and if the optical waveguide is to have a large core, that is a core diameter of approximately one millimeter, the required difference in the two indices of refraction would be even smaller; that is, on the order of 10^{-4}.

The intensity of light in an optical waveguide decreases as the radial distance from the waveguide center increases. Theoretically, of course, this intensity will never be zero regardless of the radial distance from the waveguide center the light travels. By proper selection of the core diameter, and the core and cladding indices of refraction, the decrease in intensity can be made to be so rapid, however, that the intensity can be presumed to be zero at a finite radial distance. There is no intensity distribution common to all waveguides, and intensity distribution will vary with the core diameter, the index of refraction of the core and cladding glass, and with the frequency of the light being transmitted; that is, within the same waveguide infrared light will have a different intensity distribution than will ultraviolet light.

If light intensity has not decreased to a very low level when it reaches the cladding-surrounding medium interface of the optical waveguide, three undesirable results may occur. First, light could strike said interface, be dispersed and then be reflected back into the waveguide core region. Second, some of the transmitted light could strike the interface and escape from the optical waveguide. Third, some of the escaped light could penetrate an adjacent optical waveguide thereby causing cross-talk.

(II) Description of the Prior Art A present method of producing an optical fiber is to insert a rod of glass possessing the desired core characteristics into a tube of glass possessing the desired cladding characteristics. The temperature of this combination is then raised until the viscosity of the materials is low enough for drawing. The combination is then drawn until the tube collapses around and fuses to the inside rod. This resulting combination rod is then further drawn until its cross-sectional area is decreased to the desired dimension. During the drawing process, the rod and tube are normally fed at different speeds to produce a fiber with the desired core to cladding diameter ratio. This method is normally used to produce glass optical fibers having a large core and a thin outside cladding. As an example, the total diameter to core diameter ratio of such a fiber might be 8:7.

Attempts to produce optical waveguides with a very small core diameter and a very thick cladding, that is a total diameter to core diameter ratio on the order of 100:1, by this method have been unsatisfactory because of the extreme difficulty in maintaining the core and cladding dimensions. An additional problem is often caused by numerous tiny air bubbles and foreign particles being trapped at the core and cladding interface. These air bubbles and foreign particles are a source of light scattering centers. In addition, the core and cladding materials of any waveguide must be selected so that there is only a very small, yet precise, difference between the two indices of refraction. Glass tubes and glass rods which have small, yet precise, differences in their indices of refraction are not readily available. Variations in core diameter or in either index of refraction will affect the transmission characteristics of a waveguide.

Summary of the Invention It is an object of this invention to provide a method for economically producing an optical waveguide which overcomes the heretofore noted disadvantages.

Another object of the present invention is to provide a method for producing an optical waveguide that will not excessively attenuate transmitted light, and will propagate the light along a given fiber in preselected modes.

Still another object of the present invention is to provide a method for producing an optical waveguide that will not cause excessive dispersion shift in the transmitted light and will not be subject to cross-talk from adjacent fibers.

A further object of the present invention is to provide a better interface between the core and cladding of an optical waveguide.

Still a further object of the present invention is to provide a way of preventing the formation of light scattering centers at the core and cladding interface of an optical waveguide.

Another object of the present invention is to provide an optical waveguide which overcomes the hereinabove noted disadvantages.

Broadly, according to this invention an optical wave-guide is produced by applying a film of material on the inside wail of a glass tube. This film of material may be a glass possessing the optical and physical qualities desired of the waveguide core; a doping material that will diffuse into the inside surface of the glass tube thus forming a film of glass on the inside tube wall having optical and physical qualities desired of the waveguide core; or a doped glass the dopants of which will partially diffuse into the inside surface of the glass tube resulting in a combination film of applied doped glass and doped glass formed by diffusion. This composite glass structure is then heated until the structure reaches a temperature at which the materials have a low enough viscosity for drawing. The composite structure is then drawn to reduce the diameter thereof until the film of core glass is collapsed, that is it seals the longitudinal hole to form a rod surrounded by cladding material; thereafter continued drawing of such a composite rod further reduces the diameter thereof to form a glass optical fiber which possesses the characteristics of an effective optical waveguide. That is, it transmits preselected modes of light without excessive attenuation, it is not subject to cross-talk by adjacent optical waveguides, and it does not cause excessive dispersion in the transmitted light.

These together with other objects, features and advantages of the present invention will become apparent to those skilled in the art from the following detailed description and the attached drawing.

Brief Description of the Drawings

Fig. 1 shows a fragmented elevation view partially in cross-section of an optical waveguide being formed in accordance with the present invention.

Fig. 2 shows a cross-sectional view taken along line 2—2 of Fig. 1.

Fig. 3 shows a cross-sectional view taken along line 3—3 of Fig. 1.

Fig. 4 is an illustrative graph showing light intensity distribution as a function of the radial distance from the center of an optical waveguide.

Detailed Description of the Invention Referring to Fig. 1 there is shown a thick wall tube 10 of cladding glass, on the inside wall of which a thin film 12 of core glass has been formed. This thick wall tube of cladding glass and the thin film of core glass is heated by the surrounding furnace 14 until tube 10 and core glass 12 reach a temperature at which the materials have low enough viscosity for drawing. This composite structure is then drawn until the longitudinal hole 16 collapses, that is the core glass fills the hole 16 to form solid core 22 surrounded by cladding 20, and the cross-sectional area is reduced until fiber 24 is produced which has a thick cladding surrounding a small diameter solid core. The total diameter to core diameter ratio is preferably in the range of 10:1 to 300:1, although ratios outside of this range would not be inoperable.

A thick wall tube from which suitable cladding could be produced may be prepared by core drilling solid rods of glass having the index of refraction desired of the cladding glass. The rough surfaces left by drilling may be made smooth by such methods as: starting with a larger blank and redrawing the blank after core drilling, by laser milling the inside surface, by mechanically polishing the inside surface, by fire polishing the inside surface or by washing the tube in hydrofluoric acid. A particularly suitable method of smoothing the inside wall of the tube is by first mechanically polishing the rough surface left by core drilling and thereafter flame polishing the mechanically polished surface. Since minimum attenuation of the transmitted signal is a desired feature of any communication system, an optical waveguide should be produced from a glass having minimum light absorption characteristics, and although any optical quality glass may be used, a particularly suitable glass from which to make the cladding is fused silica.

The core of an optical waveguide may also be produced from any optical quality glass having a suitable index of refraction. However, since the core glass should have a higher index of refraction than the cladding, and also have physical characteristics similar to the cladding; the same type of glass used for the cladding doped with a small amount of some other material to slightly increase the index of refraction is an excellent choice for the core glass. Therefore, if pure fused silica is used as the cladding glass, fused silica doped with a material to increase the index of refraction can be used as the core glass. There are many possible materials that can satisfactorily be used as doping material alone or in combination with each other. These include, but are not limited to, titanium oxide, tantalum oxide, tin oxide, niobium oxide, zirconium oxide, and aluminum oxide. A particularly suitable core glass is fused silica doped with titanium oxide. There are two reasons, however, for not adding too much doping material to the core glass. First, since additional doping material will cause the index of refraction to increase, the difference between the index of refraction of the pure fused silica cladding glass and the core glass will also increase forcing a decrease in the allowable core diameter of the waveguide. Second, if too large a percentage of doping material is added the fused silica Will lose its excellent light transmission qualities. The percentage of doping material that should be added to fused silica to produce a small, yet precise, change in the index of refraction can readily be determined by one skilled in the art.

To produce the waveguide core, a film of glass with the desired physical and optical qualities, such as viscosity, coefficient of expansion and index of refraction, is formed on the inside wall of a tube of the desired cladding material. The film of glass may be formed by applying a layer of glass with the desired qualities, applying a layer of doping material that will diffuse into the inside surface of the glass tube thus forming a film of glass with the desired qualities; or applying a layer of doped glass, the doping material of which will partially diffuse into the inside surface of the glass tube, resulting in a film of applied glass and a film of glass formed by diffusion the combination of said films having the desired qualities. The desired film may be applied in a variety of methods including but not limited to radio frequency sputtering, sintering a film of soot deposited by a flame hydrolysis process, chemical vapor deposition, depositing a glass frit layer, and by bathing the inside wall of the tube in a fluid that will make an "ion-exchange" with the glass. A particularly effective method is to sinter a soot layer of the desired material, deposited on the inside wall of the tube, by a flame hydrolysis process similar to that described in U.S. Pat. 2,272,342 issued to J. F. Hyde, and U.S. Pat. 2,326,059 issued to M. E. Nordberg. A modification of the Nordberg process that will deposit titanium doped fused silica on the inside wall of a glass tube follows. Dry oxygen is bubbled through a tank containing a liquid mixture of approximately 53% by weight silicon-tetrachloride, $SiCl_4$, and 47% by weight titanium-tetrachloride, $TiCl_4$. These liquids are at approximately 35°C. $SiCl_4$ and $TiCl_4$ 'vapors picked up by the oxygen are then passed through the center of a gas-oxygen flame where they are hydrolyzed to form a "soot"; that is, minute glass particles, with a composition of approximately 95% by weight SiO_2 and 5% by weight TiO_2. The glass soot leaves the flame in a steady stream and is deposited on the inside wall of a glass tube by directing the stream of soot at an open end of the tube. The thickness of the soot layer is controlled by the flow rates, distance between the glass tube and the flame, inside diameter of the glass tube, and the time allowed for deposition. Uniformity of the soot layer is greatly enhanced if a slight vacuum is applied to the end of the tube opposite the flame. The tube and the soot are then heated until the soot sinters; thereby resulting in a thin film of titanium doped fused silica glass being bonded to the inside wall of the tube. Sintering a soot layer of approximately 30 microns results in a glass film of approximately 2 microns.

The necessary ability of an optical waveguide to transmit a useable quantity of light is dependent upon the light energy not being dispersed by "light scattering centers," as well as not being excessively absorbed by the glass material. These centers are often caused by tiny air bubbles or impurities in the waveguide at the core-cladding bond. The method of this invention provides an unusually clean and strong bond, thus eliminating most of these "light scattering centers."

The intensity of light in an effective optical waveguide decreases rapidly as the radial distance from the core center increases. Therefore, if this intensity approaches zero at a distance from the core center that is less than the radius of the waveguide, then the waveguide will not be noticeably affected by cross-talk and phase shift problems. In Fig. 4, the radial distance rx represents the radius of the waveguide core, and r_2 represents the outside diameter of the waveguide. The light intensity is shown as being substantially wholly within the core, and decreases to almost zero at the outside diameter of the waveguide as is necessary for an effective optical wave-guide.

To limit light propagation along an optical waveguide to preselected modes, the core diameter, the core index of refraction, and the cladding index of refraction must be coordinated according to Equation 2. An example describing a technique that might be used in determining the coordinated values of the core diameter a, the core index of refraction "n_1" and the cladding index of refraction "n_2" follows. If fused silica is chosen to be the cladding glass, the index of refraction of the cladding would be approximately 1.4584. The number 1.4584 is normally assumed by those skilled in the art to be the index of refraction of fused silica, for sodium light $\lambda = 5893$A., and is the same value given in the *CRC Handbook of Chemistry and Physics,* page E-224, 49th edition, 1968. Further, if the weight percentage of SiO_2 and TiO_2 are selected so that the core glass has a resulting index of refraction of 1.466, the core diameter necessary to limit light propagation to the mode HE_{11} within the optical waveguide is determined by solving the equation

$$R = \frac{2\pi a}{\lambda}\sqrt{(n_1 + n_2)(n_1 - n_2)}$$

where:

R = 2.405, and thereby sets the cutoff value such that substantially only light mode HE_{11} will propagate down the fiber

a = radius of the core

λ = wavelength of the transmitted light (sodium light)
= 5893 A.

n_1 = core index of refraction = 1.466

n_2 = cladding index of refraction = 1.4584

If the above equation is solved using the parameters as listed it can be shown that the core radius a should be less than or equal to 1.5 microns if the right hand side of the equation is to be less than or equal to 2.405 thereby limiting light propagation to the HE_{11} mode.

It has been found that light absorption properties may be decreased and light transmission qualities improved in titanium oxide doped fused silica formed into optical waveguides, if the waveguides are drawn in an oxygen atmosphere and then "heat treated" in an oxygen atmosphere. The "heat treatment" referred to consists of heating the waveguide in an oxygen atmosphere to between 500° C. and 1000° C. for not less than 30 minutes; the length of treatment being related to the treatment temperature. Lower temperatures require longer treatment periods, while treatment at the higher temperature allows shorter time periods.

A specific example of a waveguide produced by the practice of this invention follows. A $\frac{1}{4}$ inch diameter hole was drilled down the longitudinal axis of a $\frac{3}{4}$ inch diameter rod of fused silica that was approximately 5 inches in length, thus forming a tube. The rough surface left by drilling was then mechanically polished. The mechanically polished surface was then flame polished to produce a very smooth surface on the inside wall of the tube. A soot of fused silica doped with titanium oxide was then applied to the inside tube wall by the heretofore described modified flame hydrolysis process. This glass soot was approximately 20 microns thick, and was approximately 94.75% fused silica and 5.25% titanium oxide. This composite structure was then heated until the deposited glass soot sintered; thereby resulting in a thin film of titanium oxide doped fused silica bonded to the inside wall of the tube. Sintering occurred at approximately 1450° C. The composite structure was then further heated in substantially an oxygen atmosphere until it reached a temperature at which the materials had low enough viscosity for drawing (approximately 1900° C.). The composite structure was then drawn to reduce the diameter thereof until the film of titanium oxide doped fused silica was collapsed, that is, it sealed the longitudinal hole to form a solid core surrounded by pure fused silica. The resulting composite rod was then further drawn to reduce the diameter thereof to a final diameter of approximately 100 microns. The core of the optical wave-guide was measured at approximately 3 microns in diameter. The core index of refraction was approximately 1.466 while the cladding material had an index of refraction of approximately 1.4584. After the fiber was drawn, it was heat treated at 800° C. in an oxygen atmosphere for approximately three hours. An optical fiber produced by the method as outlined in this example possesses the characteristics desired in an optical waveguide using sodium light, that is propagation of one light mode, namely HE_{11} mode, without excessive attenuation, and very little susceptibility to dispersion and cross-talk.

Although the present invention has been described with respect to specific methods of production, it is not intended that such specific references be limitations upon the scope of the invention except insofar as set forth in the following claims.

We claim:

1. A method of producing an optical fiber comprising the steps of providing a glass tube, forming a film of core glass, having an index of refraction different than that of said glass tube, on the inside wall of said glass tube, heating the composite structure so formed to the drawing temperature of the materials thereof, and drawing the heated composite structure to reduce the cross-sectional area thereof and collapse said film of core glass to form a fiber having a solid cross-sectional area.
2. The method of claim 1 wherein the film of core glass is formed by steps comprising depositing a film of soot of said core glass on the inside walls of said glass tube by flame hydrolysis, and heating said tube and core glass soot until said soot sinters prior to the drawing step.
3. The method of claim 2 wherein sintering the core glass soot and drawing the composite structure is accomplished by the same heating operation.
4. The method of claim 1 wherein the film of core glass is formed by radio frequency sputtering.
5. The method of claim 1 wherein the film of core glass is formed by chemical vapor deposition.
6. The method of claim I wherein the film of core glass is formed by applying glass frit to the inside walls of said tube.
7. The method of claim 1 wherein the film of core glass is formed by steps comprising: applying a layer of material selected from the group consisting of doping material and glass in combination with doping material to the inside walls of the glass tube, and diffusing at least a portion of said material into the surface portion of the inside wall of said glass tube thereby changing the composition of said surface portion.
8. The method of claim 7 wherein the layer of material is applied by flame hydrolysis.
9. The method of claim 7 wherein the layer of material is applied by chemical vapor deposition,
10. The method of claim 7 wherein the layer of material is applied by radio frequency sputtering.
11. The method of claim 7 wherein the layer of material is doped glass frit applied to the inside walls of said tube.
12. The method of claim 7 wherein diffusion of the material and drawing the composite structure is accomplished by the same heating operation.
13. The method of claim 7 wherein the doping material is at least one of the materials selected from the group consisting of titanium oxide, tantalum oxide, tin oxide, niobium oxide, zirconium oxide, and aluminum oxide.
14. The method of claim 1 wherein the cladding tube is fused silica and the core glass is fused silica doped with at least one material selected from the group consisting of titanium oxide, tantalum oxide, tin oxide, niobium oxide, zirconium oxide and aluminum oxide.

15. The method of claim 1 wherein the said cladding and applied core glass are drawn in substantially an oxygen atmosphere.
16. The method of claim 1 wherein the said fiber with a solid cross-sectional area is heat treated in an oxygen atmosphere.
17. The method of claim 1 wherein the index of refraction of the said tube is less than the index of refraction of the applied core glass.
18. The method of claim 1 wherein the step of providing a glass tube comprises: drilling a solid rod of glass to form said tube, mechanically polishing the drilled surface of said tube, and flame polishing the mechanically polished surface.
19. The method of claim 18 wherein the film of core glass is formed by steps comprising: depositing a thin film of core glass soot on the inside wall of said tube by flame hydrolysis, and sintering said soot to form a core glass having an index of refraction different than that of said glass tube.
20. The method of claim 18 wherein the total diameter to core diameter ratio is within the range of 10:1 to 300:1.
21. The method of claim 18 wherein the composite structure is drawn in an oxygen atmosphere.
22. The method of claim 18 wherein the said fiber with a solid cross-sectional area is heat treated in an oxygen atmosphere.
23. The method of claim 18 wherein the glass rod is fused silica, and the deposited core glass is fused silica doped with not more than 15% titanium oxide.
24. The method of claim 18 wherein the index of refraction of the solid glass rod is less than the index of refraction of the applied core glass.
25. The method of claim 18 wherein sintering the core glass soot and drawing the composite structure is accomplished by the same heating operation.
26. An optical waveguide produced by the method of claim 1.
27. The method of making an optical fiber comprising the steps of: providing a glass tube, depositing a thin film of doped fused silica soot on the inside wall of said tube by flame hydrolysis, heating the composite structure, sintering said soot to form a core material having an index of refraction higher than that of said tube, drawing said structure to reduce the cross-sectional area thereof and collapse said film of core material forming a fiber having a solid cross-section with the total diameter to core diameter ratio being within the range of 10:1 to 300:1, and heat treating said fiber in an oxygen atmosphere.
28. The method of claim 27 wherein the step of providing a glass tube comprises: drilling a solid fused silica rod to form said tube, mechanically polishing the drilled surface of said tube, and flame polishing the mechanically polished surface.

References Cited

UNITED STATES PATENTS

2,237,754	4/1941	Davies —— 65—60
3,275,428	9/1966	Sieground — 65-DIG. 7
3,294,504	12/1966	Hicks, Jr.—— 65-DIG. 7
3,331,670	7/1967	Cole —— 65—4

3,479,168	11/1969	Bird et al —65—3
3,481,390	12/1969	Veltrie et al —65—3 X
2,433,116	12/1947	Greenbowe et al -65—60
1,793,529	2/1931	Taylor —65.DIG. 7
3,146,082	8/1964	Hicks, Jr. et al – 65—3
3,241,934	3/1966	Granitsas et al – 65—3
3,504,730	4/1970	Dannohl —65-3 UX
3,455,666	7/1969	Bazinet, Jr — 65—4
3,481,390	12/1969	Veltri et al —65m121 X
3,614,197	10/1971	Nishizawa —65—41 X
3,607,322	9/1971	Brady —65—41 X

FRANK W. MIGA, Primary Examiner

U.S. Cl. X.R

65—4, DIG. 7, 30, 121; 106-50, 54; 350—96 B

3.9.3 Fiber Optics (Wolf, 1976)

UNITED STATES PATENT

February 24, 1976 3,940,278

Fiber Optics*

WARREN W. WOLF, REYNOLDSBURG, OHIO, ASSIGNOR TO OWENS-CORNING FIBERGLAS CORPORATION, A CORPORATION OF OHIO

Application filed June 10, 1974, Application No. 477881

Current U.S. Class: 501/37; 385/142; 385/144; 501/69; 501/903
Int'l Class: C03C 003/04; C03C 013/00
Field of Search: 106/52,50

References Cited [Referenced By]
U.S. Patent Documents

1151911	Aug., 1915	Sullivan et al	106/52.
2212879	Aug., 1940	Kalsing et al	106/52.
2664359	Dec., 1953	Dingledy	106/50.
3258352	Jun., 1966	Paymal	106/52.
3801344	Apr., 1974	Dietz	106/52.
3811853	May., 1974	Bartholomew et al	106/52.
3821004	Jun., 1974	Lagrouw	106/52.

*The entire patent (with only one claim) is presented here.

Primary Examiner: Douglas; Winston A. Assistant Examiner: Wright; William G. Attorney, Agent or Firm: Staelin; Carl G., Overman; John W., Pacella; Patrick P.

Abstract A boron-free glass composition for use in fiber optics is disclosed. In the field of fiber optics, light is transmitted from one point to another by small diameter fibers. These fibers are composed of a central core surrounded by an outer cladding or sheath. The glass compositions of this invention can be used as the core glass or the sheath glass.

Description This invention relates to fiber optics. More specifically this invention relates to improved core glass compositions for fiber optics.

In the field of fiber optics, light is transmitted from one point to another by small diameter fibers. These light transmitting fibers are composed of a central core surrounded by an outer cladding or sheath. The fibers or filaments are bundled together to form light pipes. These light pipes are flexible and can be curved around obstacles and placed in remote or hard to install areas.

Light pipes are useful as monitoring and lighting devices in automobiles, aircraft, appliances, computers and medical instruments. They are particularly useful where noncurrent carrying and therefore nonshorting or nonsparking light transmitting devices are required for safety.

Light pipes are used in some late model automobiles. They are run from the automobile headlights and tailights to the dashboard. As long as the headlight or tailight is operating, the light pipe transmits light and shows the driver that the lights are on. If one of the lights fails, no light is transmitted by the light pipe connected to that light. And the driver knows that the bulb is burned out.

Light transmitting or optical fibers are usually made of either plastic or glass. Glass fibers are preferred over plastic fibers because of their high heat resistance. The ability of glass to withstand adverse temperature and humidity conditions for long periods of time is also a factor in selecting glass fibers over plastic fibers. While plastic fibers become brittle and discolor at temperatures above about 175° F or show a reduction in light transmission when exposed to long periods of high heat and humidity, glass light pipes are unaffected by high humidity and temperatures in excess of 550° F. Individual optical fibers or filaments used in light pipes are composed of two essential parts, namely a central core and an outer cladding or sheath. The cladding surrounds the core. The index of refraction of the core is higher than the index of refraction of the cladding. Due to this difference between the index of refraction of the core and cladding, the light entering one end of a fiber is internally reflected along the length of the fiber. The principle of "total internal reflection" explains this result. This principle says that a light ray traveling from a material with a higher refractive index (core) to a material with a lower refractive index (cladding) will be reflected at the interface of the two materials.

The amount of light transmitted through a light pipe depends on the intensity of the light source, the cross sectional area of the cores of all the fibers, the losses at each end due to reflection, the absorption properties of the glass and the length of the light pipe.

In making commercially acceptable glass light pipes the chemical composition of the core and cladding is a key factor. In addition to having the proper indices of refraction the core and cladding glasses should have similar viscosities, and liquidus temperatures. These three properties control the ease with which the cladding glass can be laid over the core glass. And when conventional glass fiber forming processes are used to make optical fibers, compatibility of the core and cladding is crucial.

Conventional glass fiber forming processes draw fine glass fibers from small holes in the bottom of a reservoir of molten glass. The usual problems of forming fibers the conventional way are multiplied when making optical glass fibers. Two supplies of molten glass are needed; one of the core glass and one of the cladding glass. Then a composite fiber, 80–90% core and 10–20% cladding, is pulled from the glass supplies. To be able to do this at commercial speeds and form commercially acceptable products the composition of the core and cladding glasses must be carefully chosen.

A highly useful optical fiber known in the art is composed of core and cladding glass fibers having the following compositions:

Cladding Glass

Ingredient	Weight Percent
SiO_2	64.1
Al_2O_3	8.2
B_2O_3	11.2
Na_2O	14.0
LiO_2	2.4

Core Glass

Ingredient	Weight Percent
Al_2O_3	8.0
B_2O_3	6.1
K_2O	6.1
BaO	47.9
Sb_2O_3	0.05

One problem with both of these glass compositions is that they contain boron. This element is detrimental in many uses of fiber optics because it absorbs gamma radiation. This radiation permanently darkens the boron-containing glasses. As a result, light transmission in the fiber optic product is reduced after exposure to the radiation.

I have now developed boron-free glass compositions which can be used as the core glass in optical fibers. The glass compositions of this invention also can be used as cladding glasses in fiber optics as long as the index of refraction of the core is higher than index of refraction of the cladding.

The boron-free glasses of this invention have the following composition:

Ingredient	Weight Percent
SiO_2	40 to 60
Al_2O_3	5 to 10
BaO	0 to 50
Na_2O	5 to 40
Sb_2O_3	0 to 0.05

Preferred core glass compositions are:

Ingredient	Weight Percent
SiO_2	40 to 50
Al_2O_3	5 to 7

BaO	30 to 50
Na_2O	5 to 20
Sb_2O_3	0 to 0.05

These glass compositions have an index of refraction ranging from 1.50 to 1.65. The preferred core glass compositions have an index of refraction ranging from 1.55 to 1.65.

Two different glass compositions of this invention can be used together to form a fiber optic product. Or the glass compositions can be used as a core glass with cladding glasses falling outside the scope of this invention. Or these glass compositions can be used as a cladding glass with core glasses falling outside the scope of this invention. All of these combinations are operable as long as the index of refraction of the core is higher than the index of refraction of the cladding. Naturally, the differences in the indexes and fiber forming properties of the glass compositions also have to be considered.

Specific glass compositions embodying the principles of this invention are set forth in Examples 1 through 6.

Ingredients				Weight Percent		
Example	1	2	3	4	5	6.
SiO_2	42.1	44.5	47.1	50.1	53.5	57.4
Al_2O_3	6.0	6.3	6.7	7.1	7.6	8.1
BaO	44.6	37.8	30.0	21.3	11.4	—
Na_2O	7.3	11.5	16.2	21.5	27.6	34.5
Sb_2O_3	0.05	0.05	0.05	0.05	0.05	0.05
Liquidus Temperature, °F	2062	1909	1643	1427	1437	1535
Viscosity, Temperature, °F						
Log Poise						
2.00	2350	2405	2334	2332	2343	2380
2.25	2235	2272	2218	2208	2212	2245
2.50	2137	2153	2122	2089	2097	2130
2.75	2053	2053	2040	1978	2000	2025
3.00	1973	1953	1965		1915	1933
Refractive Index	1.59	1.57	1.56	1.54	1.53	1.51

These boron-free glass compositions are not as sensitive to radiation damage as the boron-containing glass compositions of the prior art. Light transmitted by the boron-free glass compositions was a more true white: the amount of yellow here in the transmitted light was reduced.

The glass compositions of this invention can be used as core glass with the following cladding glass composition.

Example 7 (Cladding)

Ingredients	Weight Percent
SiO_2	64.1
Al_2O_3	8.2

B_2O_3	11.2
Na_2O	14.0
Li_2O	2.4

Liquidus: No devitrification at 1455° to 1850° F over a period of 64 hours.

LogViscosity	Temp. in °F
2.0	2470
2.5	2180
3.0	1970

Refractive Index 1.50

Light transmitting optical fiber products were actually made from the following combinations of core and cladding glass compositions:

Core	Cladding
Example 1	Example 7
Example 1	Example 2
Example 2	Example 7
Example 2	Example 4

I have described this invention in considerable detail and do not wish to be limited to the particular embodiments shown. Modifications and variations within the scope of the appended claims are intended to be included.

I claim:

1. A boron-free glass composition capable of being the core glass or cladding glass in a light transmitting optical fiber product; said glass composition consisting essentially by weight of:

Ingredient	Weight Percent
SiO_2*	40 to 50
Al_2O_3	5 to 7
BaO	30 to 50
Na_2O	5 to 20
Sb_2O_3	0 to 0.05

and having an index of refraction ranging from 1.55 to 1.65.

*AUTHORS' NOTE: Silicon dioxide or SiO_2, Al_2O_3, BaO, Na_2O, and Sb_2O_3. These metallic oxides control the properties of the core and cladding for the fibers.

3.10 INTELLIGENT NETWORKS

Intelligent networks have influenced the telecommunications industry in a way that permits the network to have some distinctive, subtle, adaptive, and intelligent properties [3.10.1]. These properties include the capability for the called party to be able to pay for the call [3.10.2] and/or for the call to be automatically forwarded to an alternate location where the call may be received [3.10.3]. This capacity to dynamically change the routing of calls from one number (location) to another number (location) was the basis of the first sequence of 800-number services. Typically, the routing takes place based on the contents of a writable database (known as the service control point [3.10.4]) within the network. The signaling network is then invoked to receive the new number from the database when a call is made. The location of the database itself could be local, national, or international. In addition, the location of the caller and the called parties could be local, national, or international. In a sense, the network performs with some programmable and rudimentary intelligence. Appropriately, this type of network was called IN/1 to indicate its role as the first version of an intelligent network.

This routing intelligence can be enhanced to provide a variety of other specialized services, such as caller ID (number identification) [3.10.5], emergency number services, call forwarding, etc. However, the basis for modifying or altering the nature of call billing, setup, interconnect, disconnect, and/or service was established by Roy Weber's first patent in 1980, "Data Base Communication Call Processing Method." The traditional and very deterministic type of service provided by the telecommunications industry before this patent was changed drastically by the concepts enunciated in this patent. A variety of such services could be offered, and the numerous methods of building these networks to make these newer services possible also started to increase. Thus, at least four additional architectures [3.10.6] of networks (labeled as IN/1+, IN/2, AIN (advanced intelligent network), and knowledge-domain networks [3.10.7]) have gradually deviated from the network presented in Weber's 1980 seminal and inherent patent (see Section 3.10.1).

Computer scientists are especially alert to the fact that the true concept of artificial intelligence, or AI, [3.10.8] is not deployed in conventional intelligent telecommunications networks. Instead, it is deployed in a number of patents for knowledge-processing [3.10.9] or traditional AI systems [3.10.10]. The capacity of the network to service telephone or computer calls based on the contents of the information that is being communicated is the basis of a new group of patents. These networks may be programmed to extract patterns, derive inferences, and analyze the perceptions and goals of the user [3.10.11]. In addition, they may be blended with the network's programmable routing capability and the artificially intelligent agents [3.10.12]. These agents follow the development of the current pattern for the exchange of information or data and thus modify the network's response with many distinctive forms of intelligence.

In a sense, the networks can be programmed with numerous routing programs and with AI and thus can act as distributed information and knowledge robots. Such networks have the ability to perform a variety of very personalized and/or programmable functions, such as acting as individual instructors, isolating and identifying patterns, and performing medical diagnostics.

The seminal patent leading to a variety of subsequent patents is Weber's basic idea in 1980 for a database communication call-processing method [3.10.12]. With hindsight, a computer scientist may deem Weber's patent trivial because the concept of indirect addressing known as "indirection" [3.10.13] to fetch and execute machine-language instructions from various memory locations was practiced since the days of assembly-level programming of computers. And, in essence, this patent only asserts that a call may be completed based on a number that is stored at a different field in a database, much as one CPU [3.10.14] instruction may be replaced by another from a different memory location. However, the basic difference is that this type of function in Weber's patent is being implemented and executed in the telecommunications domain rather than in the traditional computer domain.

From the perspective of computer science and telecommunications, Weber's patent is acknowledged as one of the most significant contributions to the telephone industry. From a business perspective, the 800 number services have brought a different thrust to economic activity and to the buying preferences of both individuals and businesses.

In this patent, we have retained the most significant concepts, drawings, and claims. We also suggest that readers find in this patent more than the basic ideas documented. In particular, we suggest that readers explore the long-reaching implications that lie behind the design and implementation of intelligent networks. It starts here with this patent and progresses through advanced intelligent networks and finally to the most recent medical networks and universal services intelligent networks [3.10.15].

REFERENCES (C-3.10)

[3.10.1] Ahamed, S. V., "Intelligent Networks." in Meyers, Robert A (Editor), *Encyclopedia of Telecommunications,* Chapter 4, pp. 159–174. Academic Press, San Diego, 1989.

[3.10.2] Ambrose, W. D., Maher, A., and Sasscer, B., *The Intelligent Network.* Springer Verlag, New York, 1989.

[3.10.3] Beaty, A. Jr., "The Evolution of the Intelligent Network." *Telecommunications,* February 1989.

[3.10.4] Boese J. O., and Robrock, II, R. B., "Service Control Point: The Brains behind the Intelligent Network." *Bellcore Exchange,* November–December 1987, pp. 13–17.

[3.10.5] Miller, P., "Intelligent Network/2: A Flexible Framework for the Exchange Services." *Bellcore Exchange,* May–June 1987.

[3.10.6] Haas, J. M., and Robrock, R. B., "Intelligent Network of the Future." *Proceedings of the IEEE Global Telecommunications Conference, December 1986,* pp. 1311–1315.

[3.10.7] Ahamed S. V., and Lawrence, V. B., *Intelligent Broadband Multimedia Networks.* Kluwer Academic Publishers, Boston, 1997.

[3.10.8] Ullmam, J. D., *Database and Knowledge Base Systems.* W H Freeman & Co, New York, 1988. Also see Rich, E., *Artificial Intelligence.* McGraw Hill Series in Artificial Intelligence. McGraw-Hill, New York, 1983.

[3.10.9] Ahamed S. V., and Lawrence, V. B., "Knowledge Processing System Employing Confidence Levels." U.S. Patent No. 5,809493, issued on September 15, 1998, assigned to Lucent Technologies, Murray Hill, NJ, 1998.

[3.10.10] Garcia O. N., and Chien, Y. T., *Knowledge-Based Systems: Fundamentals and Tools.* IEEE Computer Society Press, Los Altamitos, CA, 1991.

[3.10.11] Wooldridge, Michael, *Introduction to MultiAgent Systems.* John Wiley & Sons, New York, 2002. Also see Ahamed S. V., and Lawrence, V. B., "Personal Computer-Based Intelligent Networks." U.S. Patent No. 5715371, issued on February 3, 1998, assigned to Lucent Technologies, Murray Hill, NJ, 1998.

[3.10.12] International Business Machines, *IBM/370 Principles of Operation.* Publication No. A22-6821-3, White Plains, NY, 1974.

[3.10.13] Stone, H. S., *Introduction to Computer Architecture.* SRA, Chicago, 1980.

[3.10.14] Hayes, J. P., *Computer Architecture and Organization.* McGraw Hill Book Company, New York, 1988.

[3.10.15] "Hospital-Based Integrated Medical Computer Systems for Processing Medical and Patient Information Using Specialized Functional Modules." Patent Application No. 09/152,809, filed by Lucent Technologies, Holmdel, NJ, 1998. Approved for the issue of U.S. Patent on March 13, 2001, foreign countries GB, Germany, France, and Japan. U.S. Patent No. U.S. 6,272,481 B1 issued on August 7, 2001. Also see Krol, M., "*Intelligent Medical Network.*" Ph.D. dissertation, City University of New York, 1996.

R.P. Weber

Data Base Communication* (Weber, 1980)

UNITED STATES PATENT

March 4, 1980 4,191,860

Data Base Communication Call Processing Method

ROY P. WEBER OF BRIDGEWATER, NJ, ASSIGNOR TO BELL TELEPHONE LABORATORIES, INCORPORATED MURRAY HILL, N.J., A CORPORATION OF NEW JERSEY

Application filed July 13, 1978, Application No. 924170
Current U.S. Class: 379/115; 379/211; 379/221; 379/277
Intern'l Class: H04M 003/42; H04M 007/06
Field of Search: 179/18 B,18 BA,18 EA,18 DA

References Cited [Referenced By]
U.S. Patent Documents

3189687	Jun., 1965	Miller	179/18.
3692946	Sep., 1972	Budrys et al.	179/18.
3975595	Aug., 1976	Berube et al.	179/18.

Primary Examiner: Brown; Thomas W.
Attorney, Agent or Firm: Padden; Fredrick W.

Abstract A method is disclosed for processing call data among telephone switching offices and a remotely situated data base system for deriving call service information needed to complete

*Original patent 19 pages, 23 claims, and 5 drawings; edited to 11 pages, 5 claims, and 5 drawings.

special service calls, such as INward WATS (Wide Area Telephone Service) calls to a destination specified by the derived information. A network of local and toll offices compiles a special service call message including a caller dialed 800 number and an originating area code for the INWATS service and then transmits that message to the data base system over CCIS (Common Channel Interoffice Signalling) facilities including signal transfer systems.

At the data base, the message is translated into a directory-unlisted number identifying an INWATS customer station designated to serve the call. Before the call is switched through the toll switching network under control of the unlisted number, the open/closed and busy/idle conditions of the INWATS station are checked at the data base to determine the availability of the station to serve the call. Traffic data is accumulated at the data base by originating area code and by the unlisted INWATS number. When the number of calls to the unlisted number per unit time exceeds a threshold, the data base instructs the telephone system to inhibit such calls from being served for a prescribed period.

Claims What is claimed is:

1. A method of deriving call routing information from a data base system for a special service call processed by a communication switching system characterized by receiving at said data base system from said switching system a special service call message including a code dialed by a caller and augmented by a code identifying the originating area of said call, translating said message into a call routing message at said data base system, and transmitting said routing message from said data base system to said switching system.
2. A method of deriving call routing information from a data base system for a special service call processed by a communication switching system characterized by receiving at said data base system a special service call message from said switching system, translating said message into a call routing message at said data base system, and transmitting said routing message from said data base system to said switching system, and further characterized by said call routing message comprising a station number designating a destination station and checking at said data base system the availability of said station to serve said call during said translating of said special service call message into said routing message.
3. The invention of claim 2 further characterized by controlling the sending of said station number from said data base system to said switching system only when said checking indicates that said station is available to serve said call.
4. The invention of claim 2 further characterized by obtaining at said data base system another station number designating an alternate station to receive said call when said checking indicates that said destination station is unavailable to serve said call, and sending said other station number from said data base system to said switching system.
5. The invention of claim 1 further characterized by said routing message comprising a station number designating a destination station, storing at said data base system data signals indicating the busy/idle status of said destination station, supplying at said data base system another number designating an alternate station to receive said call when the stored data signals indicate that said destination station is busy, and sending the supplied other number from said data base system to said communication switching system.

Description. *Technical Field* This invention relates to communication call routing and particularly to a method for automatically processing telephone call data from a telephone system over CCIS (Common Channel Interoffice Signaling) direct signaling facilities to a remote data

U.S. Patent Mar. 4, 1980 Sheet 1 of 4 4,191,860

FIG. 1

TELEPHONE - DATA BASE SYSTEM FOR INWATS USING CCIS

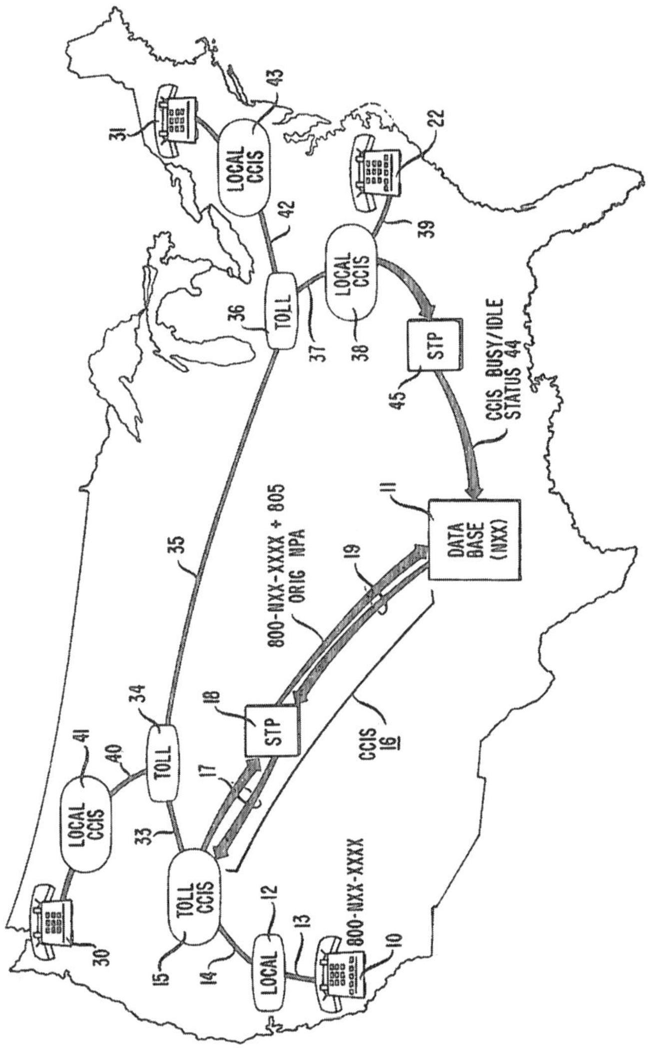

FIG. 2

ORIGINATING AREA CODE TO POTS NUMBER TRANSLATOR TABLE

AREA CODE	POTS #	COUNTER
201	212-525-3333	100
208	NONE (NULL ENTRY)	1
609	212-525-3333	50
805	919-658-3399	10
808	206-582-2044	5
919	919-658-3390	5

20; FILE 28; 27

FIG. 3

POTS NUMBER CALL PROCESSING DATA TABLE

21

POTS # DDD #	LOOP BIT	CLOSED TIME TIMES, DAYS	BUSY/IDLE BIT	ALTERNATE POTS# DDD#	NETWORK MANAGEMENT			
					COUNTER	THRESHOLD	ON / OFF	INHIBIT TIME (SEC)
FILE 29: 206-582-2044	0	NONE	0	212-525-3333	51	100	0	0
FILE 32: 212-525-3333	0	NONE	0	NONE	150	90	1	6 SEC
FILE 23: 919-658-3399	0	5 PM - 9 AM, ALL SUN	0	206-582-2044	10	125	0	0
	49	24	25	26	46	47	50	48

U.S. Patent Mar. 4, 1980 Sheet 3 of 4 4,191,860

FIG. 4

CCIS INWATS DATA BASE FLOW

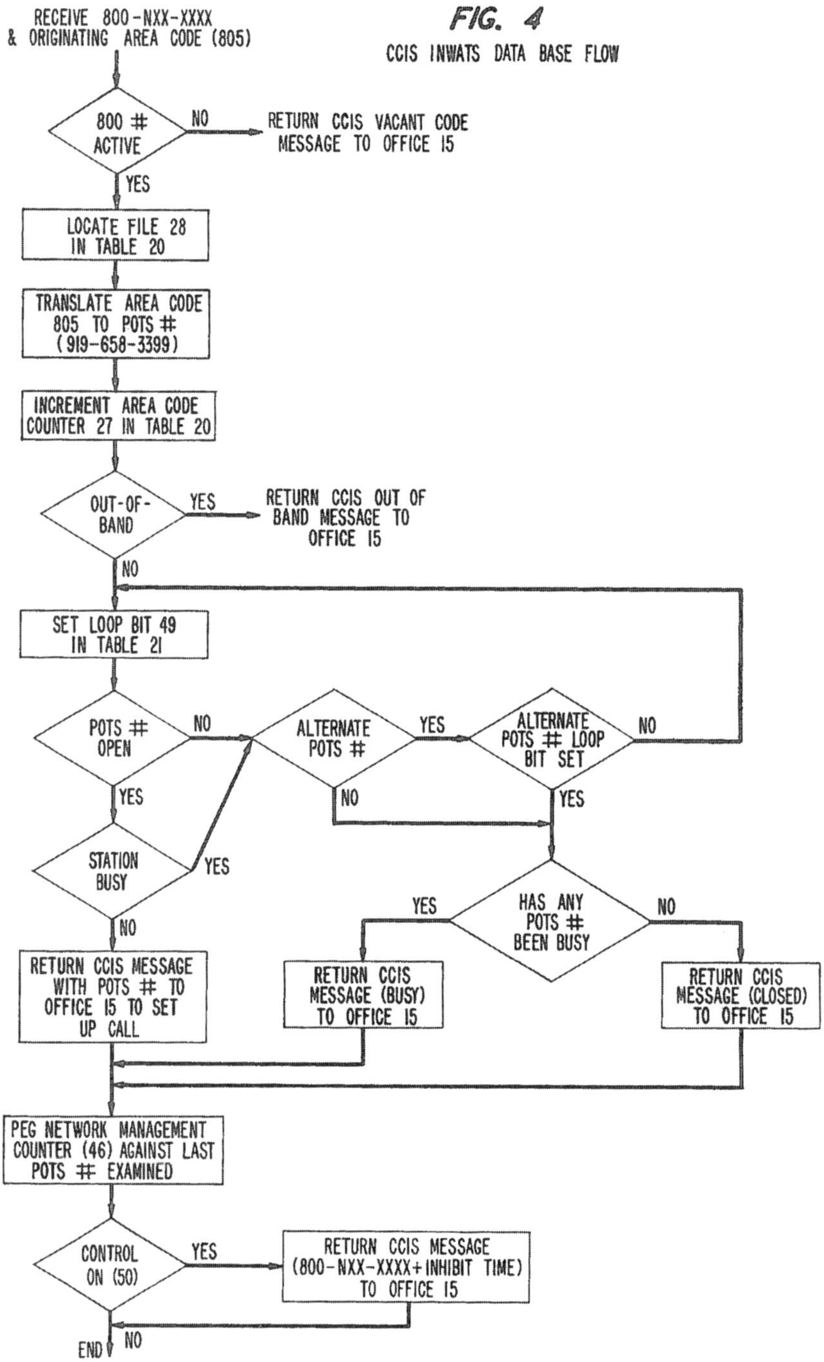

FIG. 5
NETWORK MANAGEMENT

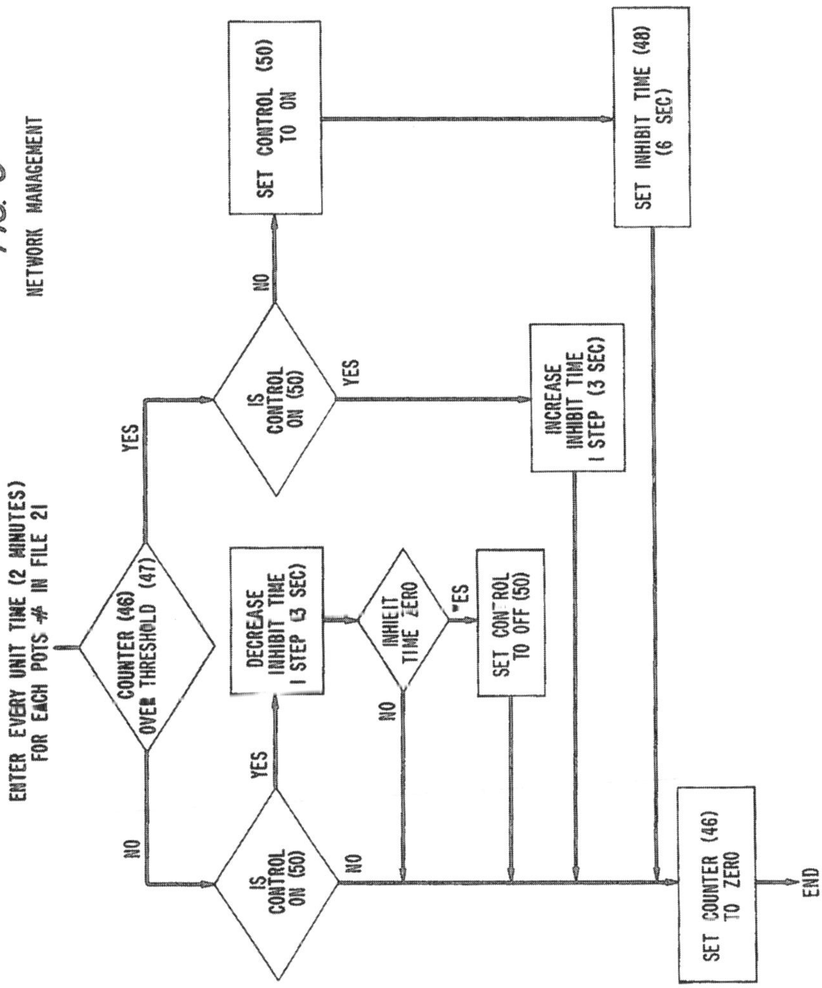

base for translation into a call control message which is then sent to the telephone system for controlling the further completion of call connections to a destination in accordance with the control message.

The invention specifically pertains to procedures for processing an INward WATS (Wide Area Telephone Service) call over the CCIS facilities to a data base system for obtaining the number of an available station to which the call is to be completed.

Background Art. INward WATS is popularly known as INWATS and is frequently referred to as the "800" number toll-free telephone service. It allows a customer to establish an area of the country from which he can receive calls without charge to the calling parties. In the United States, the service is currently available for both intrastate and interstate calls. Tariff costs for the service are based on the number of customer lines, the band of rate states selected, the total monthly hours of usage, and the total number of calls completed.

Over the last decade, the volume of INWATS calls has increased to the extent that INWATS traffic has become a substantial percentage of all toll calls served by existing telephone switching systems. The service has proven to be especially useful for business customers in obtaining travel and hotel reservations, purchase orders, and the like.

Despite the commercial success of INWATS, the ever expanding customer demands for the service and the projected extension of INWATS for international calling has presented a number of problems for the telephone industry and its customers. Prevalent among the problems are: (1) the rigid geographic bands for INWATS, (2) the requirement for a multiplicity of INWATS numbers, (3) routing and numbering inefficiencies due to service entitlement check operations of originating and terminating screening offices, (4) ineffective attempts due to all INWATS customer lines busy conditions caused, for example, by mass calling to 800 numbers advertised, for example, on television, and (5) the absence of traffic statistics for INWATS customers on the number of calls made from each area code.

To elaborate, in the United States, a customer purchases the service on intrastate and/or interstate bases and is supplied with one or more INWATS numbers. Such an arrangement is necessary because of state and federal tariffs. Interstate INWATS is currently offered for a maximum of seven geographic bands. Band 1 generally involves all states bordering the customer home state; Band 2 typically includes all of Band 1 and additional states bordering Band 1; Band 5 presently covers the continental United States; Band 6 adds Hawaii and Alaska; and Band 7 includes Puerto Rico and the Virgin Islands. In some cases, multiple bands of intrastate service are also offered. Customers purchase the various service bands on a 240 hours/month or a 10 hours/month basis.

The rigid band assignment heretofore has required a customer to purchase an entire band encompassing many different telephone area codes even though the customer may desire INWATS service only in a portion of that band. A need has therefore existed for a procedure and equipment which fulfills customer requests for INWATS service covering only desired telephone area codes.

A principal INWATS problem has been the necessity for the purchaser in many instances to be assigned a multitude of different INWATS numbers. Illustratively, several airline reservation services with multiple regionalized answering points presently are provided with between 10 and 20 different INWATS numbers to obtain reasonably economical INWATS service throughout the United States. The different numbers are required by the customer because of geographical location, business hours and the designs and operations of existing switching equipment. Such multiple number requirements have proven to result in customer and user confusion, errors in directory number listings and advertising, as well as complications in obtaining directory assistance for nonlocally published INWATS numbers and in conformity with the call originating area codes. New call processing techniques and facilities are therefore needed to eliminate the plural numbers for each INWATS customer.

An INWATS number presently includes a special area code of 800, followed by a specific NNX code for the terminating NPA (Number Plan Area), followed by a four digit number XXXX. "N" in the code refers to any digit except 0 or 1 and "X" refers to any one of 10 digits. Of the NNX codes, all NN2 codes are currently reserved for intrastate use. At the present time, each NPA in the United States is assigned one NNX code with the exception of the NPAs 212, 312 and 213 each of which have two NNX codes.

INWATS number usage is readily understood by considering an example of a typical INWATS call. Assume that a party in Indianapolis wishes to place a call to an INWATS customer with the assigned number 800 + 241 + 2312 in the Atlanta, Georgia area. After the caller dials the number, the call is routed to an originating screening office which utilizes the 800 + 241 to determine both the terminating city (Atlanta) and the band of the calling party with respect to the terminating city. The screening office then sends a number 142 + 2312 to a terminating screening office for Atlanta. The underlined "2" indicates that Indianapolis is in Band 2 with respect to Atlanta and the "4" represents the 241 NXX code. On the basis of the received 142 + 2312, the terminating screening office ascertains whether or not the call originated from a permissible rate band purchased by the called customer. If the call is entitled to be completed, the terminating screening office controls the establishment of call connections to the called customer.

A disadvantage of the foregoing INWATS call processing arrangements is that INWATS call screening, routing and service entitlement checking functions are performed by costly, complex and special INWATS facilities integrated into many individual originating and terminating screening offices throughout the nation and without call rerouting capability in such offices in the event of call overload or equipment failure. Another disadvantage is that no procedure is available for determining the busy condition of INWATS customer lines before calls are switched through the toll network. Present day switching systems determine that all INWATS customer lines are busy only after the calls are routed through the originating and terminating screening offices to the local terminating office. As a consequence, the telephone network occasionally is overloaded, particularly during heavy traffic periods by mass calling to the all busy INWATS customer lines. Such overloads obviously are undesirable and prove to be a problem for telephone companies especially because no facilities are available for dynamically identifying INWATS customer busy conditions and then precluding the toll switching of INWATS calls to that customer.

Another disadvantage is that existing telephone systems and call processing techniques have been unable to provide adequate traffic data to INWATS customers on the number of calls received from a given geographic area. The data would aid in the determination of the numbers of INWATS lines and stations as well as the personnel needed to serve the calls. It would also enable customers to determine whether the service is economically warranted for that area.

Disclosure of the Invention. The foregoing problems are solved and a technical advance is achieved by a call processing method which automatically processes prescribed communication calls from a switching network to a data base system for deriving call service information and then returning the call to the network for completion in response to the information supplied by the data base. In embodiments of the method, the information provided by the data base is either a destination number, such as for a machine message announcement or a station, or a call completion authorization code.

The method is utilized for special service calls, such as INWATS calls, served by a communication system, such as a local or toll originating screening office or a communication customer service system. It involves a process of sending a special service call data including a dialed number and a suffix code from the switching system to a data base system, translating the received message into a call control message at the data base, and returning the control message to the communication system to control the establishment of call connections in accordance with the control message. Advantageously, the control message comprises the routing message including a number identifying either a subsystem, such as a voice storage facility, which supplies machine messages to the caller, or a called party station such as an INWATS customer reservation bureau which furnishes operator services.

For INWATS service, each INWATS customer is assigned a single ten digit "800" number which is dialed by callers on all calls to that customer regardless of their place of origination. The number is used to control the establishment of connections through a telephone switching system to a centralized data base system for deriving all of the call routing, traffic measurement and network management data on calls to the INWATS customer.

The data is utilized to screen calls for INWATS service entitlement and for specifying a directory-unlisted POTS (Plain Old Telephone Service) DDD (Direct Distance Dialing) number to which entitled calls are preferably to be routed to the INWATS customer on a toll-free basis to the caller. Data is included in the data base for specifying the busy/idle status and open/closed station condition of the preferred number. An alternate unlisted number is also stored in the data base for routing calls to that number whenever the preferred number is busy or closed.

The data base is strategically located to serve a multitude of telephone offices and illustratively is one of several designed to serve as call distribution systems for the entire United States. It is advantageously arranged for traffic measurements of calls to the INWATS customer on the basis of originating area code. These measurements enable the customer to analyze traffic patterns and "800" number usage by area code, to obtain adequate numbers of INWATS lines and personnel for serving the calls, and to determine the economic need for the INWATS service in that area. The data base measures the traffic by counting each call by the originating area code and by accumulating a count of all calls to the INWATS customer. When the latter exceeds a predetermined threshold per unit time, the data base manages the number of calls routed to the preferred and/or alternate numbers in such a way as to reduce the probability of ineffective call completion attempts due to mass calling of the INWATS customer. It does so by instructing the telephone system control to inhibit calls to the INWATS number for a prescribed period of time. This action relieves the telephone voice network, data base and CCIS facilities from real time involvement on calls to busy stations and makes them available for serving other calls.

The data base comprises a first table in memory which includes a file of originating area codes (NPAs). Each such code is correlated in the file with an individual POTS number when INWATS service is purchased by the customer for that area code. In response to a receipt of a dialed INWATS number and the call originating area code from a telephone system, the data base examines its NPA file of originating area codes for screening the call to check whether the INWATS customer is entitled to receive calls from that area code on a toll-free basis to the caller. Entitlement is identified by the presence of the POTS number in the same memory file as the received area code. The NPA file also comprises a counter which is incremented on each call attempt per originating area code.

A second memory table includes an individual file for each POTS number that stores busy/idle data and open/closed hours information for that POTS number. The busy/idle data is dynamically loaded under control of the call terminating office signaling the busy and idle status of the line(s), or line group, associated with the POTS number. The data is advantageously used to reduce the probability of ineffective attempts to complete calls to all busy or closed stations and thereby reduces the unnecessary use and tie-up of toll telephone switching equipment. It also serves to signal when the station is idle and therefore available to serve calls.

Each POTS file in the second memory table also includes an alternate, or second choice, POTS number to which an INWATS call is routed when the preferred first choice POTS number is either busy or closed. The file also comprises a counter which is incremented on each call involving the preferred POTS number to accumulate a count of the total number of calls generated. The same file includes threshold and call inhibit time data for managing the number of calls that are directed to the POTS number within a given interval of time.

The procedure for processing an INWATS call comprises a calling party dialing the 800-NXX-XXXX of the desired INWATS customer. The local telephone office serving the calling station is responsive to the dialed number for routing the call to a toll office equipped with CCIS (Common Channel Interoffice Signaling) facilities and sending to that office the dialed number. The toll office routinely identifies the area code from which the call originates and then sends it, as a suffix code, together with the dialed number over the CCIS direct signaling system to the data base identified by the CCIS system from the dialed 800-NXX.

At the data base, the received INWATS number is checked to ascertain that it is an active working number. Next, the file of the originating area code is located in the first memory table and the entitlement of the call to be made from that originating area is verified by extracting the POTS number of the INWATS customer station preferably designated to serve the call. The NPA file counter is then incremented to indicate a call attempt involving the originating area code.

Thereafter, the file of the preferred POTS number is located in the second memory table and the open/closed and busy/idle status of the destination station associated with that number is examined. Upon determining that the station is open and idle, a CCIS message containing the preferred POTS number is sent from the data base over the CCIS system to the toll office for enabling call connections to be established from that office through the toll switching network to the station associated with the POTS number. At about the same time, the counter in the POTS number file is incremented to accumulate data on traffic to that POTS number and to control call routing to it during mass calling periods when the number of calls exceed a predetermined threshold.

Busy and idle data for each INWATS station is dynamically routed from an associated terminating telephone office over CCIS facilities to the data base for updating records in the file of the station POTS number in the second memory table. The busy data controls the processing of INWATS cells selectively to alternate POTS number(s) and to audible busy station messages.

The call processing procedure is arranged to return CCIS messages from the data base to the toll office to identify the dialing of a nonworking INWATS number, a call which is not authorized on a toll-free basis from that call originating area, and the busy or closed status of the INWATS station designated to serve the call. The method is advantageously useful for data base service on automated collect calling, toll call fowarding and other special service calls as hereinafter described.

Brief Description of the Drawings

> Fig. 1 shows, in block diagram form, a configuration of telephone switching offices, CCIS and signal transfer facilities together with one of several data bases illustratively for serving INWATS calls in the United States;
>
> Fig. 2 depicts an area code to POTS number listing of a data base translator and an originating area code call counter;
>
> Fig. 3 illustrates the data base translator listing of busy/idle, closing time, loop, alternate route number, and network management data for the POTS number;
>
> Fig. 4 is a flow chart of data base system operations for an INWATS call; and
>
> Fig. 5 is a flow chart of network management operations.

The drawing illustrates a number of local telephone switching offices, such as offices 12, 38, 41 and 43 which serve associated telephone stations, such as stations 10, 22, 30 and 31. The drawing also discloses toll offices such as offices 15, 34 and 36 which form a toll switching network serving the local offices. The structural details of such local and toll offices form no part of the present invention and are only described herein to the extent necessary for an understanding of this invention. Each of the local offices, by way of example, is suitably an electronic program controlled telephone system of the No. 1 ESS (Electronic Switching System) type disclosed in the *Bell System Technical Journal (BSTJ)*, September 1964, Volume XLIII, Number 5, Parts 1 and 2; and in the R. W. Downing-H. F. May-F. F. Taylor-U. Ulrich U.S. Pat. No. 3,570,008 of Mar. 9, 1971. Those disclosures and the citations referred to therein may be consulted for a complete understanding of the construction and operations of a typical local switching office 12, 38, 41 and 43.

Each toll office is illustratively an electronic program controlled telephone system of the No. 4 ESS design with CCIS facilities as disclosed in the *Bell System Technical Journal (BSTJ)*, September 1977, Volume 56, No. 7.

CCIS facilities utilized in conjunction with the local and toll offices and the data base system 11 are disclosed in the *BSTJ*, February 1978, Volume 57, No. 2 and in W. B. Smith-J. B. Synnott, III U.S. Pat. No. 3,624,613 of Nov. 30, 1971.

Detailed Description. In FIG. 1, the processing of a call to an INWATS customer is explained by assuming that it is originated from a calling station 10 in California. The INWATS customer is assigned a single telephone number which is dialed by callers on all calls to that customer regardless of their place of origination. The number comprises ten digits including an 800 area code portion followed by seven digits NXX-XXXX. The NXX digits identify the data base, such as data base 11 in Texas, which stores the INWATS customer data utilized for processing the call to the appropriate destination. The NXX-XXXX digits identify the INWATS customer within the data base. The "N" digit is any value 2–9 and each of the "X" digits is any value 0–9.

For the exemplary embodiment, call processing is described by assuming that the call may be routed to any one of three destinations, namely, first, second and third choice numbers respectively identifying stations 22, 30 and 31. Each such station is depicted as an individual telephone station for the purpose of illustration. It is to be understood that in many cases INWATS call answering facilities commercially utilize automatic call distributor systems involving numerous groups of incoming telephone lines served by a local or toll telephone switching system. A typical distributor system is disclosed, for example, in S. F. Dunning U.S. Pat. No. 3,111,561 of Nov. 19, 1963. The three destination stations will suffice to explain typical call situations encountering open/closed and busy/idle conditions and the likelihood of completing INWATS calls in geographic areas with different local business hours.

After station 10 initiates a call and dials the INWATS customer number, office 12 recognizes the 800 area code and routes the call illustratively over a trunk 14 to a toll office 15 equipped with CCIS facilities. Office 12 also forwards the dialed INWATS number to office 15 for assembly in office 15 with the area code for that portion of California from which the call originates. Office 15 determines the originating area code based on the serving of the call over trunk 14 in a known manner.

Office 15 routinely translates the dialed 800 digits and determines that the call is to be routed to a data base for deriving call routing data. As a consequence, office 15 formulates an INWATS CCIS message including an identification of the special service call, the office 15 and the originating area code as well as the dialed number and sends that message over the CCIS link 17 to a signal transfer point or system 18. The latter is responsive to the received message for translating the 800-NXX digits to determine that the INWATS customer data is stored in the data base system 11, which illustratively is located in Dallas, Texas. System 18 then forwards the CCIS message to data base 11 over a CCIS link 19. System 18 is essentially a data link, or packet, switching arrangement for CCIS applications. Packet data switching facilities are disclosed for example in A. G. Fraser U.S. Pat. Nos. 3,749,845 of July 31, 1973 and 3,979,733 of Sept. 7, 1976. Structure and operations of CCIS are described, by way of example, in the 57 *BSTJ* No. 2, page 230, et seq. Thus, the call is processed from the toll telephone switching network to a data base for deriving call service information.

Data base system 11 is, by way of example, a commercially available system, such as a Western Electric Company, Incorporated 1A processor (disclosed in 56 *Bell System Technical Journal* No. 7, February 1977), equipped with disk storage and a system of programs to establish, edit and manage information stored in memory. It is equipped via a peripheral unit bus (disclosed in 57 *Bell System Technical Journal* No. 2 February 1978) with CCIS facilities for interface communication with the signal transfer system 18 and the toll network. System 11 is responsive to the received INWATS number and originating area code for deriving all of the stored information needed for processing the call to an idle and available INWATS customer station designed as the preferred called station to serve the call. The information includes data for the active-inactive status of the dialed INWATS number, the storage files associated with the active number, the inband/out-band character of the call based on the received originating area code, and a directory-unlisted POTS number of a destination subsystem or station.

For each active INWATS number, data base system 11 is equipped with translator tables 20 and 21 of FIGS. 2 and 3. Table 20 contains a list of originating area codes and a correlated list of POTS DDD (Plain Ordinary Telephone Service Direct Distance Dialing) numbers identifying one or more called stations primarily designated to serve calls from the respective originating areas. Each listed area code is uniquely associated with an individual one of the listed POTS numbers when the INWATS customer is entitled to receive calls from that originating area on a toll-free basis to the calling party. A single such POTS number may be shared by a plurality of different originating area codes. Whenever table 20 contains a POTS number for an originating area code, calls from that area are designated as being "in-band". Otherwise, in the absence of a POTS number for the area code, (null entry) calls from that area are "out-of-band" for the INWATS customer and calls are not completed on a toll free basis to the caller.

Illustratively, on the call originating from station 10, the originating area code is 805 and table 20 lists a correlated POTS number 919-658-3399, which is a called station 22 in North Carolina, preferably designated as a first choice to serve the INWATS calls. Accordingly, such a call is "in-band". If it were not, no POTS number would be listed in table 20 for the area code 805.

Each originating area code in table 20 is furnished with a counter 27 which is incremented on each call attempt involving that area code. In this way, statistics are derived on calls by area codes regardless of whether the INWATS customer purchases the service for that area.

Table 21 contains a plurality of files each of which is associated with an individual POTS number in table 20. Each such file, such as file 23, contains eight individual items of data for each such number. These data items are utilized, as appropriate, for completing calls to the associated POTS number and for recording the number of such calls for network management purposes. The first item 49 is a loop bit used to ensure that the alternate routing does not loop on itself. The second item 24 identifies the times and dates on which the station associated with the POTS number 919-658-3399 is closed or personnel is unavailable for serving incoming calls. Closed times are, for example, nonbusiness hours, such as 5 p.m.-9 a.m., holidays, and weekends, such as Sunday. The third item 25 contains a bit indicating the current busy-idle status of the POTS number 919-658-3399. The fourth item 26 contains an alternate POTS DDD number 206-582-2044 which is used when the preferred POTS number is busy, closed or unavailable. The fifth item 46 contains a call counter which accumulates a count of the number of calls made to the POTS number 919-658-3399 for a current time interval, for example, two minutes. The sixth item 47 contains a threshold number for inhibiting calls to the POTS number. The seventh item 50 indicates whether a network management control is currently effective based on the previous time interval. The last item is a time interval 48 specifying the period for inhibiting data base inquiries to the 800 + NXX-XXXX. Entry 48 is dynamically updated by a procedure described in FIG. 5.

The utilization of data in tables 20 and 21 and INWATS information processing operations of the data base system 11 are now described for the call from station 10 of FIG. 1 with reference to the flow diagram of FIG. 4. After the tables 20 and 21 are loaded with the data as depicted in FIGS. 2 and 3, system 11 is responsive to the receipt of the CCIS INWATS message for checking its records (not shown) to determine whether the dialed INWATS number is active or inactive. If it is not validly in use and therefore inactive, system 11 effects a return of a CCIS vacant code message to the toll office 15 of FIG. 1 via link 19, signal transfer system 18 and link 17. Office 15 thereupon routinely causes an announcement to be returned to the caller for conveying dialing instructions and indicating that the dialed number is not a working number.

If the number is active, system 11 examines table 20 and its file 28 for translating the dialed INWATS number and originating area code into a POTS DDD number 919-658-3399 by a look-up operation in table 20. Next, the area code counter 27 is incremented to indicate a call from a station in area code 805. Had the look-up failed to find such a POTS number (null entry) correlated with the originating area code in file 28, the dialed INWATS number would be "out-of-band" to that call originating area. Resultingly, system 11 then effects an incrementing of a counter 27 and a return of an out-of-band CCIS message to toll office 15 of FIG. 1 via the CCIS links 19 and 17 and the signal transfer system 18. Office 15 then routinely effects the transmission of a reorder message to the caller.

System 11 determines that the INWATS call is "in-band" when its lock-up operation extracts a POTS DDD number 919-658-3399 from file 28. That number is then used to locate the file, such as file 23, in table 21 which contains the eight data items needed for further call processing and traffic management. System 11 initially checks file 23 to ascertain whether the station associated with the POTS number is indicated to be open. This is done by examining item 24 of file 23. If the station is open, system 11 next examines the busy/idle data, item 25. If the station is idle, then the POTS number, 919-658-3399 will be returned to office 15 for normal call setup. If the station is closed or busy, then system 11 examines item 26 of file 23 to obtain an alternate POTS number. The loop bit, item 49, is set to prevent looping between POTS numbers. If an alternate POTS number exists, such as 206-582-2044, and its loop bit is not set, then the process continues until either an open and idle station is found, no alternate POTS number is found, or transfer is made to a POTS number whose loop bit is set. The last two situations will cause system 11 to send a CCIS message to office 15, via the CCIS facilities, indicating an ineffective attempt. If the call results in an ineffective attempt and in the process of picking alternate POTS numbers a busy condition is encountered, then the CCIS message includes an indicator telling office 15 to return either a busy tone or audible message to the caller. If all encountered POTS numbers had been closed, then the CCIS message includes an indicator telling office 15 to transmit an audible message to the caller indicating the status of the called station.

Upon finding that the INWATS call is "in-band" and that a station, such as station 22, 30 or 31 associated with a preferred or alternate POTS number is opened and nonbusy, system 11 effects a transmission of a CCIS message including that number to toll office 15 via the CCIS facilities 16. Office 15 then controls the establishment of call connections to the respective station 22, 30 or 31 illustratively, over trunk 33, toll office 34, trunk 35, toll office 36, trunk 37, local office 38, and line 39 to station 22 or from office 34, trunk 40, local office 41, to station 30. Connections to station 31 similarly are extendable from office 36 via trunk 42 and office 43. The called station is then alerted to the incoming call and conversation ensues.

Each file in table 21 is equipped with a network management counter which retains a current count of the number of calls directed to the POTS number. For example, file 32 has the counter 51 for counting calls directed to the number 212-525-3333. The last file examined in the sequence of first choice or alternate POTS numbers increments its current counter item 46. This counter will be reinitialized periodically, for example, every two minutes. After the CCIS message is returned to office 15 with either a POTS number or an ineffective attempt, then item 50 of the last file is checked. This is an indication of whether a control is presently in progress for that POTS number. If the control is on, then the inhibit time, item 48, is returned in another CCIS message to office 15. Office 15 then inhibits all new data base attempts to that 800-NXX-XXXX for the prescribed time period. Calls to this number, arriving at office 15 during such an interval of time, are given an audible message indicating the situation. Periodically, for example every two minutes, all counters, item 46, are compared against this threshold, item 47. The threshold per unit time is set on the basis of the number of lines of the INWATS station, call holding times and similar factors. If the control is currently off and the counter is less than the threshold, then the counter is merely set to zero. If the control is currently off and the counter exceeds the threshold, then the control is turned on, item 50, inhibit time is set to an initial state, for example 6 seconds, and the counter is set to zero. If the control is on and the counter is less than the threshold, the inhibit time is decreased by three seconds, for example, while if the counter is over the threshold the inhibit time is increased by, for example, three seconds. Again the counter is set to zero. If the inhibit time goes to zero, then the control is taken off, item 50.

The result of the network management action is that mass calling attempts that are not likely to be completed are not routed to data base 11 or beyond toll office 15. This frees data base 11, CCIS facilities 17, 18 and 19 and the toll network to serve other calls and reduces the probability of call blockage due to mass calling of an INWATS customer.

Busy data for an INWATS station is written into the data base 11 under control of the terminating office utilizing the CCIS system. Illustratively, such data is gathered for the line(s) of station 22 by the terminating office 38 and busy/idle records for the one or more lines, such as line 39, serving that station. The busy/idle data is sent by CCIS message over CCIS facilities 44 and signal transfer system 45 the response to the appropriate busy or idle condition. When an all lines busy condition exists for station 22, terminating office 38 sends a CCIS message to transfer system 45 which translates it and determines that the busy data is to be routed to data base 11 for entering a busy status "1" in the busy record 25 of the file of table 21 for station 22. Resultingly, INWATS calls designated for routing to that station are blocked and alternate routing is effected as priorly explained. A similar CCIS signaling operation occurs to enter an idle bit "0" in record 25 when station 22 has at least one idle line.

The foregoing method, in addition to being utilized for INWATS service, is useful for automated collect calls, toll call forwarding and other special service calls of a type served for example by a Traffic Service Position System as disclosed in R. J. Jaeger et al U.S. Pat. No. 3,484,560 of Dec. 16, 1969. For the collect call service, the entitled customer is given a single multi-digit number, such as a 700-NXX-XXXX, number or a .phi.-700-NXX-XXXX + PIN (Personal Identification Number) code where the PIN code is a multidigit security code, which is given to anyone that the customer wishes to give a standing authorization to call him collect. In the data base, the customer telephone number is stored in a first memory table, such as table 20 of FIG. 2 in correlation with the number. When a caller dials the number, the data base receives from the communication system a special service call message including that number and selectively a suffix code, such as an originating area code; verifies the in-band nature of the call plus the security code if supplied; translates the received message into a call completion control message, such as a routing message including the customer telephone station number; and sends the control message to the telephone machine for controlling the establishment of call connections from the calling station to the customer station authorizing the automatic collect calling. Charges are then billed to that customer. Busy/idle and open/closed station data, alternate station routing data, and network management data including call counting such as are stored in the table 21 of FIG. 3, are usable for automated collect calling.

Toll call forwarding is essentially the same as described for INWATS and automated collect calling except that the forwarding is illustratively effected by allowing the called party to change the POTS number in FIG. 2 by contacting, for instance, an operator having access to the CCIS system for sending a message to the data base to update the number.

Reference is made to A. B. Mearns (Case 1) patent application Ser. No. 924,169, filed concurrently herewith.

3.11 THE INTERNET—A BREAKTHROUGH WITHOUT ONE INVENTION

The Internet has brought about a profound revolution in modern society. General-purpose computing facilitated the popularity of computer systems in the fifties and sixties, and the introduction of inexpensive personal computers in the eighties and nineties brought the revolution home. Since the early nineties, the Internet has even provided access to the information highway to the younger generation.

The Internet is a consolidation of inventions from many contributors. During the sixties and seventies, the federal government sponsored loosely related research topics as independent projects. These projects generated islands of knowledge that were founded on sound scientific bases from universities and corporations. Later, these islands were bridged and made into a large body of knowledge through additional investigations lasting from the seventies through the nineties.

The Internet is a perfect example of the evolution of this scientific strategy. Most of the significant contributions of the sixties through the nineties in the telecommunications arena have systematically cumulated in the Internet. The Internet is fueled by the economic motives of its participants. The inventions came about because the Internet has sustained its own growth. Over the last decade, the pace has accelerated, and the economic forces in a capitalistic society can only add to the Internet to reach further into remote areas.

Two such notable sponsors are the Office of Aerospace Research (SRMA) and the Advanced Research Projects Agency (ARPA). First, THE ALOHA SYSTEM (see Section 3.11.1), initiated in the late sixties, resulted in a novel way to share most transmission media. Although it was developed for the radio networks linking numerous campuses of the University of Hawaii and other research facilities on the Hawaiian Islands, it paved the way for random access techniques based on CSMA/CD.

Second, under the sponsorship of ARPA, ARPANET [3.11.1] evolved to effectively use computer resources. During the mid-seventies, ARPANET had gained full operational status, serving the needs of the Defense Communications Agency. Later, it was used by the Department of Defense (DoD) to transfer both classified and unclassified information and to meet most DoD data communication needs. The seminal blending of information processing and mass communication started in the 1970s and was funded until the concept gained maturity in the eighties. During the nineties, most of the generic concepts [3.11.2] and algorithms were firmly established. The packet network technology is now a inexpensive reality and has gained immense popularity.

The Internet has experienced a growth unlike the growth of any other invention because there is no generic patent that any commercial entity owns. Internet ownership is non-existent. In a sense, the concept freely floats from any country to another, from any social-intellectual circle to the next. When the telecommunication backbone network of any country can host the rather rudimentary requirements of packet switching (see Section 4.1), then Internet access can be readily provided with international gateway access at the national gateway from one country to another.

Numerous reference books have been written on the history and development of the Internet. We do present one classic paper (rather than a patent) later in this section on the start of public-domain packet switching, which facilitated the start of the Internet.

N. Abramson

3.11.1 THE ALOHA SYSTEM—Another Alternative for Computer Communications*

by NORMAN ABRAMSON
University of Hawaii, Honolulu, Hawaii

Introduction In September 1968 the University of Hawaii began work on a research program to investigate the use of radio communications for computer-computer and console-computer links. In this report we describe a remote-access computer system—THE ALOHA SYSTEM — under development as part of that research program [3.11.3] and discuss some advantages of radio communications over conventional wire communications for interactive users of a large computer system. Although THE ALOHA SYSTEM research program is composed of a large number of research projects, in this report we shall be concerned primarily with a novel form of random-access radio communications developed for use within THE ALOHA SYSTEM.

The University of Hawaii is composed of a main campus in Manoa Valley near Honolulu, a four year college in Hilo, Hawaii and five two year community colleges on the islands of Oahu, Kauai, Maui and Hawaii. In addition, the University operates a number of research institutes with operating units distributed throughout the state within a radius of 200 miles from Honolulu. The computing center on the main campus operates an IBM 360/65 with a 750 K byte core memory and several of the other University units operate smaller machines. A time-sharing system UHTSS/2, written in XPL and developed as a joint project of the University Computer Center and THE ALOHA SYSTEM under the direction of W. W. Peterson is now operating. THE ALOHA SYSTEM plans to link interactive computer users and remote-access input-output devices away from the main campus to the central computer via UHF radio communication channels.

*THE ALOHA SYSTEM is supported by the office of aerospace research (SRMA) under Contract Number F44620-69-CO-030, a Project THEMIS award.

Wire Communications and Radio Communications for Computers At the present time conventional methods of remote access to a large information processing system are limited to wire communications—either leased lines or dial-up telephone connections. In some situations these alternatives provide adequate capabilities for the designer of a computer-communication system. In other situations however the limitations imposed by wire communications restrict the usefulness of remote access computing [3.11.4]. The goal of THE ALOHA SYSTEM is to provide another alternative for the system designer and to determine those situations where radio communications are preferable to conventional wire communications.

The reasons for widespread use of wire communications in present day computer-communication systems are not hard to see. Where dial-up telephones and leased lines are available they can provide inexpensive and moderately reliable communications using an existing and well developed technology [3.11.5] and [3.11.6]. For short distances the expense of wire communications for most applications is not great.

Nevertheless there are a number of characteristics of wire communications which can serve as drawbacks in the transmission of binary data. The connect time for dial-up lines may be too long for some applications; data rates on such lines are fixed and limited. Leased lines may sometimes be obtained at a variety of data rates, but at a premium cost. For communication links over large distances (say 100 miles) the cost of communication for an interactive user on an alphanumeric console can easily exceed the cost of computation [3.11.7]. Finally we note that in many parts of the world a reliable high quality wire communication network is not available and the use of radio communications for data transmission is the only alternative.

There are of course some fundamental differences between the data transmitted in an interactive time-shared computer system and the voice signals for which the telephone system is designed [3.11.8]. First among these differences is the burst nature of the communication from a user console to the computer and back. The typical 110 baud console may be used at an average data rate of from 1 to 10 baud over a dial-up or leased line capable of transmitting at a rate of from 2400 to 9600 baud. Data transmitted in a time-shared computer system comes in a sequence of bursts with extremely long periods of silence between the bursts. If several interactive consoles can be placed in close proximity to each other, multiplexing and data concentration may alleviate this difficulty to some extent. When efficient data concentration is not feasible however the user of an alphanumeric console connected by a leased line may find his major costs arising from communication rather than computation, while the communication system used is operated at less than 1 percent of its capacity.

Another fundamental difference between the requirements of data communications for time-shared systems and voice communications is the asymmetric nature of the communications required for the user of interactive alphanumeric consoles. Statistical analyses of existing systems indicate that the average amount of data transmitted from the central system to the user may be as much as an order of magnitude greater than the amount transmitted from the user to the central system [3.11.8]. For wire communications it is usually not possible to arrange for different capacity channels in the two directions so that this asymmetry is a further factor in the inefficient use of the wire communication channel.

The reliability requirements of data communications constitute another difference between data communication for computers and voice communication. In addition to errors in binary data caused by random and burst noise, the dial-up channel can produce connection problems—e.g., busy signals, wrong numbers and disconnects. Meaningful statistics on both of these problems are difficult to obtain and vary from location to location, but there is little doubt that in many locations the reliability of wire communications is well below that of the remainder of the computer-communication system. Furthermore, since wire communications are usually obtained from the common carriers this portion of the overall computer-communication system is the only portion not under direct control of the system designer.

THE ALOHA SYSTEM The central computer of THE ALOHA SYSTEM (an IBM 360/65) is linked to the radio communication channel via a small interface computer (Figure 1). Much of the design of this multiplexor is based on the design of the Interface Message Processors (IMP's) used in the ARPA computer net [3.11.6] and [3.11.9]. The result is a Hawaiian version of the IMP (taking into account the use of radio communications and other differences) which has been dubbed the MENEHUNE (a legendary Hawaiian elf).

The HP 2115A computer has been selected for use as the MENEHUNE. It has a 16-bit word size, a cycle time of 2 microseconds and an 8K-word core storage capacity. Although THE ALOHA SYSTEM will also be linked to remote-access input-output devices and small satellite computers through the MENEHUNE, in this paper we shall be concerned with a random access method of multiplexing a large number of low data rate consoles into the MENEHUNE through a single radio communication channel.

THE ALOHA SYSTEM has been assigned two 100 kHZ channels at 407.350 MHZ and 413.475 MHZ. One of these channels has been assigned for data from the MENEHUNE to the remote consoles and the other for data from the consoles to the MENEHUNE. Each of these channels will operate at a rate of 24,000 baud. The communication channel from the MENEHUNE to the consoles provides no problems. Since the transmitter can be controlled and buffering performed by the MENEHUNE at the Computer Center, messages from the different consoles can be ordered in a queue according to any given priority scheme and transmitted sequentially.

Messages from the remote consoles to the MENEHUNE however are not capable of being multiplexed in such a direct manner. If standard orthogonal multiplexing techniques (such as frequency or time multiplexing) are employed we must divide the channel from the consoles to the MENEHUNE into a large number of low speed channels and assign one to each console, whether it is active or not. Because of the fact that at any given time only a fraction of the total number of consoles in the system will be active and because of the burst nature

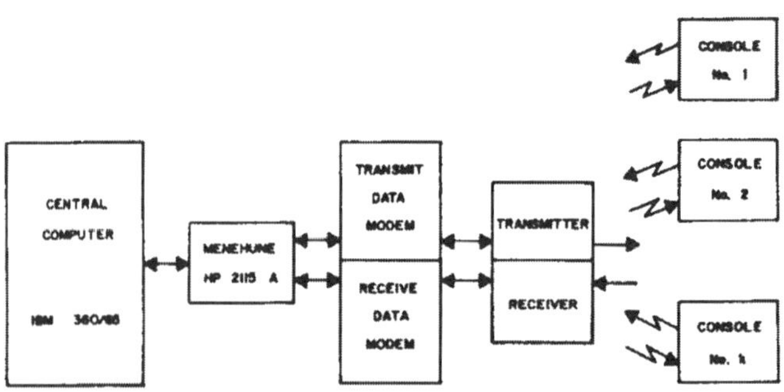

Figure 1 THE ALOHA SYSTEM

of the data from the consoles such a scheme will lead to the same sort of inefficiencies found in a wire communication system. This problem may be partly alleviated by a system of central control and channel assignment (such as in a telephone switching net) or by a variety of polling techniques. Any of these methods will tend to make the communication equipment at the consoles more complex and will not solve the most important problem of the communication inefficiency caused by the burst nature of the data from an active console. Since we expect to have many remote consoles it is important to minimize the complexity of the communication equipment at each console. In the next section we describe a method of random access communications which allows each console in THE ALOHA SYSTEM to use a common high speed data channel without the necessity of central control or synchronization.

Information to and from the MENEHUNE in THE ALOHA SYSTEM is transmitted in the form of "packets," where each packet corresponds to a single message in the system [3.11.10]. Packets will have a fixed length of 80 8-bit characters plus 32 identification and control bits and 32 parity bits; thus each packet will consist of 704 bits and will last for 29 milliseconds at a data rate of 24,000 baud.

The parity bits in each packet will be used for a cyclic error detecting code [3.11.11]. Thus if we assume all error patterns are equally likely the probability that a given error pattern will not be detected by the code is [3.11.12]

$$2^{-32} \approx 10^{-9}$$

Since error *detection* is a trivial operation to implement [3.11.12], the use of such a code is consistent with the requirement for simple communication equipment at the consoles. The possibility of using the same code for error *correction* at the MENEHUNE will be considered for a later version of THE ALOHA SYSTEM.

The random access method employed by THE ALOHA SYSTEM is based on the use of this error detecting code. Each user at a console transmits packets to the MENEHUNE over the same high data rate channel in a completely unsynchronized (from one user to another) manner. If and only if a packet is received without error it is acknowledged by the MENEHUNE. After transmitting a packet the transmitting console waits a given amount of time for an acknowledgment; if none is received the packet is retransmitted. This process is repeated until a successful transmission and acknowledgment occurs or until the process is terminated by the user's console.

A transmitted packet can be received incorrectly because of two different types of errors; (1) random noise errors and (2) errors caused by interference with a packet transmitted by another console. The first type of error is not expected to be a serious problem. The second type of error, that caused by interference, will be of importance only when a large number of users are trying to use the channel at the same time. Interference errors will limit the number of users and the amount of data which can be transmitted over this random access channel.

In Figure 2 we indicate a sequence of packets as transmitted by k active consoles in the ALOHA random access communication system.

We define τ as the duration of a packet. In THE ALOHA SYSTEM τ will be equal to about 34 milliseconds; of this total 29 milliseconds will be needed for transmission of the 704 bits and the remainder for receiver synchronization. Note the overlap of two packets from different consoles in Figure 2. For analysis purposes we make the pessimistic assumption that when an overlap occurs neither packet is received without error and both packets are therefore retransmitted.* Clearly as the number of active consoles increases the number of interferences and hence the number of re-transmissions increases until the channel clogs up with repeated packets [3.11.13]. In the next section we compute the average number of active consoles which may be supported by the transmission scheme described above.

Note how the random access communication scheme of THE ALOHA SYSTEM takes advantage of the nature of the radio communication channels as opposed to wire communications. Using the radio channel as we have described each user may access the same channel even though the users are geographically dispersed. The random access communication method used in THE ALOHA SYSTEM may thus be thought of as a form of data concentration for use with geographically scattered users.

Random Access Radio Communications. We may define a random point process for each of the k active users by focusing our attention on the starting times of the packets sent by each user. We shall find it useful to make a distinction between those packets transmitting a given message from a console for the first time and those packets transmitted as repetitions of a message. We shall refer to packets of the first type as *message packets* and to the second type *as repetitions*. Let λ be the average rate of occurrence of message packets from a single active user and assume this rate is identical from user to user. Then the random point process

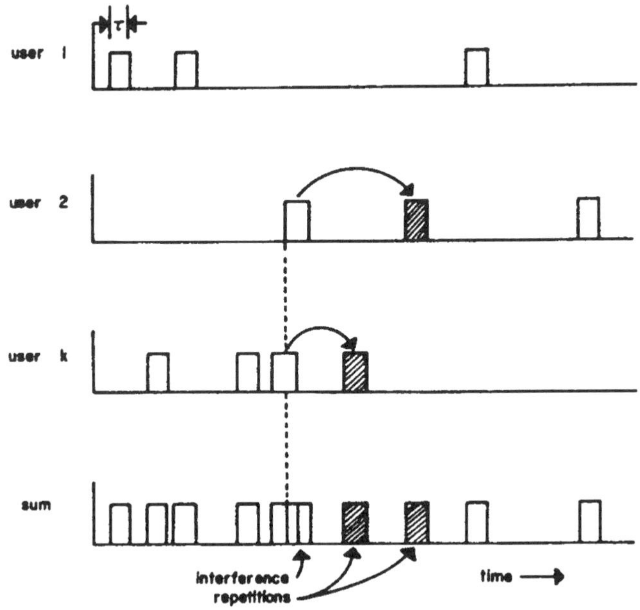

Figure 2 ALOHA communication multiplexing

*In order to ensure that the retransmitted packets do not continue to interfere with each other, we must make sure the retransmission delays in the two consoles are different.

consisting of the starting times of message packets from all the active users has an average rate of occurrence of

$$r = k\lambda$$

where r is the average number of message packets per unit time from the k active users. Let τ be the duration of each packet. Then if we were able to pack the messages into the available channel space perfectly with absolutely no space between messages we would have

$$r\tau = 1$$

Accordingly we refer to $r\tau$ as the *channel utilization.* Note that the channel utilization is proportional to k, the number of active users. Our objective in this section is to determine the maximum value of the channel utilization, and thus the maximum value of k, which this random access data communication channel can support.

Define R as the average number of message packets plus retransmissions per unit time from the k active users. Then if there are any retransmissions we must have $R > r$. We define $R\tau$ as the *channel traffic* since this quantity represents the average number of message packets plus retransmissions per unit time multiplied by the duration of each packet or retransmission. In this section we shall calculate $R\tau$ as a function of the channel utilization, $r\tau$.

Now assume the interarrival times of the point process defined by the start times of all the message packets plus retransmissions are independent and exponential. This assumption, of course, is only an approximation to the true arrival time distribution. Indeed, because of the retransmissions, it is strictly speaking not even mathematically consistent. If the retransmission delay is large compared to τ, however, and the number of retransmissions is not too large this assumption will be reasonably close to the true distribution. Moreover, computer simulations of this channel indicate that the final results are not sensitive to this distribution. Under the exponential assumption the probability that there will be no events (starts of message packets or retransmissions) in a time interval T is

$$e^{-2RT}$$

Using this assumption we can calculate the probability that a given message packet or retransmission will need to be retransmitted because of interference with another message packet or retransmission. The first packet will overlap with another packet if there exists at least one other start point τ or less seconds before or τ or less seconds after the start of the given packet. Hence the probability that a given message packet or retransmission will be repeated is

$$[1 - e^{-2R\tau}] \tag{1}$$

Finally we use (1) to relate R, the average number of message packets plus retransmissions per unit time to r, the average number of message packets per unit time. Using (1) the average number of retransmissions per unit time is given by

$$R[1 - e^{-2R\tau}]$$

so that we have

$$R = r + R[1 - e^{-2R\tau}]$$

or

$$r\tau = R\tau e^{-2R\tau} \tag{2}$$

Equation (2) is the relationship we seek between the channel utilization $r\tau$ and the channel traffic $R\tau$. In Figure 3 we plot $R\tau$ versus $r\tau$.

Note from Figure 3 that the channel utilization reaches a maximum value of $1/2e = 0.186$. For this value of $r\tau$ the channel traffic is equal to 0.5. The traffic on the channel becomes unstable at $r\tau = 1/2e$ and the average number of retransmissions becomes unbounded. Thus we may speak of this value of the channel utilization as the *capacity* of this random access data channel. Because of the random access feature the channel capacity is reduced to roughly one sixth of its value if we were able to fill the channel with a continuous stream of uninterrupted data.

For THE ALOHA SYSTEM we may use this result to calculate the maximum number of interactive users the system can support.

Setting

$$r\tau = k\lambda\tau = 1/2e$$

we solve for the maximum number of active users

$$k_{max} = (2e\lambda\tau)^{-1}$$

A conservative estimate of λ would be 1/60 $(\text{seconds})^{-1}$, corresponding to each active user sending a message packet at an average rate of one every 60 seconds. With τ equal to 34 milliseconds we get

$$k_{max} = 324 \tag{3}$$

Note that this value includes only the number of active users who can use the communication channel simultaneously. In contrast to usual frequency or time multiplexing methods while a user is not active he consumes no channel capacity so that the total number of users of the system can be considerably greater than indicated by (3).

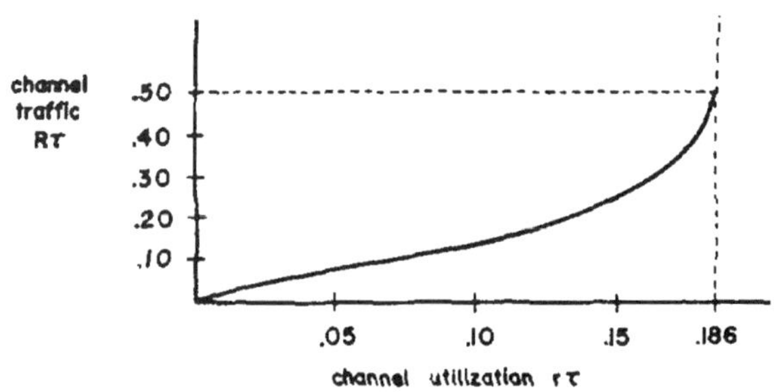

Figure 3 Channel utilization vs channel traffic

The analysis of the operation of THE ALOHA SYSTEM random access scheme provided above has been checked by two separate simulations of the system [3.11.14] and [3.11.15]. Agreement with the analysis is excellent for values of the channel utilization less than 0.15. For larger values the system tends to become unstable as one would expect from Figure 3.

3.11.2 The "Newer" Internet

LANs, MANs, WANs, and the Internet: Generic packet/frame/cell networking principles are applicable to all of these data networks. They share resources and use adaptive routing. Geographical size and expanse dictate the classification as a LAN, metropolitan area network (MAN), or WAN. Transmission paths are not dedicated, and the traffic and network conditions dictate the actual allocation of routes linking the network nodes. These network nodes work cooperatively and cogently to move data from one node to another. Most modern networks use actual network statistics to adaptively route traffic. Any node handles the traffic that it can pass without offering undue delay to local or transit traffic. The algorithms in the original/revised ARPANET or other networks (e.g., TMYNET I/II, SNA) get the packets from node to node with minimal delay.

There are only two building blocks to any network—the nodes and the links. There are abundant varieties of nodes and links. Most nodes have switching functions embedded to guide the packets to their destinations. As the header information is decoded, the routing is then performed; thus, the packets may arrive via any route, but they are sequentially assembled before delivery at the destination node.

What Is There to Lose? Most countries realize that an investment in the Internet can catalyze economic activity. For this reason, the public sector looks at the Internet the same way as does the transportation industry. However, the global impact of information highways "inter netting" the knowledge banks is significantly higher. We have every reason to expect supergeometric growth for the Internet and perhaps a slightly less dramatic growth for Internet-related businesses.

The evidence is already obvious from the growth and collapse of dot.com corporations during the last decade. However, the great competition between these startup corporations will also bring the most value to society. Scenarios of this nature have occurred in the past. The rules of economic survival of the fittest in the information and knowledge society (see Ahamed, S. V., "Social Impact of Intelligent Telecommunication Networks." Proceedings of the Pacific Telecommunication Conference PTC, Honolulu, January 18–22, 1987) may be quite different from those of a jungle society. In a very materialistic sense, the gold rush in Alaska, the wheat farmers of the Midwest, and the railroad industry in the U.S. all have left their own mark on history. The drama that is unfolding with the stampede to the Internet will change the rules of intellectual and informational Darwinism from the rules of social and economic Darwinism of the past century.

The (R)evolutionary Growth in the Internet Arena. In the major inventions of the past, society has taken its time to assimilate inventions and their consequences. The pros and cons are considered, and the absorption of the technology proceeds on a relatively slow basis. However, in the case of the Internet, the growth of its technology and its social acceptance have both been phenomenal since the eighties. Video and then digital services quickly became popular in western society. Now, the Internet and Internet services appear to be integral to the functioning of educated western societies. Users have become more information-oriented during the last two decades, and this trend is likely to gain more momentum. Our product-oriented society is shifting to an information-oriented society. Again, the economics of an information society is different from the economics of the Adam Smith society. In modern times, society is likely to be more educated and is likely to respond more quickly. These factors tend to make the inventions grow faster. By the same token these factors also make inventions extinct and obsolete more quickly. The Internet is a perfect example of this phenomenon, and it has been an eye opener to inventors that has reinforced the idea that their inventions should be valuable and timely.

Public and Social Awareness. Social awareness via the Internet is as dynamic as the Internet itself. In this context, the Internet is a vehicle for most services and needs. Illicit and pornographic material has also gained a niche here, but by and large public awareness of the Internet and Internet-related services and information causes a forceful growth of social needs, and with this growth it brings about an expression to the whims and fancies of the society.

Inventors generally do not succumb to social taste; instead, they have to convince examiners who are looking for novelty of ideas. As it is necessary from the patent examiners, these officers around the world evaluate the novelty that an invention brings in its concept or its implementation. For this reason, we are optimistic that the gadgets catering to short-lived social whims are unlikely to become major inventions—instead they will remain minor innovations, reengineering or recycled ideas that crop up as minor devices.

In a majority of cases, the excitement that a major invention brings to society exceeds the social hype that the media may create around a minor finding or a contribution. Part of the reason stems from a higher level of understanding of the public based on instant access to worldwide information and knowledge banks accessible with Internet technology.

3.11.3 The Concept of an Intelligent Internet

The anatomy, architecture, and configuration of the Internet are well-documented [3.11.16]. The most common perception of the Internet is that it can provide almost real-time access to information almost anytime and almost anywhere. From the designer's perspective, there are numerous options that can make the Internet intelligent and adaptive. Mainly, these options are documented in advanced countries. The Internet is a topic of national importance and prestige for developing countries. The topology can be quite distinctive from nation to nation, and the effects of the technology can be as dynamic as the telecommunication industry itself.

The links and the access technologies [3.11.17] are also changing quite rapidly. The next step in the life of the Internet is an intelligent Internet, which combines the features of intelligent networks (see Section 3.10) with the packet-switching concepts now implemented throughout all types of networks. This is a logical and desirable step to make one major breakthrough in society become more useful. We cautiously venture to say that clusters of smaller innovations in the three areas will make the Internet the intelligent Internet of the next generation. Together, network architecture [3.11.18], Internet billing [3.11.19] and the blending of complex service patterns is more likely to pave the way towards an intelligent Internet that can serve the information needs of future generations.

The move toward the fiber-optic backbone, or even the move toward fiber-to-the-home (FTTH) or fiber-to-the-curb (FTTC) [3.11.20] architectures, is a dominant and very established trend. This flurry of activity is likely to bring a string of inventions to the Internet. Any specialized architecture of switches for Internet traffic and Internet billing are also likely to evolve.

From the perspective of the numerous varieties of services the Internet can provide, the architecture for a truly intelligent Internet still needs to be invented and optimized. From a historical perspective, the blending of at least three types of intelligence is necessary.

First, the intelligence from the intelligent network (IN) design consideration, capable of providing a variety of programmable services is necessary. This type of intelligence can provide an impact to the IN that is analogous to the impact that INs (IN/1, IN/1+, IN/2, AIN, and USIN [3.11.20]) have brought to the telecommunications market. Routing and searching thus can be greatly facilitated.

Second, intelligence from an AI perspective (e.g., pattern recognition, expert systems, computer vision, and robotics) that is capable of performing routine and AI-based functions is necessary. This type of intelligence can provide the ease and cost effectiveness of performing routine and programmable services from the Internet Service Providers (ISPs), which can make the Internet as inexpensive as sharing a mid-size PC with dozens of users.

Third, intelligence from a knowledge-processing perspective [3.11.21] (the ability to intelligently process information that is passing through the network) can greatly facilitate the implementation of medical networks, educational networks, financial networks, etc., via the Internet.

In the Internet arena, intelligence has many directions. In the realm of telecommunications, intelligence was directed toward routing in IN/1 or in the services such as the CLASS service of IN/2 (see Bell Communications Research Reference [3.11.22]). In computer science, intelligence has been directed towards the control of hardware via control memories in processors [3.11.22] at the micro level. It is towards the control of processor executing AI type of programs doing pattern recognition in PR applications, locating expert opinions in ES [3.11.23] applications, or even controlling robotic devices [3.11.24] in robotics. In computer vision [3.11.25], other examples for personal computers [3.11.26], medical systems [3.11.27], CAD/CAM [3.11.2], etc., also exist.

In the Internet arena, the fusion of human needs (see Maslow [3.11.28]) and the intelligent steps in satisfying such needs by a single "click" will become necessary. Many legal and ethical questions [3.11.29] need to be addressed. At present, elaborate software systems can yield such solutions. However, if a form of intelligence can provide total visibility through the five layers of human needs and still not trespass on social norms, then such intelligence will complement the current role of the Internet, and we will have a brand new intelligent Internet.

3.11.4 The Success of the Internet

The downward spiraling costs of the Internet are based on its capacity to share resources. These resources are owned by the private sector on a competitive basis of free economy and are cohesively constructed to perform Internet services. The best service provider wins the most Internet customers. This strategy works for corporations as well as those supplying

Internet services. Such services can be derived from knowledge banks, medical informatics, high-speed data links, data server facilities, or distance learning centers.

The built-in feature of sharing network resources through packet-switching technology makes Internet access relatively inexpensive. Mass communication facilities of most nations are generally designed to be large enough to encourage the growth of the communication industry. This added buffer of growth fosters the Internet even in developing countries.

REFERENCES (C-3.11)

[3.11.1] Salus, Peter H., *Casting the Net: From ARPANET to INTERNET and Beyond.* Addison-Wesley Pub Co, Reading, MA, 1995.

[3.11.2] Ahamed S. V., and Lawrence, V. B., *Design and Engineering of Intelligent Communication Systems.* Kluwer Academic Publishers, Boston, 1997. Also see Ahamed S. V., and Lawrence, V. B., "Interoperability of Multiple Databases for the Design and Simulation of High-Speed Digital Subscriber Lines." *Proceedings of the RIDE-IMS Conference, Vienna, Austria, April 19–20, 1993,* pp. 262–267.

[3.11.3] Abramson, N. et. al, *1969 Annual Report: THE ALOHA SYSTEM,* University of Hawaii, Honolulu, Hawaii, 1970.

[3.11.4] Diamond, D. S., and Selwyn, L. L., "Considerations for Computer Utility Pricing Policies." *Proceedings of the Fall Joint Computer Conference,* pp 1473-1478, American Federation of Information Processing Societies, AFIPS Press, Montvale, NJ, 1968. Also see, Gold M. M., and Selwyn, *"Real Time Computer Communications and the Public Interest"* Boston University College of Business Administration document (May 1968).

[3.11.5] Fano, R. M., "The MAC System: The Computer Utility Approach." IEEE Spectrum, Vol. 2, No. I, January 1965.

[3.11.6] Roberts, L. G., *Multiple Computer Networks and Computer Communication,* ARPA Report, Washington. DC, 1967. See also Dr. Lawrence G. Roberts's website at http://www.packet.cc/published.html. (Accessed on April 4, 2004.)

[3.11.7] Kemeny, John G., and Kurtz, T. E., "Dartmouth Time Sharing." *Science,* Vol. 162, No. 3850, October 11, 1968, pp 223–228.

[3.11.8] Jackson, P. E., and Stubbs, C. D., "A Study of Multiaccess Computer Communications." *Proceedings of the Spring Joint Computer Conference,* pp. 491–504 AFIPS Press, 1969.

[3.11.9] Bolt Beranek and Newman, Inc., "Initial Design for Interface Message Processors for the ARPA Computer Network," in Report No. 1763, 1969.

[3.11.10] Binder, R., "Multiplexing in THE ALOHA SYSTEM: Menehune-Keiki Design Considerations," in *THE ALOHA SYSTEM Technical Report B69-3.* University of Hawaii, Honolulu, Hawaii, November 1969.

[3.11.11] Peterson, W. W., and Weldon, E. J., Jr., *Error-Correcting Codes,* 2d. ed. John Wiley & Sons, New York, 1970.

[3.11.12] Brown, D. T., and. Peterson, W. W., "Cyclic Codes for Error Detection." *Proceedings IRE,* Vol. 49, 1961, pp 228–235.

[3.11.13] Liao, H. H. J., "Random Access Discrete Address Multiplexing Communications for THE ALOHA SYSTEM," in *THE ALOHA SYSTEM Technical Note 69–8.* University of Hawaii, Honolulu, Hawaii, August 1969.

[3.11.14] Bortels, W. H., "Simulation of Interference of Packets in THE ALOHA SYSTEM," in *THE ALOHA SYSTEM Technical Report B70-2.* University of Hawaii, Honolulu, Hawaii, March 1970.

[3.11.15] Tripathi, P., "Simulation of a Random Access Discrete Address Communication System," in *THE ALOHA SYSTEM Technical Note 70-1.* University of Hawaii, Honolulu, Hawaii, April 1970.

[3.11.16] Images Australia Pty Ltd., *Cyberspace: The World of Digital Architecture.* Images Publishing, Mulgrave, Victoria, Australia, 2001

Also see Wodtke, Christina, *Information Architecture: Blueprints for the Web.* New Riders Publishing, Boston, 2003.

[3.11.17] Gilbert, Held, and Hundley, Kent, *Cisco Access Lists Field Guide.* McGraw-Hill Osborne Media, New York, 1999.

[3.11.18] Meyer, Eric A., *Eric Meyer on CSS: Mastering the Language of Web Design.* New Riders Publishing, Boston, 2002.

[3.11.19] Giga Information Group, Inc., Trends in Monitoring, Optimization and Billing: A Shift in Priorities: Giga Collaboration, ebook Giga Information Group Inc.

[3.11.20] Ahamed S. V., and Lawrence, V. B., *Intelligent Broadband Multimedia Networks.* Kluwer Academic Publishers, Boston, 1996.

[3.11.21] Shortliffe, E. H., and Buchanan, B. G., "A Model for Inexact Reasoning in Medicine." *Mathematical Biosciences,* Vol 23, 1975. Also see Ahamed, S. V., and Lawrence, V. B., "Knowledge Processing System Employing Confidence Levels." U.S. Patent No. 5,809493, issued on September 15, 1998, assigned to Lucent Technologies, Murray Hill, NJ, 1998.

[3.11.22] Bell Communications Research, "Plan for Second Generation of Intelligent Networks." *Bellcore SR-NPL-000444,* Issue 1, July 1986.

[3.11.23] Shortliffe, E. H., *Computer-Based Medical Consultation: MYCIN.* Elsevier, New York, 1976.

[3.11.24] Dillmann, R., Luth, T., Dario, P., and Worn, H., (Editors). *Distributed Autonomous Robotic Systems 3.* Springer Verlag, New York, 1998.

[3.11.25] Forsyth, D. A., and Ponce, J., *Computer Vision: A Modern Approach.* Prentice Hall, Upper Saddle River, NJ, 2002. Also see Ballard, D. H., and Brown, C. M., *Computer Vision.* Prentice-Hall, Englewood Cliffs, NJ, 1982.

[3.11.26] Ahamed, S.V., et al, "Personal Computer Besed Intelligent Networks," U.S. Patent 5,715,371, issued February 3, 1998.

[3.11.27] Krol, M., "Intelligent Medical Network." Ph.D. dissertation, City University of New York, 1996.

[3.11.28] Maslow, A. H., *Maslow on Management.* John Wiley & Sons, New York, 1998, Also see Norwood, George, "Maslow's Hierarchy of Needs," in *Updating Your Spiritual Vision,* on the Internet at http://www.deepermind.com/20maslow.htm. (Accessed on April 4, 2004.) Also see Maslow, Abraham Harold, *The Farther Reaches of Human Nature,* reproduction ed. Arkana, New York, 1993.

[3.11.29] Bernstein, P. A., "What's Wrong with Telecom." *IEEE Spectrum,* January 2003, pp. 26–29. See also Bernstein, Peter A., "What's Wrong with Telecom." *IEEE Spectrum Online, on the Internet at* http://www.spectrum.ieee.org/WEBONLY/publicfeature/jan03/comm.html. (Accessed on April 4, 2004.)

CHAPTER 4

Major Conclusions Derived from Socially Significant Patents

Socially significant patents have a history and use. They also have an impact. Other patents have an impact in areas other than electrical engineering, computers, or communications; however, these patents are not included in the book. Within these three technical disciplines, significant patents have altered the social norms of society. We first examine these commonalties in general and then examine them in detail. The researchers whose work we will examine rarely used a centralized and documented theme to pursue their investigations. An uncommon mixture of individual traits (e.g., curiosity, persistence, commitment, skill, observation, interpretation, detailed rigorous thinking, and reasoning and strategizing) has contributed to these singular inventions that have altered society.

In modern days, opportunities exist in a different context, and processes and microreasoning are now considered vogue by the knowledge society. However, we stand to gain significantly by adapting the mental posture of the early contributors to the tools and techniques, kinds of symbolism, and methodologies that are germane and consistent with the computer and knowledge revolution of the last few decades.

4.1 COMMON FACTORS OF SEMINAL PATENTS

Concepts, like numbers, have common factors. When we factor numbers, we can evaluate the least common multiples. In essence and in hindsight, we attempt to factor the major concepts embedded in the patents in Chapter 3 and present what appears to be the prevalent themes. In some cases, some of the steps used to reach a specific goal are intuitive. However, the earlier inventors have included their own raw thinking (see Tesla's, Edison's, and Bell's patents) in their applications. Reason and intuition are intertwined in the original writing of the inventors. Based on reason alone, a rough flow diagram of an invention may be created as a supplement to the personal information we have about these early scientists. Nonetheless, two dimensional flowcharts alone do not capture the intellectual space of the inventor.

In this section, it is not our intent to undermine the importance of these classic patents, instead, our intent is to show that some commonalties (see Sections 4.1.1 to 4.1.5) exist in most of the older classic patents.

4.1.1 Physical or Mathematical Basis

Most of the patents presented in Chapter 3 have some physical or mathematical basis. The principle could have been right in front of the inventors before they even thought of putting pen to paper. For example, the rules of electromagnetism existed long before electric machines or electrical devices came along. As another example, the basic relationship between the pressure, volume, and temperature of gaseous mediums ($PV = m \cdot RT$, i.e., the combined gas law) existed long before the Joule-Thompson valve for refrigeration came along.

This basis, first sensed intuitively by the inventor, is exploited in the invention. Then, experimenters/inventors follow a chain of experiments to prove or disprove the hypothesis. The inventors analyze this hypothesis until they can find a use for the generic idea behind the invention. On the other hand, the concept perceived by the realization of an underlying principle is relentlessly pursued by abstract thinkers and others (e.g., painters, poets, and philosophers) until harmony can be achieved between what was known and what is clear to them as a general principle. A completely new string of inventions thus can be derived.

In a specific sense, the microphone that Bell used in 1876 is based on the law of electromagnetism ($e = n \cdot d\phi/dt$) to generate voltage. The voice, encoded as electric signals, travels down a pair of wires ($V_r = V_s - I \cdot Z$, of [4.1]) in accordance with the laws of signal propagation over wire pairs. In hindsight, this concept of converting voice signals to electrical signals and propagating them over wire pairs may seem trivial, but the invention has been significant.

In a specific sense, the kinetographic camera that Edison built in 1897 is based on the synchronization of speeds of a recording camera with the forward speed of the projection system. This rule of maintaining synchronism to project the image is an underlying concept so well known that it is not even written down as a rule. The projection of images and shadows on surfaces is a concept that was well known even to the cavemen. In hindsight, this concept and its basis may seem trivial, but the invention has been significant.

In a specific sense, the CRT that Zworykin proposed in 1938 is based on the fact that electron beams carry current from the anode to the cathode by depleting the negative charge at the cathode and that, as such, these beams are susceptible to the electromagnetic and electrostatic forces that deflect the beam. In hindsight, this concept and its basis may seem trivial, but the invention has been significant.

In a specific sense, the first few computers that processed numbers and data are based on the rules of binary arithmetic and the representation of numbers as entities, such as voltages, currents, charges, etc. Binary arithmetic existed long before the Electronic Numerical Integrator and Computer (ENIAC) project, and the work of von Neumann, Konrad Zuse, and others. In hindsight, this concept and its basis may seem trivial, but the invention has been significant.

We can extend this type of dissection of the genesis of each of the inventions in Chapter 3 and tie the invention (in hindsight, of course) to a physical principle, an equation, or a well-established fact. However, we have made the point that there is a universal truth that is more enduring than the patent and that the patent is a mere manifestation of this underlying truth.

With a reasonable blending of research and foresight, any new concepts (or their combinations) proposed now as inventions can become the basis of future devices. The truth or the principle embedded in the invention now can become the basis of a new technology of the future. Hence, the seed of intuition needs human labor in order to develop an invention that has a practical use. Such inventions may be accomplished by the original inventor or the next generation of inventors who rework prior inventions into new breakthroughs. For example, antibiotics were discovered (by Sir Arthur Fleming) long after they were commonly used in laboratory environments and before their potential was discovered.

4.1.2 The Intuitive Content

In the classic works, ranging from the earliest works of Edison, Tesla, Marconi, Ampere, etc., to the patents of Shockley, Bardeen, and Brattain, the proportion of intuitive content in older patents is much higher than the intuitive contents in more recent patents. Conversely, the scientific and mathematical content is much higher in more recent patents. It almost appears as if intuitive content (i.e., the product of human imagination) is being replaced by the analysis and recombination (i.e., a product of human rationality) of preexisting variables and symbols. Such symbols and variables are now being borrowed from a preexisting platform of knowledge. Suffice it to say that newer patents are resulting from applied knowledge rather than taking the form of new or incremental knowledge.

The most alarming of such highly intuitive patents (and perhaps the most influential) is based on Tesla's concept of the rotating magnetic field (this patent is not discussed in this book), which has brought about the most generic type of electric motor (i.e., the induction motor). The genius of Tesla lies in his discovery that high-frequency electromagnetic radiation can propagate even without magnetic materials. These patents (No. 645,576 and No. 649,621, in 1900) have earned him a place as one of the earliest inventors of microwave communication systems. To affirm our earlier comment, some of these classic patents are not based on any formal mathematics or equations. Instead, the notion is conveyed in a few figures, and the generalization of the observed result is stressed in the claims.

Through the last century, the emphasis on mathematics and precision has been dominant and noteworthy. In fact, this emphasis has helped to make the imagination far more disciplined and scientific. Nowadays, we have a much finer and more exact basis on which to prove or disprove an invention. Similarly, we have a greater basis on which to optimize an invention. The rigor of scientific thinking provides deep inroads into intuition and creativity. It would be more appropriate to brand optimized re-inventions as innovations rather than as inventions.

4.1.3 Reality, Imagination, and Possibilities

The early inventors have been extremely imaginative and have extended the realm of scientific possibility. The tools and approaches they used were limited by the extent of their personal education and academic training. Academic training was restricted to the elite, some of whom perhaps compromised the imagination to the rigor of learning. This was not always the case, however, since a significant number of musicians, writers, artists and astronomers in the past were responsible for starting completely new schools of thought. However, inventions must have a scientific basis in order for them to be used to solve social problems. Newton, Maxwell, Einstein, and others deployed such mathematical approaches. In a sense, imagination is only one of elements that factored into an invention.

From a slightly different perspective, usually, the invention has to be supported financially, politically, and/or socially before it can gain any prominence. An example of this—lack of rigor and lack of external influence—is evident in Tesla's work (see Section 3.3). Raw genius is evident in Tesla's two patents (see Section 3.3). In 1901, Tesla was trying to investigate logic circuits, pulse-length modulation techniques, and communication of a signal over microwave links (also his own invention in 1900). Even today, to invent as Tesla was proposing to invent to make such invention a true contribution to society and to the stability and security of typical modern research institutions (e.g., Lucent Bell Labs, IBM Research Laboratories, or Stanford Research, Inc.) would be most desirable.

However, by itself, Tesla's superb imagination was not sufficient in 1901, just as it would not be sufficient today. As a result, these particular inventions were never implemented. From a modern perspective, convincing other scientists of the validity of his inventions would be a challenge, since his inventions do not contain mathematical and circuit analyses. Within this context, the expectations set for today's scientists are far more demanding. Scientists during the 1930s through the 1960s attempted to prove the validity of an invention before it was developed, (see Farnsworth's and Zworykin's patents during the 1930s). These scientist's outspoken approach to research is evident in the following statements: "Work, Finish, Publish" (Faraday), "Ninety-nine percent perspiration, one percent inspiration" (Edison), and "Read not to contradict and confute, nor to believe and take for granted....but to weigh and consider" (Bacon).

In professional circles, this trend toward rigorous scientific investigation is much more dominant today than in the past. In the fabrication of VLSI circuits, for example, there is no room for error. It is as important to prove the validity of established principles in civil engineering (e.g., bridge, dams, and skyscrapers), in space missions (from satellites to Venus probes), and in modern medicine (from antiseptics to laser optical surgery) as it is in proving an invention.

In academic circles, the student and the mentor may take it upon themselves to prove the ideas included in a dissertation in order for that dissertation to be considered convincing by their colleagues.

4.1.4 The Basic Social Need and Driving Force

Some of the patents in Chapter 3 describe novel methods for satisfying basic social needs. In fact, some of the methodologies are as generic as the paper on which the patent is written. The invention is only a declaration of one of the possible ways to satisfy a social need, such as the need for improved transportation or the need for improved communication, etc.

In a specific sense, the basic need of communicating over long distances was one of the major reasons for the invention of the telephone. The need to entertain the public was one of the reasons for the invention of the kinetographic camera. The need to duplicate the effects of migration of electrons and holes (and thus make electronic gadgets smaller) in semiconductor devices was one of the basic reasons for the invention of the transistor (see Chapter 3). The need to imitate the usefulness of the telephone and copy the concept to communicate over longer and longer distances was one of the basic needs for inventions of the microwave, satellite and fiber systems.

In a specific sense, the computer revolution was fueled by society's need to process large amounts of data quickly, efficiently, and economically. The computer race between countries was in progress, and the desire to hold the most useful patents became a national obsession. As a result, many scientists gained personal recognition and a sense of personal triumph during the early part of the computer era.

In a specific sense, the need to reach higher and higher frequencies for communicating devices (and thus to increase bandwidth and enhance the amount of data communicated) is one of the reasons for the maser and the maser-communications system patent. Similar social needs may be traced back to most of the patents presented in Chapter 3.

In a specific sense, the need for higher and higher frequencies in the optical band is accommodated by changing the basic structure and the innate properties of the optical waveguide (see the three related sequential patents in Section 3.9). Examples of this trend and of the use of innovative methodologies involved in creating a breakthrough include the migration from vacuum tubes to transistors (Section 3.6), from transistors to integrated circuits, from RF oscillators to lasers/masers (Section 3.7), and from multimode fibers to single-mode fibers (Section 3.9) from WDM to DWDM systems.

In a specific sense, the need to vastly enhance communication, reduce costs, and meet a large and growing set of fundamental social needs has brought about the Internet revolution. This revolution has resulted in major social change for the present generation of knowledge workers, and the impact of this development on society is as major as the impact of computers in the 1960s and 1970s.

4.1.5 The Opportunity for Originality

The lack of organized literature in any particular field of invention related to the earlier patents is clear because there was no Internet and no search engines that facilitated the immediate retrieval of relevant information. There were no firm boundaries between subjects in academia. Wisdom was the trailblazer then. Mathematicians and logicians were not necessarily known as inventors. Hence, inventors of technological devices had quite a bit of freedom to extend the frontiers of knowledge in a very dramatic way (see the contributions of Maxwell, Faraday, Tesla, Marconi, etc.) even though some would classify these men (especially Maxwell) as very astute mathematicians and logicians rather than as creative scientists.

4.1.6 Observation, Solution, and Generality

Most of the earlier patents were based on the observation of natural phenomena. Such observations were then exploited to meet a personal, social, or global need. Next, possible cause-effect relationships and generalizations were examined and a string of important patents thus resulted. This type of investigation is common even today, but now scientists are rehashing patents/innovations rather than making the kinds of major breakthroughs that resulted 40 to 100 years ago. We support this conclusion by citing the following specific cases.

In a specific sense, the observation that an electric current would deflect a magnetic needle (a natural phenomenon) was indeed the basis for generating a force field (exploitation), namely the electromagnetic force. Many significant patents have resulted, thus leading to the foundation of electrical power systems. In fact, the concept has been generalized many hundreds of times in building electric motors (e.g., induction, synchronous, single-phase, polyphase, and split-phase motors) and electromagnets (from cranes to space applications) of all shapes, sizes, and configurations. The concept also has been turned around to build electrical (e.g., synchronous, induction, transformer) generators, and so on.

In a specific sense, when it was determined that the listener does not have to actually see the person speaking, then the telephone (equipped with a traditional electromagnetic speaker) was developed. The telephone thus carried the distant sound signal described in Bell's patent of 1876. A string of other generalizations has also resulted by using different media (e.g., free space, coaxial cables, and fiber) for transmission of signals.

In a specific sense, the observation that the human eye could be fooled (a natural phenomenon) by rapidly superposing images on each other to generate a sense of movement (see the sections on the motion-picture camera and television systems) has resulted in a number of significant patents and breakthroughs in the realm of cinematography, television, computer-generated visual effects, animation, and multimedia communications.

In a specific sense, when it was noted that the levels and the direction of signal voltages and currents could be manipulated by other signal voltages and currents in a deterministic, logical, mathematical, and predictable way, the concepts for the processing of data by using operation code and operands in the central processing unit of a computer started a revolution that is still strong today.

In a specific sense, when it was observed that electrons and holes in solid-state devices at room temperature could also conduct or block the flow of current (an observation of a natural phenomenon), an entire industry based on semiconductor technology resulted. In addition, the control on this flow of current with an intermediary voltage source (akin to the grid in the electron-tube plasma environment) has also influenced inventions by Shockley, Bardeen, and Brattain. Both analog and digital semiconductor devices have now become a firm reality.

In a specific sense, the reflection and refraction of light in gems and glass had been observed for a long period in history (see Chapter 16 of [4.2]). In addition, the capacity to pull the glass strand as very fine optical fibers was observed during the latter part of the last century. Further, the capacity to alter the refractive index of glass with a suitable composition of metallic oxides (see Section 3.9) was also observed and exploited in the 1970s. Compound glass (i.e., core and cladding suitably and geometrically positioned) could also be pulled dependably as fiber. In a sense, these combined observations started the fiber-optics revolution during the latter part of the last century. Fiber-optics technology is a perfect example of a complex invention that is revolutionary in its own right because of the many interdependent observations that had to blend together to fill social, industrial, and technical needs.

It is possible to perform an exhaustive analysis by tracing the observations, solutions and generalizations, strategies or patterns in all of the patents shown in Chapter 3, (see Sections 3.6, 3.7, 3.9, and 3.11). However, in this section we reiterate the idea of a triangle of social need, scientific collaboration, and individual creativity as it applies to most of the significant patents that have emerged during the last few decades.

4.1.7 The Quest for the Best

The "quest for the best" is a fundamental drive present in creative people. This personal trait is as forceful today as it was in the past. This driving force is as evident in recent patents as it was in the patents of the last century (see Chapter 3). The "best" of earlier generations may not be the best from today's perspective. However, these patents have left clear footprints in the sands of time. However, we have the luxury of hindsight to comment and analyze!

Firmly grounded in reality, scientists realize that the quest for the best can be illusive. The process of searching for the best solution involves an expenditure of resources, and marginal gains should justify marginal costs. Hence, the inventor has to realize when the quest for the best should end in view of the overall context of and constraints on the solution of the problem. To a large extent, the true personality of the inventor reveals itself in the quality of the solution proposed in the invention, despite such constraints. In many cases, the direction of search crosses the realm of rationality and starts to become an inspiration or an insight, which leads to the best solution without the traditional cost attached to reaching the best solution. Research thus becomes almost all consuming for inventors until they find the solution that is the best. In many instances, both ingenuity and persistence become apparent as the final solution unfolds. For example, Einstein's general theory highlights the very essence of his thinking which involved both imagination, or unstructured creativity (see Section 4.3.1) and long years of work, or structured creativity (see Section 4.3.2).

In a specific sense, Edison sought the best filament material for his light bulbs. Edison's long and tedious experiments with all types of natural elements and other materials without actually seeking out the material with the highest melting point is an example of seeking the best solution to make his invention work. The nature of the inventor is brought to bear on this search process*. Being an experimentalist, Edison was without too much recourse to scientific methodology. Nonetheless, pure persistence provided him with the best solution for his skill set and for the time in which he lived. Thus, while he did not seek the materials that could emit light without becoming hot (gaseous plasma or light emitting diodes), he did invent the electric light.

*In the early days when the search process was manual and there was no Internet, nor any search engines readily available, searches were limited to the mental process the scientist's immediate personal experiences, and the experiences of other colleagues or friends. This practice is documented again and again in the works of Ampere, Volta, Gauss, and Oersted (see Chapter 3)

4.2 THE CRUCIAL DIFFERENCE

The flavor of the older breakthrough patents is reflected in the observations made in Sections 4.1.1 through 4.1.5. In more recent patents, the opportunity to develop such classic inventions is narrower. Most of the current inventions have precursors, and most recent patents can be tied to older patents or to generic, documented investigations. In today's environment of intense specialty in every scientific/technical area, the burden of proof of total originality weighs heavily on the inventor. First, the topic being resolved or studied should be unique. Second, the methodology and the approach for the solution should never have been attempted by any other scientist.

This explosion of knowledge in the fields of study of advanced research students can open the innermost frontiers of their personal thinking. We refer students to Sections 2.4 to 2.6 to re-examine their commitment to the modern scientific and creative climate in light of their own personal drives. As we have suggested before, inventions are sometimes a by-product of the social setting and the personality traits of the inventor. However, more than that these inventions have to become symbols of personal achievement and a source of joy to the inventor. In addition, they should be a means to providing a service to or satisfying a need of the society in which we live. It is for creative minds to visualize a need before that need leads to a crisis and produces a solution or a device ready for the market.

To some extent, the modern climate for basic and generic inventions is more demanding than the older climate. The balance between drive, frustration level, and achievement level for the next generation of scientists is more delicate than that for earlier generations. However, these challenges need not concern truly creative students.

4.3 SPECIFIC CONCLUSIONS FOR STUDENTS

Creativity and invention have defied formulation and programming. In perspective, we know that these are neither totally random processes nor are they totally programmable events. In this book, we have attempted to reexamine both aspects in order to take a heuristic approach by learning from the great inventions of the past and applying that information in the present with some amount of educated extrapolation. What we propose will be totally ineffective in some cases, but it is our contention that, in a majority of cases, what we say can be generalized to enhance the cumulative index of creativity (CIC) for any entity. (The CIC is defined as the average number of inventions per unit array of resources spent in the research environment.) We feel fairly confident that the suggestions will be valuable to students.

Broadly categorized, there are two methods of inventing: the unstructured approach and the structured approach. In practice, neither approach is complete in its own right. Therefore, a harmonious and adaptive composition of the two approaches helps to convey the basic traits and personality of mature scientists.

4.3.1 Unstructured Creativity

In the unstructured approach, the individual scientist depends on possessing the unique gift of a mental process that is able to sift through many layers of reasoning, association, graphical/object analysis, and logical linkage and derivation, which the mature and seasoned scientist would have done at a conscious level. If the mental processes can be labeled at a rudimentary level, then these scientists would let problems simmer in their mind for a period of time and would wait for a "miracle" of a solution to appear, even though they may not understand this solution at first. Again, without calling this a miracle, the subconscious mind then weighs and considers all of the possible solutions and performs a validity/optimality check on the most reasonable solution. Discrete mathematical formulation is not needed since the subconscious mind also performs the programming and the algorithmic validity/optimality check, sets the boundary conditions, and assesses the practicality of the solution. The gift could be at a genetic level, or it could be nurtured into a full-fledged sequential/shared activity over time.

Nevertheless, some early inventors (e.g., Edison, Tesla, and Marconi) had a facility for sharpening these subconscious functions, thus enabling them to make significant contributions over a sustained period of time. This feature becomes apparent by comparing the early patents of these inventors with their later patents. As an example, the connection between the long years Einstein spent in proving his general theory of relativity (structured creativity) and the few short moments that led him to enunciate his special theory of relativity (unstructured creativity), as he was strolling through the gardens, will remain obscure for some time. Whether an ounce of inspiration can be distilled from a sea of perspiration may remain a mystery, but to us it is evident that the highest quality of effort (with computer-assisted creativity and knowledge processing) will intercept the lowest rung of genius. For this very reason, we propose the second category of creativity, structured creativity.

4.3.2 Structured Creativity

Structured creativity is a product of directional problem solving intended to generate an optimum solution by using all of the scientific tools at a conscious human level or by using a computer in some programmed way. At the lowest level, this activity becomes computer-aided design. At the highest level, the scientist uses the computer to verify a hypothesis that the scientist has reached as a result of the unstructured creativity in the human mind.

In structured creativity, the human mind can investigate and challenge each step that leads to the end result and then reconstruct such steps to attain the desired objective. This task is routinely performed in the synthesis of new drugs, petrochemicals, new fertilizers, etc.

It is not our intent to argue that all creativity can be totally structured; instead, we hope to explain that some structure is involved in being creative. Highly disciplined people generally do not ignore their insight; instead, they use their sense of discipline to remain within the confines of structured creativity. Discipline helps them to focus the search and to generate the highest level of awareness of the best solution to the problem that they are attempting to solve.

If an invention is the result of pure logic and no intuition, then it is a derivation rather than an invention. Conversely, if an invention is the result of pure intuition and no logic, then it is also not an invention, but a fantasy. Between these two extremes lies the interface between the rigor of induction and the insight of the mind. Trained scientists have the skill to switch between these two modes of thinking under two conditions: (1) when an approach is leading to a dead end and (2) when an approach has led to some microinvention, but the validity of this microinvention needs substantiation.

Over the last century, logicians, mathematicians, and pure scientists have developed the capacity to be predictive as well as retrospective. Some segments of communications and computers can greatly benefit from these derivations. In developing structured inventions, the tools and techniques of the early mathematicians have a different use. Whereas mathematicians wrote their theorems in abstraction, computer and communication scientists have tools to verify data and signals in a time domain or in a spatial domain as these data and signals are being constantly manipulated through novel devices, microchips, or complex switching systems. The application of theorems and their validity to the objects of knowledge, information, and communication systems is a doorway to new devices and processes that scientists may not yet have explored. Such processes are likely to become the basis of inventions still to be made.

4.3.3 The Practice of Invention/Innovation

We present these conclusions in two figures, Figures 4.3.1 and 4.3.2. Figure 4.3.1 addresses the issues germane to students who are going through the process of earning an advanced degree. In Figure 4.3.2, some of the issues of scientists who are employed in a research and development environment are presented. It is obvious that Figure 4.3.2 can be logically derived from 4.3.1, but we have attempted to highlight the logical differences.

It is contradictory to write a flow diagram for a process that has no well-defined pattern. To the limited extent that some pattern exists, we present that pattern as a flowchart. But by the same token, we would suggest that readers attempt to comprehend

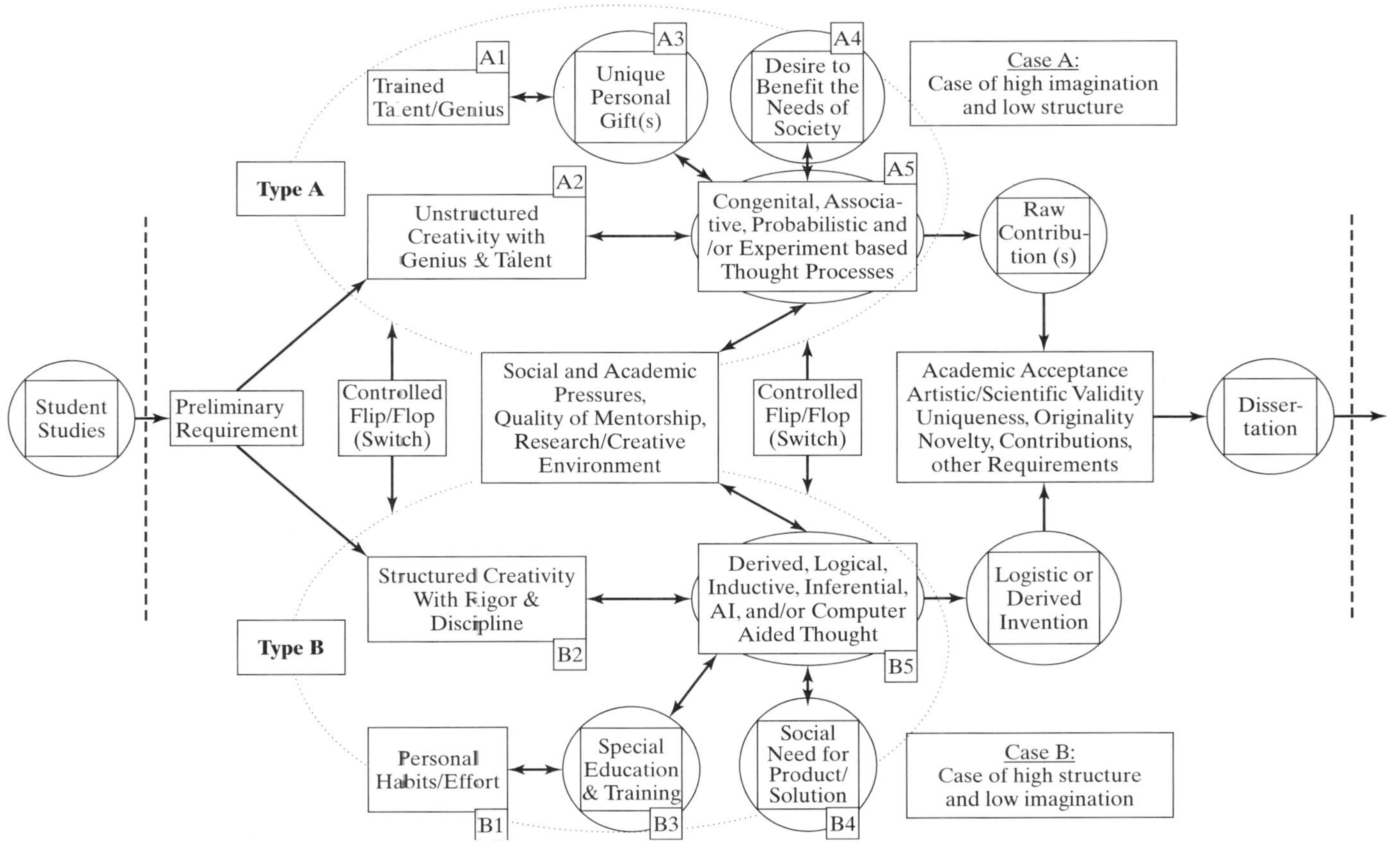

Figure 4.3.1 A possible scenario for research students to use to keep track of the course of events during their academic tenure at the university or during their career. Creative blending of these boxes and events is as much the research material as the creative blending of the topics and subject matter for the dissertation. The elements in the two ellipses are fully and adaptively connected (e.g., neural net) rather than exactly what is shown.

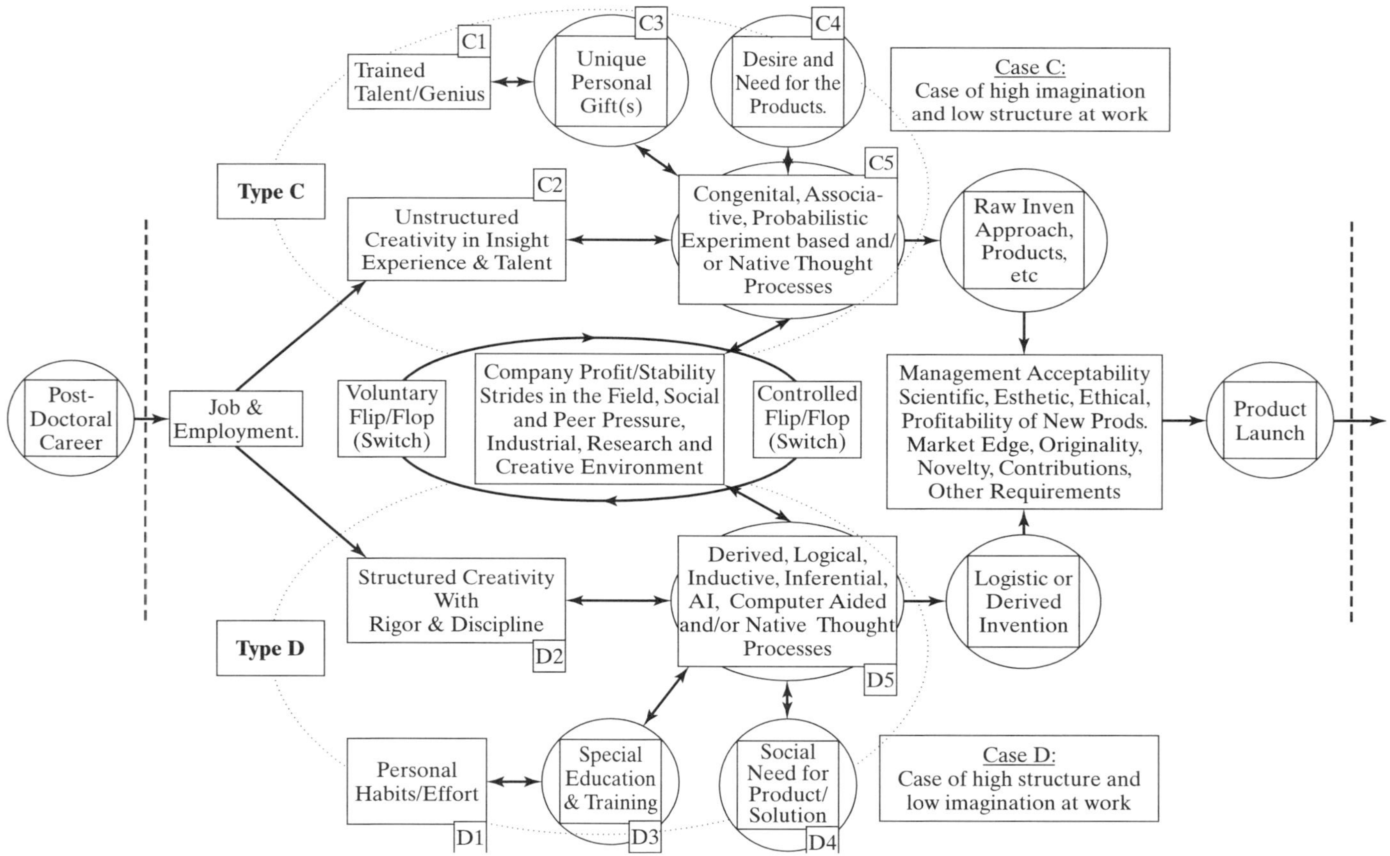

Figure 4.3.2 A possible scenario for scientists to use to keep track of their daily activities in industry/academia. Blending of the five elements in the ellipses is an adaptive and life-long process. It can change from project to project, time to time, and company to company. It is desirable that these elements be fully and adaptively connected as neural nets.

the broad constructs or blocks in Figures 4.3.1 and 4.3.2 and then flow freely and move from block to block as "the eye would see," rather than follow the sequencing of the arrows in the flowchart. The two acts of being disciplined and the act of being free to enhance and modify the contents of the diagrams could co-exist.

Student Perspective The two philosophically different approaches discussed in Sections 4.3.1 and 4.3.2 are addressed in some detail so that an optimum and adaptive blending can be suggested at a perceptual level to advanced research students. For this reason, we propose that the practice of invention depicted in Figure 4.3.1, be used to help target and focus the orientation of activities of entry-level research students in a university setting.

During the second levels of the doctoral program, it is up to individual students to evaluate their personal orientation towards creativity. The process of introspection is akin to that of a musician selecting an instrument to play in conjunction with the natural dexterity of the fingers to play the instrument and the inclination towards the music that the instrument creates. Most third-level students should be able to recognize their strengths and weaknesses. Time spent in evaluating the strengths to harness and the weaknesses to supplement (from the mentor's participation or group activities with other students) is likely to produce the genesis for a significant contribution.

Figure 4.3.1 has two major parts: a top ellipse and a lower ellipse. The diagram is almost symmetrical at the central X-axis. As we have observed, many students, traverse through this diagram many times throughout their research career. The activity is sometimes ill-conceived, without an overall idea as to the progress they make at each "visit."

However, the mentor has the responsibility of tracking the incremental changes that each student makes during each student-mentor session that takes place as the student pursues doctoral degree.

The Unstructured Approach. The unstructured approach of some students* (Type A) who are more relaxed in completing their work tends to follow the upper part of the diagram. To compensate for their less rigorous attitude, they depend on various personal skills, such as quick (but accurate) inference gathering, sharp observation, quick recall, intelligence, causal analysis, flash association, and native perception. In many cases, mentors are amazed at the raw genius of such students that, by itself, seems sufficient** to allow the student to get an advanced degree.

The ellipse in the top half of the diagram combines the students' training with their own personal attributes/abilities. It permits the student to become cognizant of the needs of society on the one hand and the requirements necessary to satisfy the examining committees on the other hand. The previous training and talent that they bring as students play a role in the thought processes that lead them to make a raw contribution. Students use all of their congenital, associative, probabilistic, and/or experimental observations to trigger the thought process that leads to the core of an invention/innovation. This contribution can be more than the basic innovative content of the dissertation. It is for the mentor to nurture gifted students and train their raw genius.

---

*In reality, students exhibit a predisposition towards being structured (painstakingly precise) thinkers or unstructured (spontaneous, free) thinkers. What can really be observed by the mentor is that the student often leans towards the Type A attributes (i.e., the congenitally gifted) as the student progresses through the research career.

**We strongly oppose such overreliance on raw genius because training and rigor should combine with the native abilities of students earning an advanced degree.

The five elements shown in the diagram hinge on the student's own attributes:

1. A1 - Training, talent, and ingenuity in the discipline wherein the advanced degree is attempted
2. A2 - Unstructured creativity combined with genius and talent in the direction most relevant to the solution of the problem at hand, as well as the ability to realize and suggest a solution within the context and parameters of the problem
3. A3 - Unique personal gifts, such as quick reflexes, the ability to make pointed observations, or the ability to interpret the key words and phrases that are germane to the optimal solution of the dissertation problem
4. A4 - Desire to benefit the needs of society such as realization of a type of gadget, invention, or a product that improves quality of life and makes day-to-day activities in society easier and/or more enjoyable
5. A5 - The ability to make congenital, associative, probabilistic, and/or experimental observations to establish the causal and interdependent relationship that leads to the manipulation of parameters or "symbols" in the scientific process, which in turn leads to a superior and/or more general/unique solution

The five elements (A1 through A5) in the top ellipse of Figure 4.3.1 play an interactive, interdependent, and adaptive role to balance out a generic invention or raw contribution. In fact, all of the skills of AI programming are practiced by the human mind in order to learn and modify the contents of these five elements and to strengthen the interactive feedback between them. Some gifted students have no formal AI training in using their raw talents, but to the mentor it can appear that the students are actually subconsciously practicing this process of finetuning of their capabilities as they talk or present their perspectives regarding the dissertation problem. This balancing act is largely a very individualistic and very human process.

Almost all students respond to the challenge of their dissertation in an open classroom discussion. An agenda of such problems should be presented and discussed by the mentor. The response varies dramatically. The mentally agile and astute catch on to the interactive process by being optimal in their response.* The mentor should listen out for the most logical, the most optimal, and the most ingenious solutions in order to identify the Type-A students. These students can also show their strengths in certain broad specialties (e.g., mathematics or programming skills as compared with systems engineering or computer-aided design skills). Observations by the mentor can offer clues to the personality types of students.

It is not our intent to quantify or streamline the process of unstructured creativity; instead, we urge students to approach this process with caution, dignity, and the realization that the process of observation of the phenomenon is likely to alter the process (akin to Heisenberg's principle). In the context of research, the process of inventing/creating remains unseen and undocumented, but the result can surpass initial expectations and sometimes lead to an entirely new frontier of science. For the highly gifted people in this particular direction, it is best to leave a beautiful thing alone, though not all to chance. Students with this rather rare type of talent should be nurtured to help them master their own gifts,

*Such student responses are akin to the response of innate organisms in nature. For example, the bee has an optimal shape for the cell (hexagonal) to provide the most room for the least amount of wax, and yet lie alongside other cells. The shapes and colors of many flowers are the result of evolutionary developments (e.g., to attract the most suitable pollinating agent). The shape and size of hummingbirds are adapted to the shape of the flowers that provide them with nectar, etc.

though too much micromentoring from the mentor can be counterproductive. In a great many cases, it takes an ingenious mentor to recognize a genius student.

The Structured Approach. The structured approach of students who are driven to rigorously complete every aspect of their work (Type B) tends to follow the lower part of the diagram. To compensate for their less daring and imaginative capabilities, they depend on different types of personal skills, such as reasoning, painstaking attention to details and differences, methodical deductions, an almost obsessive study of cause-effect relationships, conscious pattern recognition, and precision in their (object-oriented) analyses. In many cases, mentors have to provide additional help to such students and spend time in skillfully suggesting the next step in the creative process. Even though the mentor may "see" the mental struggle of the student, we propose that the mentor not directly contribute to this internal struggle, but let the student find the recourse to polish and optimize this conscious activity.

The ellipse in the lower half of the diagram shown in Figure 4.3.1 combines the student's training, habits, and attributes with the student's own specialized education and with the student's cognizance of society's need for the product or device to be invented. The five elements in the lower ellipse of Figure 4.3.1 are restated as follows:

1. B1 - Personal habits and sustained efforts are as important to the student as training, talent, and ingenuity (see A1 in the top ellipse). The willful commitment of effort is a necessary step in the field wherein the advanced degree is attempted.
2. B2 - Structured creativity, or willful approaches to derive a solution (whatever it may take to infer, derive, or deduct the solution or a part thereof), is a rigorous and disciplined process that involves a search for the most suitable clues and relationships. In a great many cases, the many dedicated weeks that Type B students spend searching can at times be replaced by a more logical and insightful remark from Type A students (see A2 in the top ellipse). This trait of some students is also recursive in their deployment of the approach. The mentor sometimes can recognize when the student takes an approach to the solution of a microproblem that is similar to the solution of the macroproblem that contained the microproblem. The approach can be painfully tedious, but the outcome can be rewarding, especially when the student has partially tailored the solution to the specific dissertation problem.
3. B3 - Special education and training such as subject-matter expertise in a related field or derivation of a similar equation or of the linkages to similar or related objects become one of the strongest key elements of disciplined, logical, and deductive students. Such students usually develop a winning strategy as compared with the quick reflexes, pointed observations, and quick interpretations of Type A students (see A3 in the top ellipse).
4. B4 - An awareness of a social need for products and solutions is the motivation for making the contributions in the first place. This Type B approach calls for logical convergence to a solution or the point-by-point reasoning of a seasoned attorney. The flash realization or intuitive leaps of the imagination typical of the Type A student is absent (see A4 in the top ellipse). However, a social need still initiates the drive for invention and innovation in both types of personalities (Types A and B).

Thus, it is the methodology used to incorporate this need into the research that is slightly different. In the next section, we demonstrate that the capacity to switch between the two modes (A and B) can blend the best of both worlds.

5. B5 - The derived, logical, inductive, inferential, AI, and computer-aided thought processes are the final stages of reaching a solution or making an invention in the context of Type B students. However, if the student does make a connection between the somewhat different approaches to being creative/"innoventive" (halfway between being inventive and innovative), then the activities of A5 can be harmoniously blended into B5 and vice versa. In fact, to be the dualist in one personality of scientist is like a single musician playing the sitar (of the Indian musician Ravi Shankar) in a musical dialogue with the violin (of Yehudi Menuhin). The impact could bring about a splurge of synergy of unprecedented intellectual dimensions, much more than anticipated or foretold. True innoventions can be (semi-) consciously made.

In an intellectual setting, the importance of the problem facing Type B students is painstakingly articulated by the mentor as a motivational tool. Sometimes, it can become an ordeal to convince students that their capacity to function in the B mode will bring some order and discipline to the hasty approach and shortcuts taken in the A mode. In many cases, the raw invention of the A types needs the optimization of the B types. For example, the development of the DWDM fiber is based on the prolonged and sustained efforts of optical-fiber technologists, even long after multimode and single-mode fibers were "innovented" (invented or innovated), installed, and deployed. As another example, compound semiconductor (electronic and optical) devices evolved long after semiconductor and optical technologies had matured.

The entire drama leading to the solution or invention of the third level B-Type students is generally the origin of a slow and tedious climb. The main steps involve the deductions, the inferences, the AI techniques, and the conscious level logic that lead to the core of their invention. The core of their dissertations for Type B students is usually based on a chain or chains of deductions or derivations. Many times, this struggle involves the 99 percent perspiration that Thomas Edison prescribe for his inventions. Sometimes it is perceptible to the mentors that there is less than 1 percent inspiration. It could be that there is also a flash of genius in the student, or it can be supplemented from the mentor's own thought process. Ideally, the bridge between these microinventions is (or should be) an intuition or a highly directional thought process. Sometimes the mentor can become overly helpful* in this creative content of a student's dissertation.

For students who follow an unstructured approach, the five elements in the lower ellipse play an interactive, interdependent, and adaptive role. This act of balancing is more drawn out and painful for this group of students, who are essentially less daring in their outlook. In fact, we have observed some of our students using the systems approach to string their painstaking inferences into one major conclusion or invention. The approach is then optimized in order to help clarify the conclusion that is drawn or the basis of the invention that has been made. For deeply ingrained Type B students, the mentor should offer the

*We strongly discourage this type of help on the part of the mentors because the student's insight into the solution of the problem or the invention should be an integral part of the process students follow to earn an advanced degree.

approach of the A-type students, without giving away the intuition.

However, if the mentor is persistent, then the student can learn to harmoniously blend effort (hard work and painstaking deductions) and deeper insight. The microcode of AI programming is practiced by such students who tend to be very rigorous in their approach. Sometimes it is painful for mentors to see students struggle to become a mere extension of AI machines. Again, this balancing act is an individual and very human process.

The Flip/Flop (Switch) and Its Conscious Control. Students may perhaps start to realize that both approaches (structured and unstructured creativity) discussed earlier in this section are essential to become a notable scholar or even a researcher. It is therefore proposed that Type A students be assigned by the mentor to work in conjunction with Type B students, and vice versa. Such unison of the thought process should enrich the research methodology of both types of students. However, it is also the responsibility of the mentor to supplement either type if the student is lacking in one area—without actually solving the core research problem. If other students of the opposite type are not available, then we suggest that the mentor at least offer the student enough reading material from the lives of established scientists to get a good reason to use a blend of the two methodologies as seen fit.

The Core Scientific Community

Research Staff. Low-level scientists frequently have narrowly defined job descriptions. However, this situation need not prevent them from being creative in their own right. Personal motivation and adherence to the highest levels of research ethics (as presented in Chapter 2) are applicable to this portion of the research community. The doctoral degree is not necessary if one wants to become a noteworthy scientist, but it can help one become an established scientist sooner rather than later. Historically, the role of technicians has proven to be invaluable in keeping records and documents detailing experimental results. These results often become the cornerstones in established scientists' navigation of the thoughts leading to an invention or a breakthrough.

Established Scientists. Postdoctoral research and/or employment may bring about considerable freedom in the research one chooses to perform. When the choice exists, every venue of investigation needs to be explored. Usually, the mature scientist uses these techniques in completing projects and in making contributions and inventions.

In the corporate environment, scientists' challenges are different from the challenges of doctoral students. Accordingly, the differences between the flowcharts shown in Figures 4.3.1 and 4.3.2 reflect the nature of the thought process for the student and the career scientist. In the corporate world, the scientist supports numerous projects in different stages of completion. Hence, the scientist's job demands far more flexibility and the ability to immediately switch between projects. In a sense, the scientist traverses the block diagram of Figure 4.3.2 every day, many times and in many ways.

Each time, the incremental move is toward corporate profit rather than toward invention. The ellipses in Figure 4.3.2 are activated as a result of the dynamic and variable business goals that drive the corporation. The boxes located near the horizontal centerline of the diagrams shown in Figures 4.3.1 and 4.3.2 differ significantly. Of course, scientists bring their own personality attributes (see Sections 2.5.1 and 2.5.2) to the job. However, the environments (academic vs. corporate) are different, thus calling for a modified modus operandi in the corporate world. The adaptations that the scientist makes in the social and corporate world may involve a scientific or ethical price that is generally not a factor in advanced studies. Hence, the ever-shifting dynamic balance that a scientist sees in Figure 4.3.2 can be significantly different from the long-term balancing act that a student sees in Figure 4.3.1.

In an intellectual sense, the two ellipses shown in Figure 4.3.2 balance out as a well-structured neural net. Learning, adaptation, and autoreprogramming are all built into updating the neural network to indicate the truest match between human response and external social and economic settings. In fact, we suggest that every mature scientist develop a (mental or documented) database of neural nets that have proved sound in the past. In most cases, the solution is to match, or incrementally change, the models in the ellipses to suit the application. To this extent, we are suggesting that scientists practice the art of invention at the highest level. Such invention would constitute the correct blending of past training, existing skill sets, and insight in light of the most recent technological developments.

To some extent, the challenge is significant because the neural net contained in the two ellipses shown in Figure 4.3.2 changes every time it is used. Keeping track of causal changes and/or slow and subtle changes in a correct or error-proof context is a creative exercise in maintaining a healthy attitude towards research. Most scientists today take time for introspection and have the most valuable set of tools and programmable techniques in both the social and the technical/scientific domains.

4.3.4 Diagrams for Four Other Environments

The flowchart shown in Figure 4.3.2 can be extrapolated to fit the four models for innovation (Section 2.5.4), adaptation (Section 2.5.3), improvisation (Section 2.5.2), and reengineering and CAD (Section 2.5.1). We have not presented four separate diagrams since the models and the corresponding diagrams can be derived from Figure 4.3.2. The complexity of the derived models decreases gradually. The contents and the feedback in the neural nets in the two ellipses reduce gradually from the diagram for invention (Figure 4.3.2) to the one for reengineering. In a sense, the top half of the diagram starts to fade gradually as the fourth model (reengineering and CAD) is approached. Whereas the objects and their interconnection in the lower ellipse of Figure 4.3.2 are more amenable to programmability, the objects in the upper ellipse resist such programming.

The basis of CAD is to exploit previously learned knowledge and expertise and then design a specific system to satisfy a new set of social needs. Non-AI-based CAD programs are used repeatedly to design a circuit, a system, or an object. Human designers usually supplement and verify CAD-generated designs. Verification of the new designs is based on intuition or on common sense. When the computer enters the realm of common sense, AI concepts start to surface. When learning and the opinions of experts are also included, then the slightly more specific topics in neural nets and expert systems become relevant. All of these approaches are programmed activities and by definition belong in the lower ellipse of Figure 4.3.2. However, if the design from the CAD-based programs leads the design engineer to find the mental counterpart (A1 through A5) based on the design process (B1 through B5), then the process of innovention (invention and/or innovation) might have already begun.

The extent of mixing and merging the mental activity of the human with the processing of the machine is a very personal experience, much as reprogramming an old application for a new compiler, or playing an old song with a new band, or a rewriting an old poem for a new culture. However, if the convergence of the mind and machine were a duet of social benefit and AI based novelty then the perfect union of thought and computation leads to an invention. A total absence of the mental activity and only AI-based novelty leads to routine CAD. These are the two extreme boundaries for merging the Type A and the Type B approaches.

It is our contention that students should reassess what they read in literature in the same vein as an engineer would assess the results from a CAD program and then critically modify the core concept to make an innovation or an invention. All of the tools and techniques of A1 through A5 become applicable at the entry point (Box A2), to the top ellipse in Figure 4.3.1. The depth of the student's penetration into the content of pertinent literature usually is a good precursor to the quality of the contribution. We would like to suggest that students approach this literature with an open mind. They should neither accept nor reject what they read (and see), but instead they should weigh and consider it carefully.

The ability of the human mind to be spontaneous, insightful, and genuinely creative at the higher levels has the ultimate edge in the architecture of the objects shown in the upper ellipse. The mystery of invention is thus sustained through the instantaneous adaptation (Type A) and rigor (Type B) of the flowchart shown in Figure 4.3.1 to the particular invention that is being made. In a sense, the concepts we present here are recursively used upon themselves.

The upper section of the diagram does not need to be present in the model for reengineering. Total rigidity of every element in this model is the representation of a robotic response in a physical environment. In the absence of a robotic environment, the situation starts to change from robots to humans, from CAD-type engineers to designers, from designers to enhancers, from enhancers to improvisers, from improvisers to adapters, from adapters to innovators, and from innovators to inventors. At the top of the skill chain is the true inventor who may wish to use the diagram we propose in Figure 4.3.1.

Alternatively, if a creative individual attains the status of the "gifted," then one may decide to live and die as a poet-philosopher in the top ellipse of the diagram. The extreme freedom of thought still needs the slight moderation of realism to walk the halls of academia as a successful student. Sometimes the poet-philosopher refuses the rigor to earn an advanced degree.

For those persistent students who would like to practice invention as a conscious activity, we suggest the following:

1. As a first step, we suggest that students should to be able to divide their mental capacities into two rather opposite directions: (a) by being creative at the highest level, which involves the manipulation of concepts and methodologies and the restructuring of the steps and procedures in the solution or investigation, and (b) by being able to reassemble the concepts, methodologies, and steps or procedures in simple and logical ways, so that most of the fundamental laws of nature would be directly applicable to the invention, solution, or investigation. This strategy involves a practice involving a sort of divide and conquer at a mental level, where exacting truth and scientific accountability are two prime concerns. In a sense, this is the activity that occurs within the upper ellipse of Figure 4.3.2.
2. As a second step, we suggest that any novelty in the solution or invention in the first step be scrutinized and re-optimized to satisfy the abstract constructs that call

for elegance and simplicity in the solution or invention. And if any complexity is encountered as a necessity, then one should be able to justify the marginal gain of the invention with respect to the marginal cost of the complexity. This is this final application of Marshall's law of microeconomics in the realm of invention. Today, mathematical tools for evaluating the invention and/or any minor variation of the invention are readily available (see the patents shown in Sections 3.6, 3.8, and 3.9). Consequently, the cause-and-effect relationship between the microsteps in the inventions is correlated to the gain in the performance of the device invented or the solution proposed. In a sense, this activity (or process) shown within the lower ellipse of Figure 4.3.1 is practiced on the contents of the upper ellipse shown in Figure 4.3.1. As a second-order perturbation in the chain of inventing, we suppose it is also possible to become creative with the objects of the upper ellipse, but within the context of the structure reinforced in the lower ellipse.

Finally, our suggestion to any group of research students is to find a comfortable and cooperative balance between being too rigorous and being too creative. Generally, we encourage the conscious practice of collaborating reality with complementary creativity and with cognition of the concepts that have given rise to this current reality in the first place.

REFERENCES (C-4)

[4.1] Miller, M. J., and Ahamed, S. V., *Digital Transmission Systems and Networks*. Vol. II, Computer Science Press, Rockville, MD, 1988 (or see any elementary text on electrical engineering).

[4.2] Anderson, A., *The Raman Effect*. M. Dekker, New York, 1971–1973.

CHAPTER 5

EPILOG

One problem facing universities and corporations is the alarming disharmony between the total intellectual commitment to the pursuit of scientific truth and the possible financial rewards associated with making an invention. There is good reason to believe that such a dichotomy existed during the time of Tesla [5.1], who introduced the concepts germane to the functioning of the induction motor, the hydroelectric generator, some aspects of radio, etc. More recently, during the time of von Neumann (see Section 3.7), as stored program control for computers was being implemented, a number of commercial suitors dominated the computer age. In this context, the relationship between curiosity and the mind on the one hand becomes intertwined with the relationship between greed and personality on the other hand.

Historically, scientists, poets, inventors, and artists have won a special place in society. However, attention tends to be slanted towards the scientists because their inventions generally bring social benefits, and society often rewards them for making noteworthy, significant inventions. Sometimes, the possible commercial value of an invention starts to distort the perception of the scientist making the invention. The temptation of riches can corrupt the joy of invention to the extent that business ethics start to take the place of scientific ethics. In recent years, large-scale unethical practices in business have been well documented. In this section, we present some precursors that the college may adapt to counter this trend.

Ethical and Moral Leniency towards Some Graduates. To some extent, students' behaviors and tendencies carry through into their future professions. If the tendency of the student has been to outwit the proctor, cheat the examiner, and outsmart the mentor (see Section 2.1.3), then society and the nation pay the price in the long run. Starting in the late 1990s, numerous high-profile cases involving unethical business conduct and illegal actions have been uncovered around the globe.* As educators, it is our contention that such tendencies should have been corrected at the colleges and universities. Proctors, examiners, and mentors have the responsibility for helping to eliminate the upcoming breed of unethical executives from classrooms, colleges, and the doctoral programs without hesitation.

*For an exposé of unethical business persons in the telecommunications industry in 2002, see IEEE, *Spectrum*, January 2003, pp. 26–29.

Bugs and viruses in computer and networks systems are handled in a similar way. Early detection of bugs, viruses, and Trojan horses can greatly minimize the damage to the host organisms. This social analogy provides a preventive and precautionary strategy that can help to safeguard society. In our experience, the inactive role of the schools, colleges, universities, corporations, societies, governments, and nations during the last decade or so creates a time bomb that is likely to blow up anyplace, anytime, anywhere. This passive tolerance comes with a hefty price tag.

Of course, it is impossible to totally eliminate all instances of unethical behavior. Nonetheless, university research should offer no refuge for some individuals' ill intentions during their college years. What is most alarming is the amount of damage that these societal bugs, viruses, and Trojan horses cause to the network and the fabric of society. During the last decade, scores of politicians, lawyers, accountants, and executives have fallen prey to greed, hypocrisy, and criminal behavior in the hopes of attaining wealth and power. We do not have a quick-fix solution to this problem. Instead, we present the methodology for making genuine inventions in such a way as to make the scientist aware of the extraneous forces that may interfere with their pursuit of scientific truth. This temptation to stray from strict ethical boundaries must be consciously checked. The forces of personality (see Section 2.4.1) and character of the individual (see Section 2.4.2) dominate the behavior of advanced students.

In the student-mentor relationship at the university, the role of the mentor is critical. In the corporate setting, the role of the corporate ethics officer (if there is such an individual) also becomes critical. In this section, we emphasize the corporate environment for two reasons. First, corporate ethics can slip rapidly when the commercialization of an invention becomes the inventors' claim to fame or makes their fortune. For example, the magnetic domain devices [5.2], charge-coupled devices [5.3], linear motors [5.4], etc., did not have an impact on society. Second, the possibility of hiding behind the legal system after having made inaccurate claims and building a personal fortune (e.g., the recent Enron-like swindles [5.5]) from an invention or an unethical corporate or political move can be quite easily disguised. There are many examples of such illegal or unethical maneuvers in corporations and governments.

5.1 THE COLLEGE PERSPECTIVE

The college life is rarely spent in isolation. Groups of research students usually congregate in academic and social settings. Study groups are formed to prepare for the qualifying and the area examinations. In most instances, the senses are heightened and the psyche is tuned. As opportunities arise, mentors and students should make the most of genuinely educational and creative experiences.

In a college setting, many such opportunities can be created and exploited by students and their mentors. In a multi-student gathering, the leadership of one or more students becomes crucial in order to channel the direction of thought from the mundane to the creative. One or two visionary students can bring the practice of creativity to life here.

In student groups, the group's acceptance of one or more student leaders is not easily achieved. Sometimes the mood of the group can become combative. In such delicate situations, it is critical that the group's leader find a way to spark the group's collective creativity while still emphasizing the importance of scientific and intellectual analysis. If such leadership does not evolve by itself, then the role of the mentor becomes essential. In this section of the Epilog, we present a case study of a classroom for doctoral students where the mentor plays the essential role of teacher, leader, and facilitator of group discussion and communication.

The entire paper[1] involves the participation of seven doctoral students (Galathara Kahanda, Tasneem Kazmi, Lin Leung, Nazli Mollah, Norongrit Waraporn, Marcus Pinto, and Tarek Yahya) at the Graduate Center of the City University of New York and one of the authors (Syed Ahamed) as their mentor. It is replicated here to indicate that inventions can be planned and executed as an organized and dedicated activity.

5.1.1 Abstract (Blending Intelligences)

In this paper, we report a case study to invoke the creative talent of seven graduate students through the use of conscious concept-mining techniques during a doctoral course. The first half of the semester is dedicated to the discussion of the most noteworthy inventions in the field. During the second half of the semester, the mode of the class is first changed to foster a more creative and independent role for the students and then is changed further to foster creative/collaborative thinking. In this case study, the topic addressed is telecommunications, including new optical-transmission techniques and optical interconnects [including Add Drop Multiplexers (ADMs)]. The classroom format reported here attempts to provide a friendly and open environment that encourages students to share their different experiences after they listen to the mentor present a current technical problem specific to a narrowly confined technical area. Moreover, this classroom environment is also intended to foster students' willingness to explain their personal convictions regarding the kinds and levels of creativity they discover as a result of the 40 minutes (or so) of classroom discussion. During the second half of the semester, both the faculty member and the students finetune the novelty of the contribution that may result from the 2-hour class.

[1]"Blending Natural Intelligence with Artificial Intelligence." Paper published in the *Proceedings of the 2002 International Multiconferences in Computer Science* (14 Joint International Conferences), Monte Carlo Resort & Casino, 3770 Las Vegas Blvd. South, Las Vegas, Nevada, USA. June 24–27, 2002.

5.1.2 Concepts of Group Creativity

In the first half of the course, a broad introduction to major inventions is presented to students to help the students understand the mindset of the inventors and to make the students aware of the inventors' contributions to society. The ground rules for the second half of the course are very simple. Any of the seven students can talk about a new direction of thought (e.g., associative, inductive, logical, inferential, or creative) related to telecommunications at any time. The methodology for the mentor is to take a mental note from one or many creative contents of the students perspective and supplement it from a broad background of what the major scientists had contributed in the last 10–12 years. These core contributions are taught during the first half of the semester. During the later part of any lecture during the second half of the semester, the faculty member would propose a new device, an algorithm, approach, or any invention that threaded through the students' perspectives made during the discussion. The whole class and the mentor then enhance, improvise, synthesize, and reconstitute the idea(s) so as to generate a new invention or a variety of dissertation topics for the students to investigate.

The chalkboard or whiteboard is used to provide a visual focal point of the main theme under discussion (e.g., the source of intelligence in networks or the role of doping for optical amplification), the crucial elements or objects in any invention and their interrelationships, the novelty in older inventions, and the contributions of individual students during the class. The faculty member will also enter the visual objects (e.g., drawings, equations, linkages, and forward and backward pointers) on the board to emphasize the known part of the older inventions, algorithms, equations, etc., as well as the novel part of the student discussion.

The class becomes fully cognizant of the mentor's role in helping them to generate an invention during the 2-hour class and generally shows great enthusiasm in attempting to make the most contributions to the group's success. The main stimulus on the part of the mentor is to make creative and constructive linkages between what was known and what was not known at the time of the invention. In fact, this approach helps the more reserved and quieter students in the class to contribute actively. The atmosphere is supportive and congenial, and the mentor never allows students to become critical or hostile toward each other.

5.1.3 The Mining of Generic Concepts

From the mentor's perspective, each class needs an initial list of basic concepts, an organization, and a set of possible relationships between these basic concepts that have made past inventions in the field of telecommunications or computers highly successful. For the mentor, it is suggested that their role as inventors be invoked to gather this list of generic concepts, their organization, and their interrelationships, even though it is not totally essential that the inventions be derived from the mentors' own research. Instead, classic inventions like the basic telephone [5.6], concepts involving the experimental fabrication of the transistor to control the flow of electrons and holes in semiconductors [5.7], the classic development of optoelectronic materials [5.8], the doping of optical fiber materials [5.9] to realize optical amplification, etc., should suffice. These concepts serve well to be planted as the intellectual mines in any classroom.

In this specific case of the last seven classroom sessions during the latter half of the semester, three generic concepts selected from a list of classic inventions, such as the telephone (Bell), the transistor (Shockley, Bardeen, and Brattain), microwave transmission (Tesla or Marconi); phonographs and movies (Edison), and optoelectronic materials and optical amplification (Bell Labs Innovation) selected from a classic list of significant inventions; two patents [5.10, 5.11] on knowledge processing; and two topics from advanced-level students [5.12, 5.13] nearing the completion of their research constitute the seed topics for discussion. Most of the sessions led to new patent-worthy ideas or research topics for other students in the course to pursue.

5.1.4 Group Effort and Resulting Innovations

The exact outcome of the class, in this mode of searching for new patent-worthy ideas, is not totally predictable. If the conditions are perfect, then the mentor and the students should conceive a new machine, a new algorithm, a new approach, or a new theory at the end of the class. The centralized and key role of the mentor is somewhat important because most students want to present their ideas to the mentor. In essence, the mentor plays a dual role of a teacher/motivator who mentally stimulates the class to think coherently and also as an architect who fuses bits and pieces of the students' thoughts into generic concepts that have an academic, scientific, or social value. In the rest of this section, five contributions of the group effort are presented.

Two Novel Network Concepts. New architectures for the two specific networks resulted from the discussions in the class. The first concept led to the organization of electronic governments (EGs) based on the combined architectures of the intelligent networks [5.14] (discussed during the first half of the semester) and the discussion of the addressing of subject matter using the Dewey Decimal System. The role of Internet Protocol (IP) addressing capabilities became important since the addressing of EG functions can be based on the IP mode of addressing. The second concept led to the possible architecture of a new medical network. The functions are commanded by the use of entire asynchronous transfer mode (ATM) [5.15] cells, and as machine-executable code. The network under the physician's direction initiates specific medical procedures (e.g., X-ray diagnostics, bloodwork, and physiotherapy). In a sense, such cell sequences become the complete care and treatment of the patients (subject to the approval of the physicians). The ATM header block (5 bytes) thus handles network functions, and the payload (48 bytes) handles medical function(s). Each of the modules of communication (e.g., networking and content delivery) in the new medical network is handled by the individual cells of any typical ATM network. Both the first and the second venues of investigation are being currently pursued by the second-level students.

One Extension to the Modeling of Networks. Typically, when networks are simulated the program generates congestion points, link-congestion statistics, delays, and the overall response time even as the network is performing complex knowledge-based functions. When the intermediate results are saved and analyzed, then user functions that need complex network functions can be charged based on a tariff structure that can be readjusted to maximize an economic or an opportunistic goal in addition to satisfying a quality of service (QoS) goal. This approach levies the highest tariff based on the QoS and the address locations. In a sense, these network simulators can also serve as revenue projectors based on different tariff structures. Since the data traffic is readily available from the simulation models, the revenue stream can now be predicted (quite accurately) from the traffic studies.

Intelligent CPU. During the class discussion (about 45 minutes long), the notion of an intelligent CPU was conceived. The architecture of this new CPU consists of a program cache that would receive segmented executable instructions to segments of the problem. In an execution mode, a series of program segments (from a database machine) would be used to solve the many smaller problems that constitute a complex problem. The next solution strategy would be based on the success of the previous set of executable codes, and the degree of success would be monitored to help select and either intensify or abandon a particular strategy.

If one looks back on the newly conceived CPU, three elements of associative or derived thinking become apparent. First, the concept of the intensification of the solution strategy was borrowed from optical fibers. Thus, the original concept was altered to suit this intelligent CPU because optical fibers have fixed gain Erbium doped fibers rather than variable gain fibers. In fact, the CPU strategy used an equalization technique [5.15] rather than an amplification technique. Second, the trial-and-error strategy is borrowed from the human approach in solving new problems. Finally, the resolution of a complex problem into finite and manageable subproblems was taken from the pipeline CPUs that use special hardware to perform very finely segmented and very specific hardware functions.

Generic Microprograms for Intelligent CPUs. In the architecture of processors, the concept of pipelining is used to fragment one long OPC (operation code) into a series of smaller and finite operations. In the propagation of optical signals through a long-haul fiber, the concept of optical amplification is used to recover the optical signal and regenerate it by optical amplification. In solving tedious and unstructured problems, human beings use the segmented-solution approach in order to string the entire solution together. These three basic and proven methodologies were primed as concepts, and the challenge was to determine if a new CPU could be designed that could solve complex and knowledge-based problems.

5.1.5 Differences Between Brainstorming and Concept Mining

The conventional approach to brainstorming is narrowly to constrict the thinking of the members to a particular scientific or corporate problem. The nature and extent of immediate resources and the period for the solution also become crucial issues. In some corporations and organizations, a team starts with a well-known solution to a corporate/organizational problem and re-designs that solution and its strategy using new and immediately available resources to come to a new and improved solution. In research laboratories, the constraints are relaxed, and a more long-term solution is sought by both teams and by individuals.

In the process of concept mining, the faculty member spends about half the semester priming the students' thinking and background knowledge base. The classroom environment is never threatening to the students. If there is a basic flaw in a student's reasoning, then the mentor may enter into a one-on-one dialogue with the student to help clarify that student's thinking. The aura of academic freedom sails through the classroom quite well because most of the students are already teaching in universities. The relationship between the mentor and the students is far more collegial. The span for the sustained intellectual and personal relationship among the participants becomes longer. If the mentor monitors the group dynamics in an effective manner, the benefit to the students can be greater than that derived from a typical classroom experience.

5.1.6 Conclusions

As a result of our experiences with this group of doctoral students in a relatively stress-free learning environment, we report that the traditional ideas of data mining can be translated as concept mining and can be implanted in the minds of PhD students. Students were not instructed in the practice of AI, since all of the students on this team already had some exposure to the basic tools and techniques of AI. During the first half of the semester, we suggest that the mentor highlight the circumstances that have led scientists to make major inventions in the past. During the second half of the semester, it becomes necessary to reformat the class to encourage all five (associative, inductive, logical, inferential or creative) forms of human thought. These thought processes become the seeds of student contributions to the classroom discussion. Although, they start out as routine thinking, they become the nodes in a flow graph of an invention.

The level and homogeneity of the class are crucial to the group's ability to develop slim threads of inductive thinking into creative thinking. Perhaps the hardest challenge for the mentor is to document consciously the logic embedded in the ideas of the students into a cohesive blueprint of an invention. In hindsight, the mentor's role is akin to that of a systems designer who is actively building a flow graph for a project in which the goal is not specified. More than being a systems designer, the mentor also vaguely evaluates the coordinates of the nodes as they are being actively contributed by the students in the classroom.

In a sense, the mentor has to demonstrate skill in the scientific and the social areas in order to help the class apply its collective creativity towards the development of a new entity. In this particular class of seven students, the foundation was already built by promoting a supportive and congenial classroom atmosphere.

Sometimes the group's synergy and momentum of thought need refocusing. Such refocusing can occur through restructuring the content shown on the (black) board, where the students' path of developmental thinking had been delineated by the mentor. As a result of constructive group interaction, students perceive that the latter half of the semester involves rearranging various chunks of knowledge rather than to focus on particular information at a very detailed level. We have not tried to implement this approach to enhance or rebuild basic mathematical equations or methodologies. From our experience with this class, however, it appears reasonable to conclude that the methodology is feasible for use in other disciplines.

5.2 THE CORPORATE SETTING

There are two kinds of corporate research: fundamental and applied. The basic difference between the two kinds of research is the extent of articulated managerial goals. The long-term individual responsibility of the scientist is higher in fundamental research, and corporate rewards are expected during a longer timeframe. The extent of investigative freedom is more restricted in applied research, but research managers share in the success or failure of that research. When an ingenious idea is conceived as a result of either type of research, a flurry of developmental activity follows, and corporations sometimes keep such activity confidential to make their product more marketable and to thus gain a larger market share and larger revenues. This effort is hardly conceived as research, although many notable inventors (e.g., Edison, Bell, Marconi) practiced both research and development in their own personal laboratories and went on to establish major corporations.

5.2.1 A Historic Perspective

Typically, the development of new products and enhancement of existing products are essential for corporate survival. In the early days, fundamental research was rare in corporations until DuPont initiated a formal research laboratory in 1902 in the United States. By 1920, the DuPont research laboratory employed about 300 engineers. In contrast to this approach to fundamental research, specific (product- or service-oriented) research is necessary for almost all products and services introduced into a society. For a long period in the late nineteenth century, discipline rather than innovation was the preferred method for gaining cost reductions, greater profits, and increased marketability. However, many major contributions were made based on the ingenuity of individual inventors.

Through the early part of the twentieth century, as many as 3,000 research laboratories were established in the United States. Other noteworthy research contributions came from General Motors for the invention of ethyl-based antiknock fuel and of synthetic rubber for the tire industry. Today, many millions of major and minor innovations are developed every day. Growing amounts of technical information and growing numbers of information networks are both responsible for the rapid development of numerous recent innovations (rather than inventions).

Traditionally, innovation has been nurtured [5.16] in the realm of industrial research. However, the stiff competition that has emerged during the 21st century poses a serious threat to the funding and survival of corporate research. In a sense, the globalization of trade and commerce is undermining the creative potential of corporate scientists. The emphasis in research is likely to shift to academia, where the genius of a few becomes a pathway to invention and recognition for those individuals. Fundamental and basic research in corporate [5.17] entities, in most cases, is too expensive to support. In the final analysis, funds for research come from society, either as accumulated profits/monopolistic gains of corporations or as additional taxation by governments that offer research grants to academic institutions.

5.2.2 A Current Corporate Perspective

Corporations sometimes attempt to foster creativity in small groups by inviting executives to retreats or other stress-free locations in order to pose challenging problems. Executives thus can take the opportunity to discuss their concerns regarding problem areas and to suggest remedies and solutions. Such meetings may help corporations address problems that exist; however, these meetings may not provide the ideal setting for fostering genuine creativity. Leadership is also another essential consideration, since the boss–employee relationship is already implied in the corporate environment.

To some extent, corporations discourage the development of breakthrough inventions for two major reasons. First, the hierarchical power structure of corporations tends not to foster multidimensional freedom of thought. Whereas the intellect has no bounds while exploring the truth, the unwritten code of conduct for executives encourages these individuals to maximize personal and corporate gain and hide many of their most creative thoughts from what is said in the open. This lack of freedom of thought and speech can create a sort of intellectual prison.

Second, the role of image building in corporations is important, and many people strive for recognition, essential to being the most promotable [5.18], even if they must behave dishonestly at times. Whereas intellectual honesty is a critical trait of scientists, such honesty is often less than desirable for executives. In some research organizations, administrative and research careers are deliberately separated into two distinct groups. However, an underlying struggle for power frequently exists, and many businesspersons focus on gaining greater prestige than their peers rather than channeling their creativity in the workplace for the betterment of society.

5.3 CREATIVITY FOR PROFIT-MAKING

There is little wrong in profiting from creativity. Only ethics and long-term social values need realignment. Outrageously, many gifted scientists opt to climb the corporate ladder rather than to seek truth. Financial gain can seem more attractive than the corporate ladder. Yet, many poets and philosophers hide away in some remote cliff off on a corporate island. Highly creative people oftentimes do occupy prominent corporate positions; as such, the rest of this section relates to those who wish to contribute to both corporate profits and creative goals.

5.3.1 Corporations, Innovations, and Social Progress

Corporations need financial resources to survive. A very rudimentary diagram showing the relationship between motivation (social needs), corporate action, and social progress is depicted in Figure 5.3.1. Typically, corporate wealth and resources are either borrowed or accumulated over a period of time. As we discussed earlier, corporate progress can occur as a series of

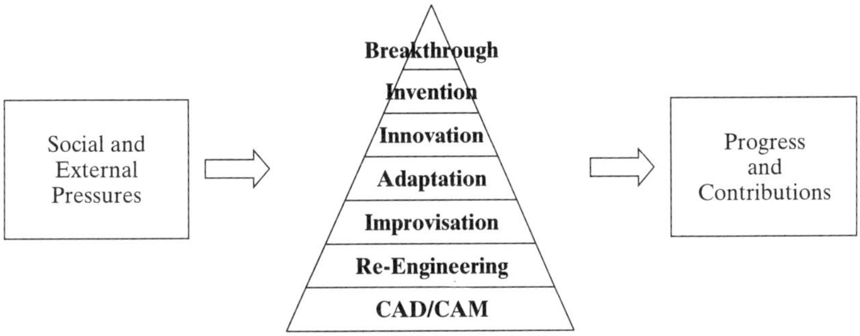

Figure 5.3.1 The major strides of scientific creativity are generally evident in history as breakthroughs (e.g., transistor, laser, and optical fibers). Inventions are documented as patents, and innovations are also evident in the industrial arena to solve problems of an industrial or commercial nature. Adaptations are routine in most corporate environments. Improvisations and re-engineering are expected from the engineering staff as a part of their duties, and CAD/CAM is also incorporated in most system designs to enhance existing products and reduce their cost of generating new products in accordance with slightly different specifications.

breakthroughs, inventions, innovations, adaptations, improvisations, re-engineering efforts and CAD/CAM efforts. These activities typically bring in sales and thus generate corporate revenues. In large part, the role of creativity in corporations is not limited to scientific creativity, but instead incorporates all aspects of making social progress while also assuring the financial stability of such corporations. Thus scientific creativity is highest at the top three layers of the pyramid (i.e., at the breakthrough, invention, and innovation levels shown in Figure 5.3.1).

The basic differences between corporate creativity and scientific creativity start to become clear as profiteering takes on an important role in purely business environments. However, when corporations encourage the development of genuinely new products, scientific creativity also begins to play a role. A tiny niche for scientists thus can appear. Modern corporations that seek to derive wealth based on the novelty of their products employ scientists whose role is akin to that of a financial officer.

The wealth of technological corporations may be traced to the intellectual few who create rather than to those who copy. Like the wealth of nations that may be traced to those nations' natural resources, corporate wealth may be linked to creativity and the scientific collaboration of the corporation's creative people who are assets rather than just employees of the corporations. Such scientists understand business as much as executives understand technology, and a new blending of industrial talent starts to take shape. The dynamics of such corporations are shown in Figure 5.3.2. Scientific research generally resides in a dynamic-loop environment where social needs and social rewards for corporations are intertwined. National and economic forces in society also contribute to the cycle. Sometimes the quality and quantity of inventions are both influenced by the extent of government support and the national need for products that have been conceived but not yet invented. Wartime inventions and space-age technology are examples that have shaped the profile of American creativity [5.19]. In a strict sense, only the top three layers shown in

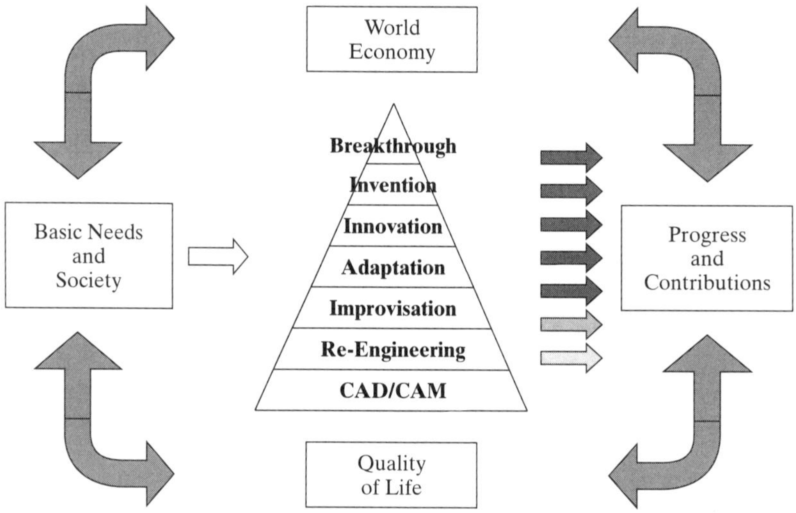

Figure 5.3.2 The environment for scientific activity within a corporate environment. Both scientific and business aspects become important for technology-oriented corporations.

Figure 5.3.2 can be classified as creative. At the lower levels, the job of the employee is to perform a scientific or a managerial task.

Corporate creativity and innovation are not unrelated to the academic pursuits of students. Many industrial scientists and managers have earned advanced degrees. The difference in the approach is in the concentration towards the goal. For many students, the advanced degree is a career goal. For many scientists in the industrial community, however, contributions to corporate profitability can advance their career goals. These two sets of goals become closely integrated over the span of a corporate career. The tools and techniques to pursue both goals are essentially the same. The flowcharts shown in Figures 4.3.1 and 4.3.2 need to migrate in the scenarios depicted in Figures 5.3.1 through 5.3.5. The student may apply the tough marketability tests from industry to academic inventions. The industrial scientist, in turn, may be rewarded for the rigor of scientific thought taught at the universities. Opportunities for creativity never end for the researcher; instead, they flow into new veins of exploration and discovery.

5.3.2 Personal Attributes, Group Dynamics, and Corporate Funding

When external factors such as social demands and governmental or political pressure that contribute to the corporate research and development activity are isolated from internal factors, then two dominant forces driving the research teams become apparent. Both the personality attributes (see Section 2.4) of the team members and the group dynamics (see Section 2.3) govern the daily activity and thus the overall effectiveness of the team. Whereas absolute rigidity of the structure (e.g., as in the military) helps to ensure that the job will be done, absolute latitude of thought can lead to significant contributions. In a sense, a blending of both extremes, intellectual freedom and a procedural infrastructure would be desirable in corporate research. Selecting the right personnel for the team becomes a social process rather than a strict guideline. Some of the issues that influence corporate research are represented in Figure 5.3.3.

In the university environment, student-mentor teams play a crucial role. Even though teams may consist of small numbers of researchers in university environments, these relationships can become quite intense. It is not fair to expect students to understand the teamwork necessary to make major inventions in modern society. If a mentor deems it necessary for student commitment to exceed normal and expected levels, then it seems fair to offer sufficient financial rewards to the student. Funding for research has made inroads even in academia.

In Figure 5.3.4, the interplay between the role of researchers' personality attributes, group dynamics, and the productivity of the entire team is taken one stage further. Slightly outside the core team of researchers and technicians is the outer layer of management that needs to monitor the more latent and hidden assets and setbacks of the research teams. An unobtrusive form of mentorship exists in corporations that have mature and stable research teams. In some cases, the administrative role of the manager of the research department is made into an elected and/or rotational position, thus eliminating a sense of professional competition. Such highly collegial and stable research environments lead to classic breakthrough discoveries or inventions, such as Hamming Codes [5.20], satellite communications [5.21], semiconductors (see Section 3.6), and masers (see Section 3.7) conceived at Bell Laboratories.

For the development of silica glass (see Section 3.9) specially doped and fabricated to the H_{11} mode for transmission of light through the fiber, the collaboration of scientists R. D. Maurer, P.C. Schultz, and D. B. Keck at Corning Glass is a classic example. These recent inventions are based on a rich history at Corning Glass of the development of original work that has helped to define and refine the composition of glass.

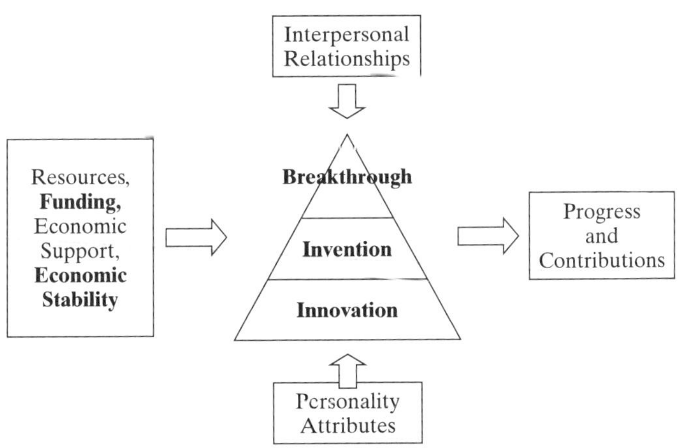

Figure 5.3.3 If the external forces that contribute to corporate research and development activities are isolated from the internal forces, then two dominant forces driving the research teams become apparent: the personality attributes (see Section 2.4) of the team members and the interpersonal relationships (see Section 2.3) or group dynamics that govern the daily activity.

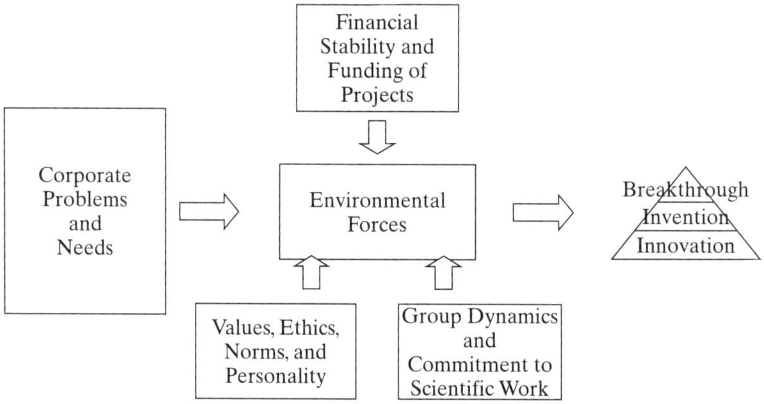

Figure 5.3.4 The problems and needs of society that are generally brought into the corporate environment may consist of problems with a particular product line or ideas for technological innovations. If a short-term solution is necessary, then the engineers working at the lower level of the seven-layer invention pyramid (see Figure 5.3.2) will find a quick fix for the problem or need. However, if the problem needs the attention of scientists, then appropriate financial resources are allocated and factored into the financial stability and funding of research projects. Uncertainty of the outcome generally dampens the flow of funds. All the same, many major inventions are made in corporate research laboratories.

In offering these aforementioned cases of collaboration among scientists, we allude to the fact that such a collaboration between the managers of corporations can also lead to breakthrough industries. Within any corporation, such collaboration can lead to new or novel products to serve unique corporate needs. We suggest that the methodology of research be practiced in the managerial rungs and vice versa. The resources among the scientists are generally shared within corporations, and management skills and techniques may also be shared in a congenial and collegiate fashion, and vice versa.

In the rest of the chapter (Sections 5.4 and 5.5), we propose that the creative talents and functions of scientists and managers should be blended. Managers can thus incorporate scientific inventions of devices and principles into corporate practices and profitability goals. When such a seamless overall corporate functionality is achieved, a second level of managerial revolution may be expected based on the crosspollination of two of the necessary elements (innovation and profit motive) of social change. In Figure 5.3.5, the underlying forces leading to creativity are represented. The interaction between money, management, and personalities (depicted in the triangle shown in Figure 5.3.5) influence the work environment, and vice versa. The interplay between the four objects in the triangle and the external factors thus becomes continuous and pervasive.

5.4 THE MANAGEMENT OF RESEARCH

Management and creativity conflict at times in certain key ways. Even though they could conflict, they can both be concurrent, coherent, cogent, and consistent. The term

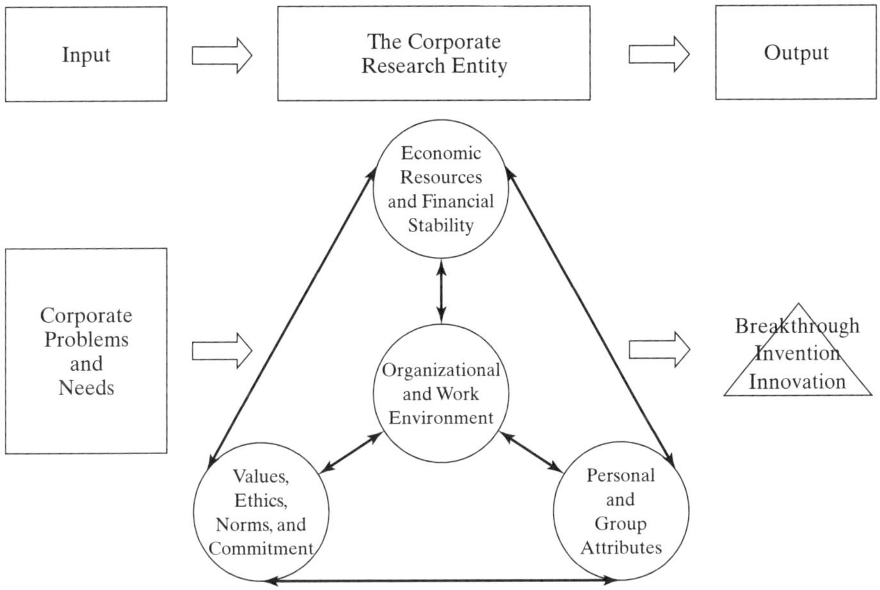

Figure 5.3.5 Diagram to illustrate the role of corporate and social pressures that drive corporate breakthroughs, inventions, and innovations. The bottom four layers shown in Figure 5.3.2 are not shown here because these lower-level functions are routine and are handled by the bottom engineering staff rather than by scientists.

management implies constraint, whereas the term research implies an absence of constraint. Total freedom would be as extreme as total slavery in research thus leading to the familiar bell shaped curve for the distribution of the fruitful mix of the two extremes. The management of research deals with the administrative functions in support of scientists who are (reasonably) free in their intellectual pursuits. This freedom does have bounds, however, and such bounds should be respected. Once again, the makeup of the team and the role of the manager become crucial in maintaining the balance between corporate structure and creative freedom. Mature managers and scientists tend to avoid conflicts in order to build an atmosphere of collaboration.

The interpersonal relationships of the research team form a complex system of checks and balances between the needs and the strengths of the members. In an ideal setting, their contribution is comparable to the symphony of many musical instruments each played in harmony and rhythm directed by a conductor. Both masterpieces and chaos may result. The manager thus plays the delicate and sensitive role of a leader who balances his self interest, the company's interests, and the long-term rewards for the members of the team.

These three different sets of interests can suffer as a result of forces such as corporate changes, the drying up of funds, personal conflicts, etc. In such cases, cohesion can turn into chaos, the harmony can turn into bitterness, and symphony can turn into noise.

In many instances, the code of ethics is neither documented nor articulated since the personalities of the team members and the leadership style of the manager are both highly variable. For the true executive, the art of managing the human team involves a combination of intuition and reasoning. It is easy for a research team to lose its research edge through mismanagement, misunderstandings, or personal biases. On the one hand, the policy of total nonintervention is exploited by immature scientists. On the other hand, a policy of a military style of dictatorship leads to the desert style instinct of the killer-hunter rather than the dedicated poet-philosopher who seeks the truth and the sophistication that lies hidden in the investigation that is being pursued.

5.4.1 The Individual Scientist

The individual scientist becomes the cornerstone of the research team in most corporate environments. The scientist is constantly examining ideas that may lead to novelties, inventions, or innovations that could (by an estimation of probability) bring a profit to the corporation.

A representation of the model for the management of creative processes shown in Figures 4.3.1 and 4.3.2 is simplified in Figures 5.4.1 and 5.4.2. In Figure 5.4.1, the basic attributes of the students and their mode of activities appear as four lines in the flow graph: A (the gifted), B (the meticulous), A/B (the very gifted/less meticulous, who have a conscious capacity to flip-flop (switch) between the two, as described in Section 4.3.3), and finally B/A (the very meticulous/less gifted, who have a conscious capacity to flip-flop (switch) between the two, as described in Section 4.3.3).

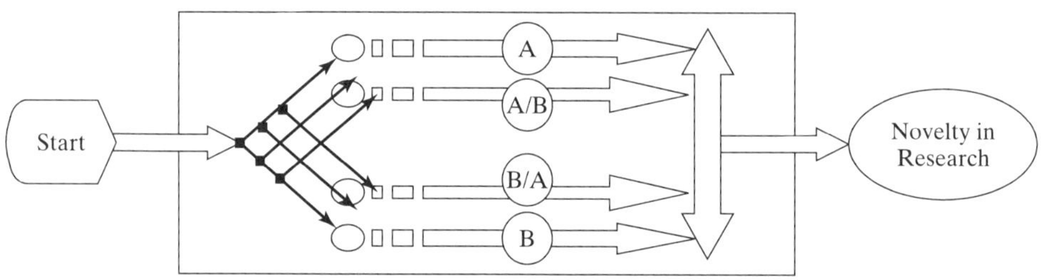

Figure 5.4.1 A simplified flow graph for searching for novelty in the academic world. A, A/B, B/A, and B are explained in Figure 4.3.1 and in Section 4.3.3.

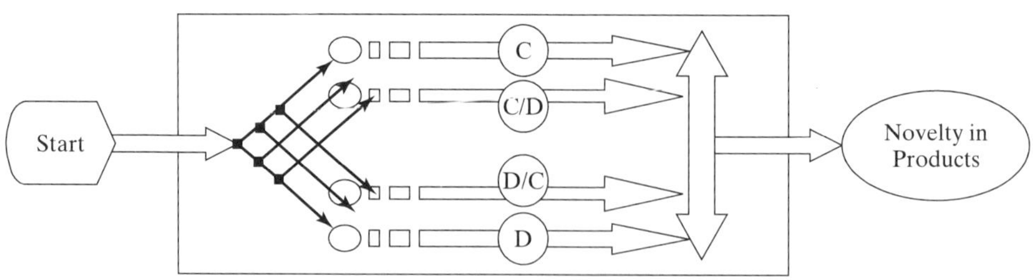

Figure 5.4.2 A simplified flow graph for searching for novelty in the corporate world. C, C/D, D/C, and D are explained in Figure 4.3.2 and in Section 4.3.3.

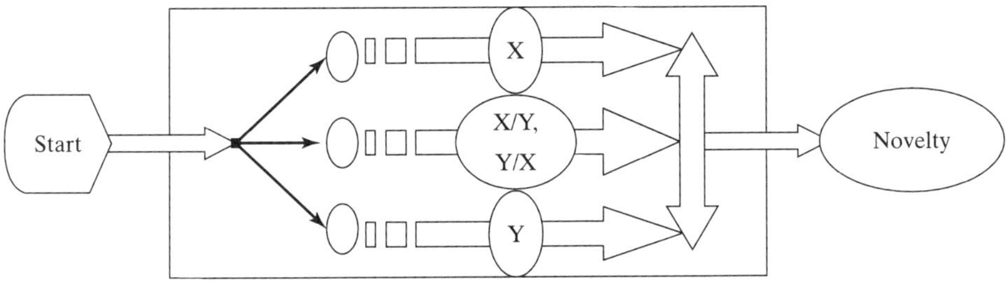

Figure 5.4.3 A simplified flow graph for searching for novelty in the corporate world, where the new employees come from academia. X, X/Y, Y/X, and Y are momentary assumptions of A, A/B, B/A, and B and/or C, C/D, D/C, and D roles and attributes in any combinations and proportions.

An analog of Figure 5.4.1 is also possible for professional corporate scientists. As shown in Figure 5.4.2, there are four lines—C, C/D, D/C, and D—where C indicates the activities of the most gifted scientists and D indicates the activities of the most meticulous scientists.

To construct a managerial model, it becomes necessary to consolidate these two figures into one and rename symbols A, B, C, and D as X (for A or C) and Y (for B or D*). In the real world, human beings do not follow this type of machine-based logic; instead, they can freely migrate their thought processes from the academic domain to the corporate domain, and vice versa. In addition, they may blend the two thought processes (as a member of an academic team or as a member of a corporate research team) to evolve their own way of tackling the problems associated with creativity. When this level of intricacy in the thought control starts to surface, then the practice of creativity appears to have become recursive to the extent that the scientists are being creative in the practice of their creativity. In such cases the symbolism, though slightly more complex, may still be represented.** A simplified diagram is shown in Figure 5.4.3.

Figure 5.4.3 thus becomes the foundation of the rather elaborate model for the management of creative people. As we discussed earlier, the predictable part of the practice of creativity is (to some extent) outlined in Sections 2.5.1, 2.5.2, and 2.5.3.

Figure 5.4.3 can be further condensed using the concept of a simple switch that connects a series of inputs at "Start" to a series of outputs at "Novelty." The scientist can be an intelligent switch that connects any, some, or all inputs to any, some, or all outputs. The attributes of performing the switching function are the contents of the activities symbolized as X and Y in any combination. Each scientist thus can be viewed as a functional block of a corporate research entity, as symbolically represented in Figure 5.4.4. A series of such switches (researchers) acting in sequence or in parallel with in a corporate structure would depict a corporate research laboratory. The functions of management of the researchers in a flexible and adaptive fashion is represented at S in Figure 5.4.4.

*Symbolically, [X = A .or. C] .and. [Y = B .or. D].

**[$X_i = A_i$ (and/or) C_i] and [$Y_j = B_j$ (and/or) D_j], where i and j are the components in the ellipses shown in Figures 4.3.1 and 4.3.2 or the attributes of the researchers described in Section 2.5. The representation of X/Y becomes more complex, but the bounds may be extrapolated.

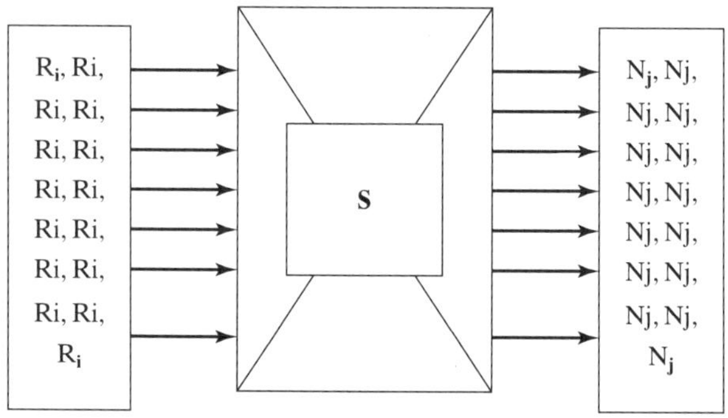

Figure 5.4.4 A symbolic representation of the creative role of corporate scientists who switch, recombine, blend, and mentally process an array of individual resources [R_i, which are their own attributes (X, X/Y, Y/X, Y)] to create an array of novelties (N_j) in order to facilitate the achievement of corporate goals. The role of the manager/administrator is designated as S.

Two major functional blocks arise: basic research activity and management of such activity to realize the group synergy in order to harness a corporate goal. Now the problem of management does not have to enter into the private domain of the individual researcher. This private domain of the researcher is linked to all aspects of learning how to perform creative research. It is immaterial whether it is derived from the mentor, an associate, a conference, or a casual remark from a colleague, the boss, or an inspiration.

In most corporate research laboratories, numerous scientists work together, and both individual and group contributions are common. Whereas the emphasis thus far has been on reinforcing the contributions of the individual scientist, it is also possible to address the role of management in allocating individual resources, group resources, departmental and corporate resources, as well as the role of management in unifying and integrating their contributions as individual novelties (e.g., inventions and innovations), group novelties (e.g., processes and products), departmental novelties (e.g., devices and systems), and corporate novelties (e.g., major systems and socially desirable and financially rewarding products). Thus, the continuum and the momentum of corporate research may be continued and expanded. We examine the stability or instability of such conditions of slow divergence, unstable dissonance, or convergence in the next section.

5.4.2 The Management of Scientists

In the corporate environment, the modern trend has been to pay scientists for their inventions rather than to reward them for their research. Although rewards are often nonmonetary, they can offer sufficient motivation for scientists. In most cases, when the scientist or the team is mature, no management is better than over management. Nonetheless, management often needs to direct research pursuits in order to align research contributions with corporate goals.

Many inventors and contributors seek out novelty and elegance in their research rather than monetary gain. From Veblen's contribution [5.22] to the scarcity theory, the most valued is the rarest of the rewards. As such, the onus now shifts to the research manager, who must reward the most valued scientists with the rarest of the gifts within the group that is most likely quite affluent and obviously well recognized. Two major resources exist: genuine appreciation without excessive flattery and genuine sincerity and concern. Such rewards should be individually tailored to each scientist. Some examples of such individualized rewards would be the distribution of a small book documenting the contribution and its relevance to corporate goals, contributions to honor societies, establishment of a website, an international 800 number for call-access for spreading the contributions, etc.

In a sense, the management of creative people is a creative activity in its own right. It can become challenging and rewarding to socially creative people. Again, the goal is to seek a creative balance between socially creative people, who can have a scientific mindset, and scientifically creative people, who can have a social mindset. The areas of overlap become more sparse at the highest levels of performance in both of the domains (the social sciences and the physical sciences). The help of management consultants for this very specialized function of avalanching creativity within the group in any discrete direction that the corporation expects the maximum financial return is another alternative. In many instances, scientists have business skills that may be explored in order to allow the scientist to make scientific, social, managerial, and financial contributions to the corporation and thus to ensure the continuation of ongoing research.

We base our overview of the management of scientifically creative people on two broad concepts: unity of purpose and freedom of thought. The concept of unity of purpose works in favor of corporations that finance, run, and manage research laboratories, and the concept of freedom of thought works in favor of each individual scientist.

This type of partitioning of functions was practiced even in the days of Adam Smith, who argued for unity of control (in the hands of management or capitalists) and division of labor (to exploit the manual or dextral skills of laborers). Although the context is not exactly the same, we may elevate and reconstruct this approach to the management of scientists.

At the corporate level, the unity of purpose to make a dramatic financial return on investment becomes the dominant goal. At the individual scientist's level, the freedom of thought that is documented in Figures 4.3.1 and 4.3.2 dominates the investigations of physical phenomena.

5.5 TWO MAJOR MANAGEMENT FUNCTIONS

On the left side of Figure 5.4.4 lies the first managerial task of resource allocation, and on the right side lies the second task of unification and blending of novelties. At the macro level, all resources come from corporate funding (typically 10% of corporate revenues), grants, and other donations. Again, at the macro level all unification involves the painstaking and tedious process of product integration based on the four groups (i.e., individual, group, departmental, and corporate) of novelties and inventions.

These two tasks of management are represented in Figures 5.4.5 and 5.4.6. In Figure 5.4.5(A), a typical decision tree involving probabilities of success is shown. The actual use of such decision trees is not a pure science, even though the derivation and the implementation are topics in operations research taught at engineering schools or topics in management science taught at business schools. To a large extent, the practice of

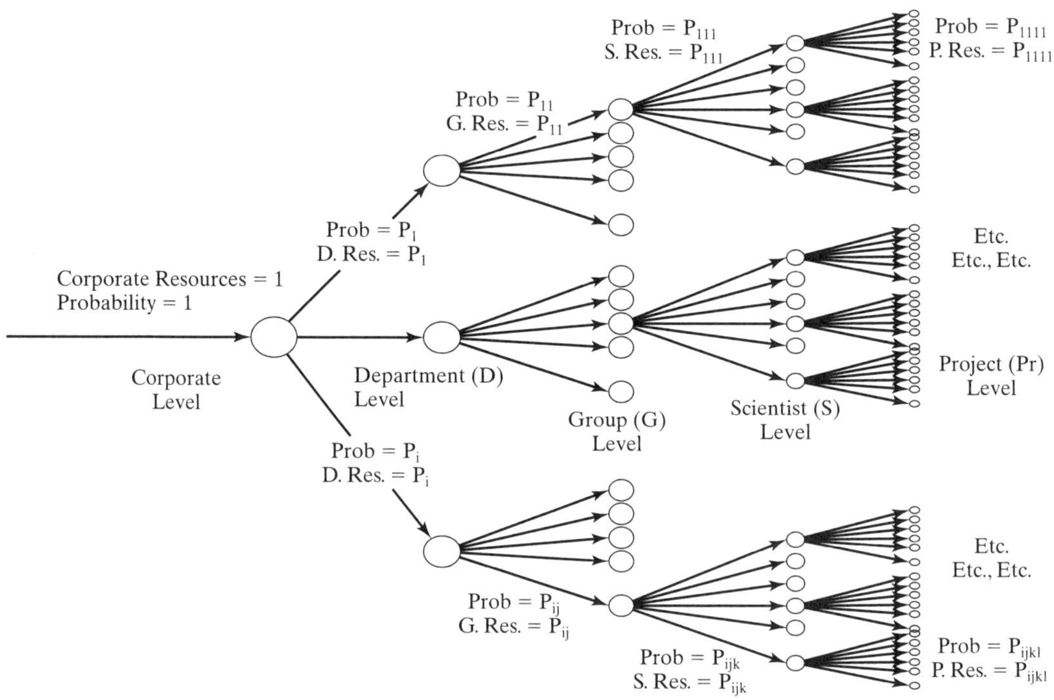

Figure 5.4.5(A) A decision tree approach to the allocation of resources in order to maximize the expected return. P designates probability of success of a project. D, G, S and Pr stand for department, group, scientist, and project. Res designates the allocated resource

management demands some skills that are beyond the academic training of business schools, and we leave it to seasoned managers to practice these acquired skills in the area of resource allocation.

When managers prefer to use judgment and skill in allocating an array of (limited) corporate resources, then the managerial task also becomes creative in its own right. Much like scientists who blend their resources (both corporate and personal) with their own personal and academic skills, managers often blend and distribute resources rather than simply allocating them. To this extent, managerial creativity can become highly specialized and personal. The diagram for this type of allocation is presented in Figure 5.4.5(B). Of course, a certain amount of bias and subjectivity can creep into the allocation process. However, in most instances those projects with the highest probability of producing significant profits receive funding, and scientists with the strongest track records receive support based on a combination of rationality and need.

Product integration and blending the products for marketing can be creative tasks in their own way. However, business tools and methodologies can be the routine practice of management. At the outset, the expected market demand for and scientific/social acceptability of the product become crucial. These expectations can soon become subjective, and many unproven inventions can drain large amounts of corporate research funds. Extreme objectivity coupled with sound reasoning and caution are in order to promote the invention at individual, group, department, and corporate levels. It is almost like walking a decision

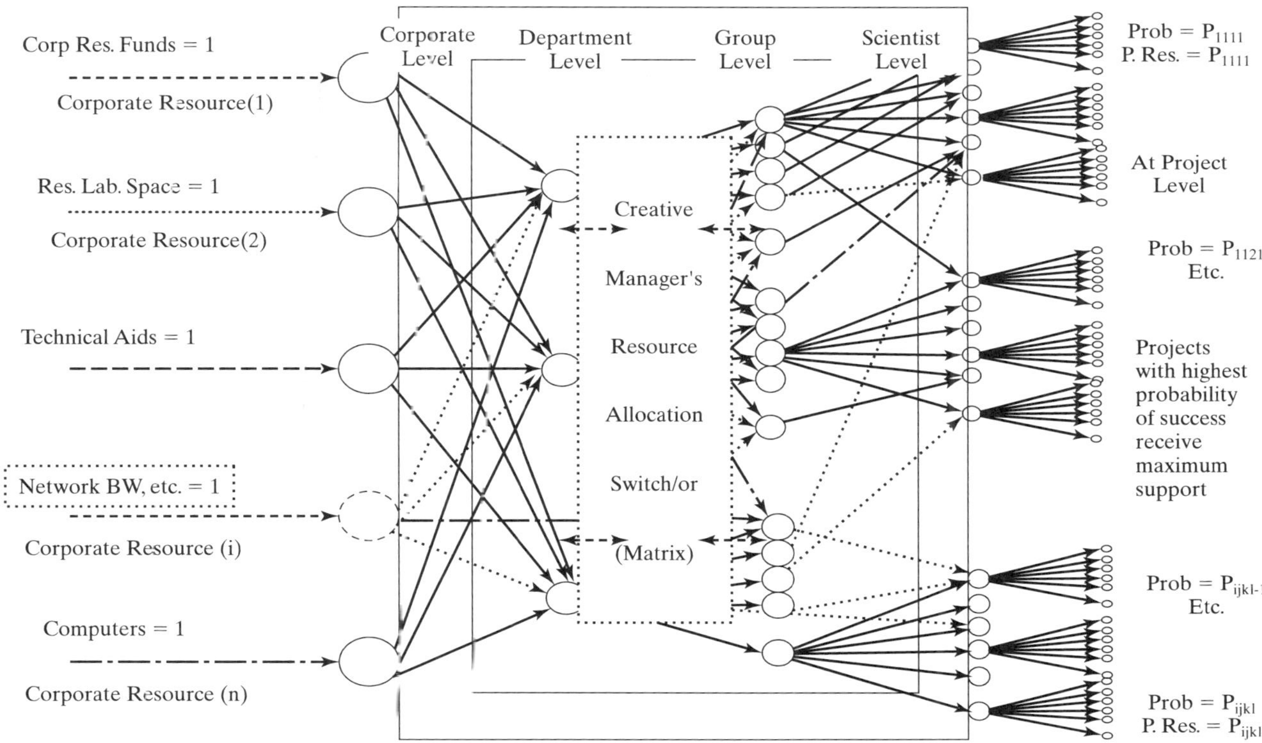

Figure 5.4.5B A creative manager's approach to the allocation of an array of resources to maximize their overall utility in research.

tree backwards, leading to one or more products at each level. In this chapter, we will not discuss such routine functions of management, since they are only tangential to the management of research; however, if there is a creative component in the process, then the flow graphs of inventions shown in Section 5.4 apply. For this reason, Figure 5.4.6 is divided into two sections. In Figure 5.4.6(A), the role of plain integration is emphasized. Combinations and permutations of individual contributions are cascaded and arranged in different architectures to offer new products. This type of activity corresponds to production innovation rather than to an invention or a breakthrough.

In Figure 5.4.6(B), the blending of inventions is depicted. The corporate team now alters the inventions of individuals, groups, and departments to suit the needs of different markets. It also investigates its own inventions in the hopes of making them more universal. This feedback loop generally lasts much longer in order for substantial inventions and breakthroughs to be made. Corporate, economic, and product stability drive corporate inventions and contributions in this direction. From a historic perspective, Bell-Laboratories has satisfactorily portrayed this type [Figure 5.4.6(B)] of blending function, whereas product-driven industries like IBM and the automobile and aircraft industries are driven towards the integration mode depicted in Figure 5.4.6(A).

5.6 THE OVERALL CORPORATE RESEARCH MODEL

It is now possible to combine Figures 5.4.4 to 5.4.6 into a set of corporate management models for research that honor both of the two essential roles: the role of scientists and the role of managers.

Typically, the role of managers can be purely integrative (e.g., somewhat autocratic, military-like, or X theory of management) or creative (e.g., democratic, laissez faire, or Y theory of management) within one dimension (the F dimension, which stands for functionality). The former role can be intrusive (typical of a production environment), or the later exclusive role (typical of the research environments) in the other dimension (the E dimension, which stands for environment). Then, within the F and E two-dimensional space {F,E}, four corners (0,0; 0,1; 1,0; and 1,1) exist.

When managers enter the domain of creativity and work with established and inventive scientists, the modus operandi is ($F = 1, E = 1$) or (1,1). Likewise, the other extremes of the F, E two-dimensional space are (0,0), (0,1), and (1,0). The space near the (0,0) region is a totally robotic environment, the region between (0,1) is unstable since absolute rigidity in research is bound to fail, and the region near (1,0) is possible for high productivity, but congenial for the highest level of creativity from managers in both the managerial and the technological domain because they are concerned with the financial returns of invention for the corporate revenues. Near corner (1,0), very talented managers "program" their slave scientists to invent according to a schedule.

The most highly desirable region [approximately (0.7–0.9, 0.7–0.9) for the (F, E) space] may also be approached by scientists who have acquired a flair for business in dealing with research. Conversely, managers may initially lack technological skill sets, but these skill sets may be acquired from their scientist colleagues over a period of time. When such collaboration is desired, then human interaction between managers and scientists starts to become crucial. Contrary to usual management and research styles, the distinct characteristics of responsibility, trust, concern, and creative partnership need to be developed between the management and the research

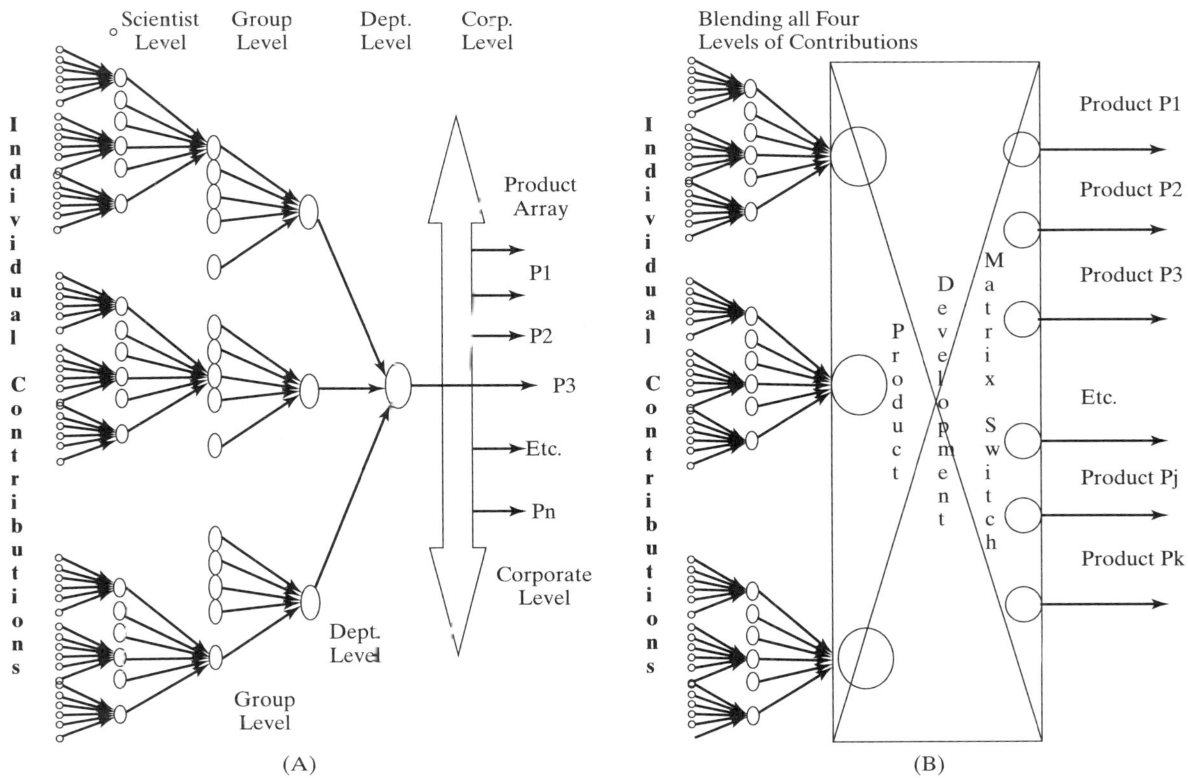

Figure 5.4.6 (A) Direct integration of individual contributions into different marketable products. (B) Creative blending of individual contributions into different marketable products.

teams. In a stable and cooperative mode, such mutual respect and interdependence may evolve between the scientists and the managers over time.

The two-dimensional {F,E} space does not have an independent existence. For the proposed model, if the {F,E} dimensions are considered as x and y dimensions, then the corporate resources {R} play the crucial role of continuity for the {F,E} space. In a realistic sense, the square located within the four corners (0,0; 0,1; 1,1; and 1,0) in the {F,E} space assumes the position of a base of a pyramid, with the financial resources {R} of the corporation along the z dimension. The classic research pyramid starts to take shape in the corporation, with scientists, managers, and corporate resources forming its three-dimensional existence. We use this concept in Sections 5.2 to 5.5.

Another dimension of freedom also exists for managers. Whereas some managers may prefer to depend more on a theoretical economic approach for dealing with the allocation of resources, other managers may take a more liberal approach in considering an entire array of resources (e.g., funding, office and laboratory space, technical aid, computers, and instrumentation) and depend on a creative way of blending the resources allocated to individual scientists. When managers practice this mode of resource allocation, the block for resource allocation starts to more closely resemble Figure 5.4.6(B), except that the arrows are reversed. Groups of resources are creatively distributed by managers to a group of scientists who carry out research on different projects. There is considerable amount of skill and managerial creativity in this function as well.

There is no reason to assume that a highly creative team of scientists will be led by a group of well-balanced managers. The two critical functions of managers (resource allocation and product integration) can have a significant effect on the overall stability of the corporate research team and on the corporate profitability needed to sustain the research team. The same conditions are also true on a national basis whether research can be sustained in a country or over a period of time.

If a group of creative scientists is formed in a corporate or a university setting, the delicacy of their existence can be gradually threatened by managerial neglect of the two critical functions, resource allocation, and product integration. We depict four possible scenarios in Figures 5.4.7 to 5.4.10. The community of scientists (on a corporate, academic, or national basis) forms the central creative and inventive core of personnel responsible for the breakthroughs, inventions, and innovations that drive corporations, academia, or the country. This group of creative individuals works closely with the two groups of managers. The first group provides the resources and facilities for the scientists to create novelties and inventions, and the second group exploits the novelties and inventions for social and economic benefit.

If the allocation is practiced based on a mechanical decision-support-system type of methodology, then the left-hand side of Figures 5.4.7 and 5.4.8 become similar to Figure 5.4.5(A). If the unification of novelties or inventions is practiced on a mechanical product integration, then the right-hand side of Figures 5.4.7 and 5.4.9 becomes very similar to Figure 5.4.6(A).

In Figure 5.4.7, the overall representation of a managerial model of the separation of managerial tasks from the tasks of scientists is shown. Resource allocation takes place on the left-hand side, and integration of inventions and novelties takes place on the right-hand side. Both of these managerial functions are routine and lack the imagination necessary to make them most valuable to the long-term survival of research scientists. The scientists work free from the corporate pressures and can focus on inventing.

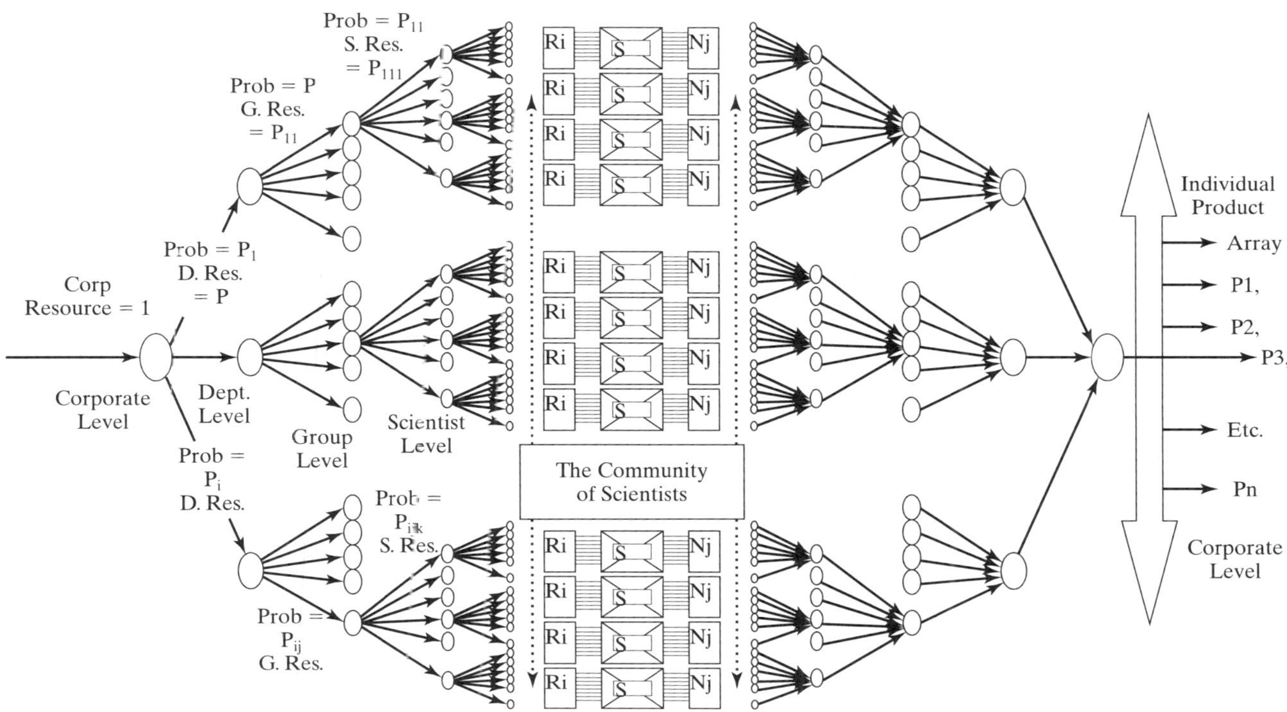

Figure 5.4.7 Overall representation of a managerial model of the separation of managerial tasks from the tasks of scientists. Resource allocation takes place on the left-hand side, and integration of inventions and novelties takes place on the right-hand side. The scientists are free from corporate pressures and can focus on inventing.

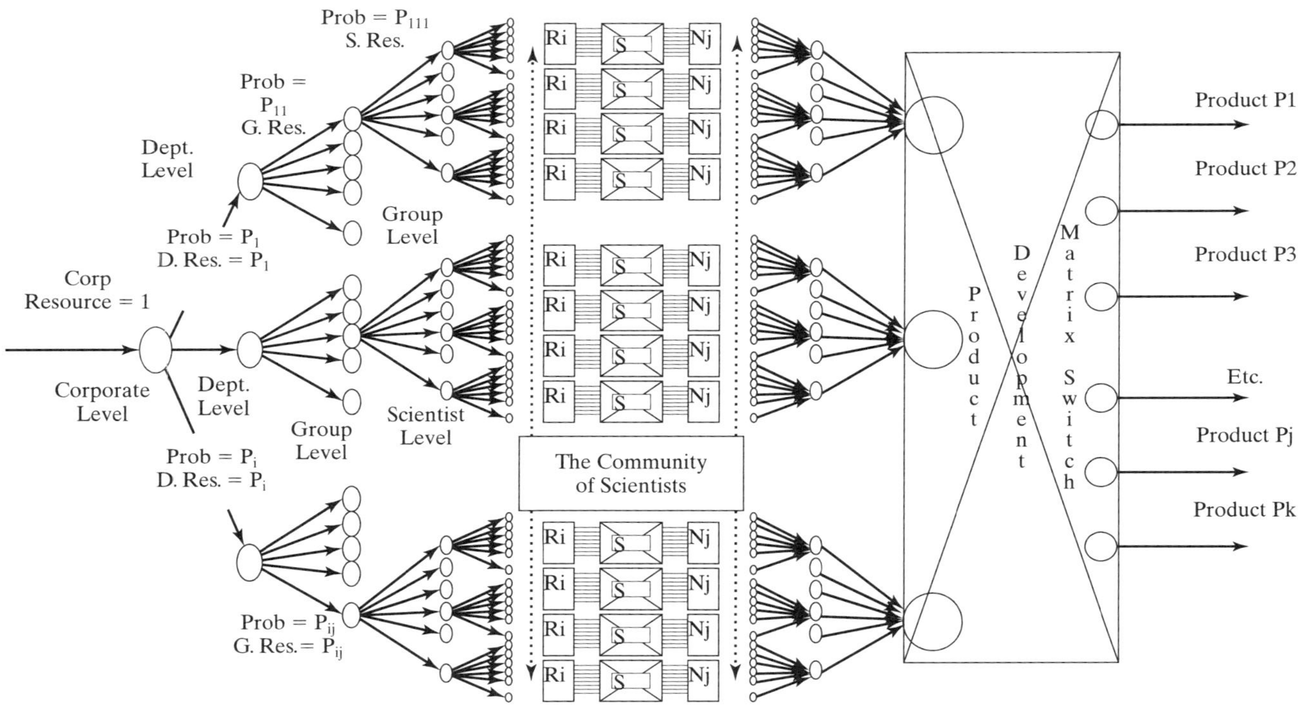

Figure 5.4.8 Overall representation of a managerial model for the partial separation of managerial tasks from scientific/research tasks. Resource allocation takes place on the left-hand side, and the blending of inventions (of individuals, groups, departments) now takes place on the right-hand side. Scientists are free from corporate pressures until the creative managers redefine the type of inventions that will satisfy the product needs. This may be a small price to pay to blend the creativity of managers and scientists.

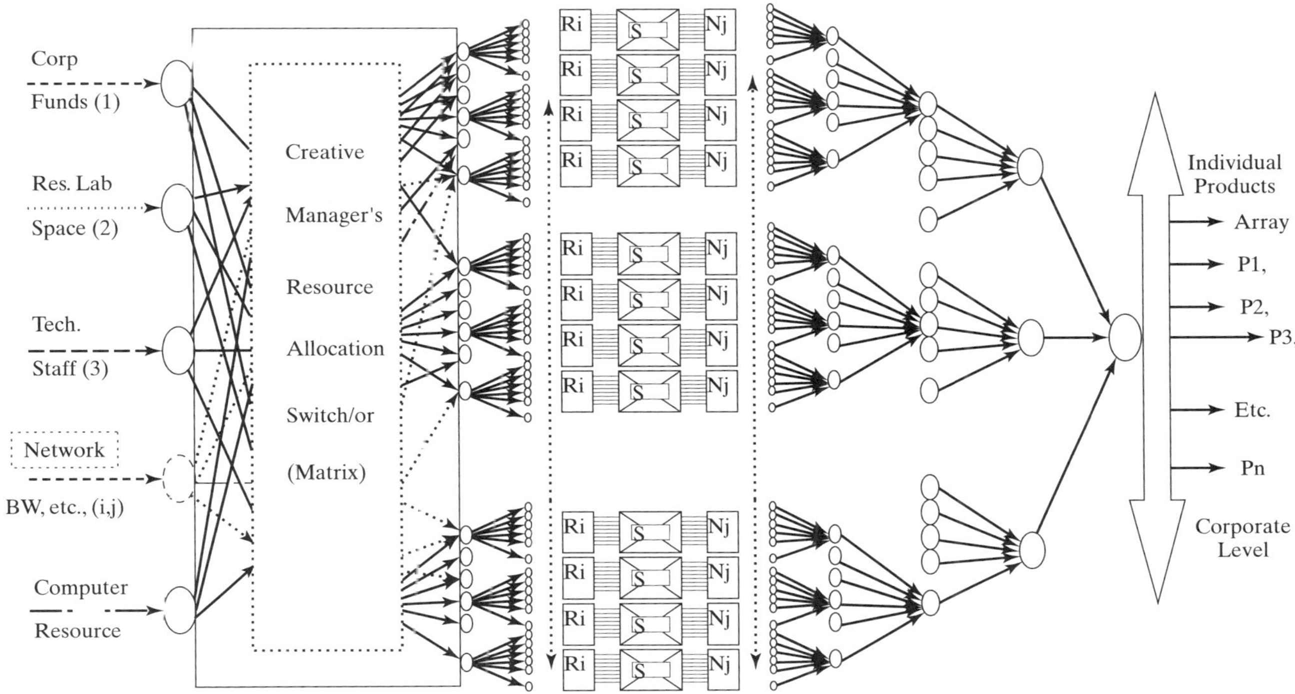

Figure 5.4.9 A case of an unbalanced managerial team. Corporate resources are appropriately allocated, but the product development team fails to use the invent.ons of the community of scientists to enhance corporate profitability.

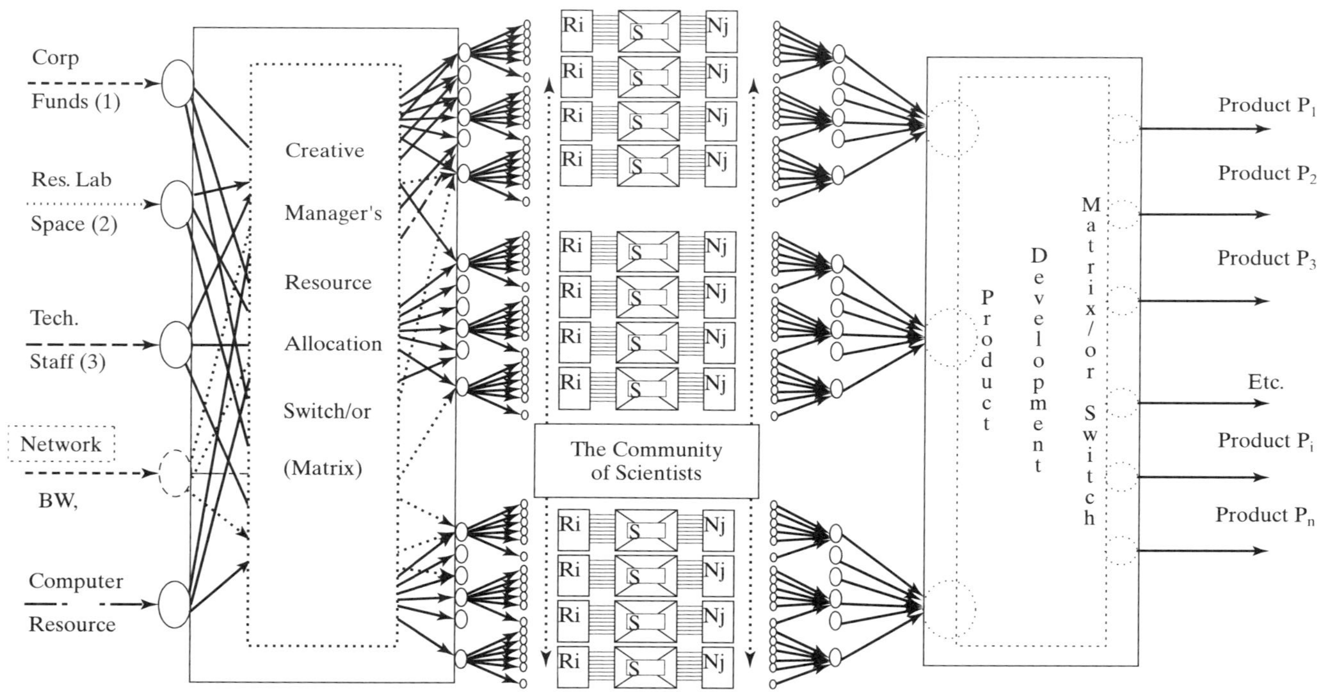

Figure 5.4.10 Overall representation of a creative management team that surrounds a community of creative scientists. Managers, like scientists, are also creative, and they use their talents to manage resource allocation and use in order to maximize the output from the community of scientists. On the product and marketing side, management also exercises its creative talent in blending each of the inventions and contributions of the scientists to maximize the potential for profits. This particular scenario maximizes the creative output of managers and scientists and hence maximizes the collaboration between the two groups.

In Figure 5.4.8, the overall representation of a managerial model for the partial separation of managerial tasks and scientific/research tasks is shown. Routine resource allocation takes place on the left-hand side, and the blending of the inventions (of individuals, groups, and departments) now takes place on the right-hand side. Scientists are free from corporate pressures until the creative managers (on the right-hand side) redefine the type of inventions that will satisfy the product needs. This may be a small price to pay to blend the creativity of managers and scientists. The condition is generally somewhat stable and is associated with the growth part of the corporate cycle or the expansion phase of the national economy.

In Figure 5.4.9, a case of an unbalanced managerial team is depicted. The corporate resources are appropriately allocated to the scientists, but the product development team fails to use the scientists' inventions to enhance corporate or national profitability. The corporate or a national technological edge may start to disappear, which is generally associated with the condition of the inevitable downturn of corporate and national economic cycles.

In Figure 5.4.10, an overall representation of a creative management team that surrounds a community of creative scientists is depicted. Managers, like scientists, are also creative, and they use their talents to manage resource allocation and use in order to maximize the output from the community of scientists. On the product and marketing side, management also exercises its creative talent in blending each of the inventions and contributions of the scientists to maximize the potential to reach marketing goals related to product volume or profits.

This particular scenario maximizes the creative output of both the managers and the scientists and hence maximizes the collaboration between the two groups. This condition is generally associated with the stable part of corporate life cycles and national economic cycles. Historically, such conditions are ideal for the development of a stable community of managers and scientists. All of the products are used to satisfy the social needs of the present society, and most current inventions are used to satisfy the needs of future generations. The former function is accomplished by the marketing chains that supply society with the economic goods that corporations produce. For the same reason, the latter function is accomplished within the research communities of universities, corporations, think-tanks, etc., that society can sustain.

In this stable mode (shown in Figure 5.4.10), it is desirable that managers use both sets of refined managerial skills (resource allocation and product integration), which are consistent with the significance of the scientists' inventions. Unfortunately, this is not always accomplished. Other dimensions, such as human skills also enter the overall picture. The demands on managerial creativity become as intense as the demands for scientific creativity. Together, social and scientific progress can both be achieved. This utopia of marriage between social and scientific talent calls for a society where (scientific) truth, virtue, and (mathematical) elegance can coexist. The present environment of corporate scandals (see Sections 2.1.1 and 2.1.5) forms the antithesis of a perfect union of scientists, managers and social leaders. Some stable cultures fostered in the Ancient empires (the golden years of Chandargupta, The Buddhist years of Ashoka, the Ottoman Empire, etc.) have caught a glimpse of this harmonious trilogy of truth, virtue, and beauty in the daily lives of common people.

From our experience, it seems the more mature scientists and novice managers engage in situations akin to the scenarios shown in Figures 5.4.7, 5.4.8, and 5.4.9. When the work experience is limited for managers, then the effect can be quite disruptive for the scientific community and unproductive for the corporation. Over a period, teams of scientists migrate to more balanced environments, if such environments are available. If such a climate extends over a generation or two on a national basis, then the decay and death of a scientific, artistic, literary, business, etc., culture starts to becomes evident. Everybody become a wage earner in a downward spiral of creativity.

When the converse scenario (i.e., mature managers and novice scientists) exists, then the managers must help the new scientists much like a mentor would help research students. In a sense, the relationship can be long lasting and rewarding only if corporate and financial stability is also present. When a short-term financial goal is of primary importance, this scenario is unlikely to succeed, except by threatening the scientists who must invent. This strategy has worked in some warlike situations around the world, where scientists have known to be abducted so that they can perform military research. Even some democratic and free societies have developed weapons of mass destruction (e.g., nuclear devices and ICBMs) and new strategies (e.g., fully fueled airplanes as missiles and robotic war planes) for destructive purposes.

REFERENCES C-5

[5.1] Seifer, Marc J. *Wizard: The Life and Times of Nikola Tesla: Biography of a Genius,* reprint ed. Citadel Press Book/Carol Publishing Group, Secaucus, NJ, 1998. Also see Pate, Richard, "Table of Contents Page," in *Rich's Tribute to Nikola Tesla,* on the Internet at http://home.earthlink.net/~drestinblack/teslatoc.htm. (Accessed January 2003 and April 1, 2004.)

[5.2] Bobeck, A. H., *Magnetic Bubbles.* North-Holland Pub. Co./American Elsevier Pub. Co., New York, 1975.

[5.3] Melen, Roger, and Buss, Dennis (Editors), *Charge-Coupled Devices: Technology and Applications.* IEEE Press, New York, 1992.

[5.4] Laithwaite, E. R. (Editor), *Transport without Wheels.* Westview Press, Boulder, CO, 1977.

[5.5] Bernstein, P. A., "What's Wrong with Telecom." *IEEE Spectrum,* January 2003, pp. 26-29.

[5.6] Bell, A. G., "Improvements in Telegraphy." Letter Patent No. 174,465, United States Patent Office, March 7, 1876. (Available from the United States Patent and Trademark Office.)

[5.7] Shockley, W., "Semiconductor Amplifier." U.S., Patent No. 2,502,488, 1950 (application September 24, 1948). Also see Bardeen, J., and Brattain, W.H., "Three-Electrode Circuit Utilizing Semiconductive Materials." U.S., Patent 2,524,035, 1950 (application June 17, 1948). (Available from the United States Patent and Trademark Office.)

[5.8] Dutta, N. K., "III-V Device Technology for Lightwave Applications." *AT&T Technical Journal,* 1989, pp. 5-18.

[5.9] AT&T Internal Communications, 1992. Also see "Undersea Communication Technology." *AT&T Technical Journal,* Vol. 74, January-February 1995, pp. 75-82. Also see "Erbium Doped Fibers and the Next Generation of Lightwave Systems." *AT&T Technical Journal,* Vol. 71, No. 1, February-March 1992.

[5.10] Ahamed, S. V., and Lawrence, V. B., "Knowledge Processing System Employing Confidence Levels." U.S., Patent No. 5,809493, 1998. (Available from the United States Patent and Trademark Office.)

[5.11] Ahamed, S. V., and Lawrence V. B., "Hospital-Based Medical and Patient Information Using Specialized Integrated Medical Computer Systems for Processing Functional Modules." U.S., Patent No.6,272,481, 1988. (Available from the United States Patent and Trademark Office.)

[5.12] Kazmi, T. S., "Simulation and Design Studies of Knowledge Processing in Intelligent Network Environments." PhD dissertation, City University of New York, 2002.

[5.13] Goldbert, N., "Financial Intelligent Networks." PhD dissertation, City University of New York, 2004.

[5.14] Weber, Roy P. "Data Base Communication Call Processing Method." U.S., Patent No. 4,191,860, 1980. (Available from the United States Patent and Trademark Office.)

[5.15] ATM technology is documented in most modern telecommunication textbooks-for example, see Ahamed, S.V., and Lawrence V. B., *Design and Engineering of Intelligent Communication Systems.* Kluwer Academic Publishers, Boston, 1997.

[5.16] Rosenbloom R. S., and Spencer, W. J. (Editors), *Engines of innovation: U.S., Industrial Research at the End of an Era.* Harvard Business School Press, Boston, 1996.

[5.17] Roussel, P. A., Saad, K. L., and Erickson, T. J., *Third Generation R&D, Managing the Link to Corporate Strategy.* Harvard Business School Press, Boston, 1991.

[5.18] Pachter, Barbara, and Brody, Marjorie, *Climbing the Corporate Ladder : What You Need to Know and Do to Be a Promotable Person.* Skillpath Publications Inc., Mission, KS, 1995.

[5.19] Michalko, Michael, *Cracking Creativity: The Secrets of Creative Genius.* Ten Speed Press, Berkeley, CA, 2001.

[5.20] Sloane, N. J. A., and MacWilliams, F. J., *The Theory of Error-Correcting Codes.* North-Holland, New York, 1998.

[5.21] Roddy, Dennis, *Satellite Communications,* 3d ed. McGraw-Hill Professional, New York, 2001.

[5.22] Diggins, John Patrick, *Thorstein Veblen: Theorist of the Leisure Class.* Princeton University Press, Princeton, NJ, 1999, Chapter 4. Also see Blaug, Mark, Great *Economists Before Keynes: An Introduction to the Lives and Works of One Hundred Great Economists of the Past,* reprint ed. Cambridge University Press, New York, 1989.

Glossary

Academic Research environments: Research facilities and management teams at universities and colleges, meant primarily for teaching and other educational activities. The flavor of academic research changes if the projects are funded by external agencies. The conflict is heightened if the project goals are not consistent with the social or cultural goals of the student populations.

Academicians: Faculty members and administrators who mentor student research or who participate in the administrative duties of the university or college.

Accountability: A term generally used in the context of research staff, teams, and students when they publish or make overly general claims about their findings. Largely, it is the responsibility of research staff to be accountable and validate its findings and intermediate steps in relation to the environments where the results were found or derived.

Adaptation: The practice of altering the basic concepts of earlier inventions or innovations to solve a localized problem. In many cases, as technology changes, adaptations become imperative in order to make manufactured products optimal and less expensive to manufacture.

Algorithm: A step-by-step specifiable procedure for solving a problem or achieving a goal, often (though not always) by computer. The design of algorithms generally implies that the entire process offers the best results or the most rapid convergence to the solution for an entire set of problems.

ALOHA systems: An approach to communicating information by parceling it into packets to be distributed through a shared network. This system is discussed in detail in the Chapter 3 of the book.

Analogize: To draw parallels (or comparisons) between different concepts, processes, approaches, or solutions with the purpose of learning whether certain knowledge or strategies can be derived or applied between solutions. Analogizing is an established way of finding new solutions to problems.

Artificial Intelligence (AI): The capability of a machine programmed with intelligent functions to solve a class of sophisticated problems, often dealing with pattern recognition, robotics, the opinions of experts, or the embedding of intelligent agents. Generally, AI programs are triggered by inputs and events in the environment and lead to unique and precise results that are outside the scope of solutions of routine program execution. A considerable amount of inferential decision making in portfolio management, optimal design, education, medicine, etc., can be handled by AI systems and networks.

Brainstorming: The practice of discussing the relevant details of a problem at hand amongst a group of skilled technicians to achieve an immediate or a short-term solution. In corporate settings, brainstorming moves a project from a dead end or a slow pace and blends the perspectives of many engineers, scientists, and technicians to move the project onto a fast track.

Breakthrough: An invention that significantly alters the course of a society in a positive direction. Generally, this type of invention exploits a radically new concept or a law of nature to benefit the society and has a long-term effect toward the betterment of humanity. It is also possible to conceive of negative breakthrough inventions that deter a society by perverted, though intelligent and creative, people.

Bubble devices: A group of magnetic devices investigated during the 1960's to explore their possible adaptation as memory and logic devices in computer hardware and subsystems. Because bubble devices were inexpensive, their initial impact was significant, but the devices lost the appeal very quickly because of imperfections in the orthoferrite materials they were made of and in their technology. Poor propagation (in both the tee-bar and chevron varieties), interaction effects, and imperfections in these "bubble" or magnetic domains were the major reasons for the demise of the devices and technology. Problematic size, power consumption, and temperature sensitivity were the other contributing factors. Still, this pseudo-invention contributed to the exploration of the more promising VLSI technologies of the 1970's.

C+ and C++: Object-oriented programming languages that evolved after the C language proved viable as a general programming language for a number of applications. Being inherently powerful and close to the actual CPU functions, the C language and its enhanced versions provide great flexibility in the computationally intensive functions encountered in AI and object-oriented applications, as well as in graphical functions (C++). A vast amount of raw computational power is available through the elegant and optimal use of the syntax embedded in these newer languages.

CAD/CAM: Computer-aided design (CAD) and computer-aided manufacturing (CAM) bring the computer closer to engineering environments. These techniques are routinely practiced in most design and production centers in order to reduce the costs and improve the accuracy, productivity, and documentation of product changes. Coupled with numerical programming of the controlled machines, CAM is used in most robotic manufacturing plants.

CADD: Computer-aided design and drafting, used to design and produce drawings specific to the requirements for transferring information from design houses to production houses.

Categorize: To classify objects or activities on the basis of their attributes. In the innovation domain, categorizing concepts and subconcepts on the basis of the modularity and the universality of these components of invention can be very fruitful in synthesizing inventions or subinventions to suit a particular social or device need.

CAT SCAN: A term coined after the invention of MRI techniques to scan specific regions of a body or the entire body in order to explore abnormalities. The term *scan* is retained because a CAT scan is generated after the system examines the properties of small segments of the body or the entire body systematically. The overall picture is generally assembled by the piecewise accumulation of the results for each individual segment.

CCD, charge-coupled devices: These devices were explored soon after bubble technology started to lose its glamor. The objective was to retain some of the sequential access inherent in the bubble technology and yet take advantage of the semiconductor device technology. In contrast to bubble devices, CCDs carry analogue information encoded as the measure of the charge in each "bucket" or cell. Major drawbacks in CCD technology are dark currents, residual charge, and charge dissipation. Some current systems still deploy CCD components.

Classic Creativity: This type of creativity has yielded classic and socially significant contributions. The inventor, artist, poet, or philosopher finds a linkage to some basic, universal, or inherent law(s) of nature or a deeper truth in the contribution that is made. For this reason, and for the social value, the contribution is used by a large cross section of humankind and lasts a long time.

Codec: A device that converts an analogue signal (such as speech, voice, or video) into its digital or binary equivalent, to be transmitted, processed, enhanced or refined by modern digital techniques.

Commitment: The predisposition of students and researchers to be devoted to the subject matter being investigated.

Concept: A basic fragment of an invention that is mentally processed, enhanced, refined or integrated by the scientists using an ill-defined practice of creativity to generate a concept that can become a new invention. This new invention is fragmented into smaller and modular concepts in the society over a period of time and is itself mentally processed, enhanced, or refined by the next generation of scientists. The process expands at an exponential rate for those who can practice creativity in this structured mode (e.g., Edison and Tesla to some extent)

Concept Mining: The practice of (or the art of) implanting incomplete or partially evolved concepts or fragments of various inventions from a mentor into the minds of a group of doctoral students in order to bring about a synergy of thoughts leading to a full-fledged invention. It is not necessary that the fragments of inventions be derived from the mentor's own contributions, but it is desirable to implant the fragments of concepts that are aligned towards the solution of an innovative or still unsolved problem, but based on the contributions of classic inventions.

Concepts, fragments of: The modular building blocks of solutions to problems. In a normal sense, the accepted solution to any given problem stems from concepts connected together in a coherent and an elegant fashion. In the specific sense, the concepts in invention stem from solutions that are not documented as the working or optimal solutions. For this reason, the flow diagram of an invention needs to attempted with an iterative, recursive, modifiable, and adaptive approach. Some of the possible blueprints for the approach to a solution are presented in Figures 4.3.1 and 4.3.2.

Control of a Creative Process: The control an individual inventor, artist, poet, or philosopher exerts over a subconscious creative process that may start to unfold quite unexpectedly. This process of spontaneous or unstructured creativity (Type A), when supplemented, augmented, and, to some extent, controlled by structured creativity (Type B), is defined (by us) as control over a creative process. The converse (i.e., Type B controlled by Type A) is also an accurate and valid definition. However, this later control is learned by great exercise of

Type A creativity controlled by Type B creativity. It teaches the individual to make Type A creativity control Type B creativity and vice versa. The process is made feasible by deliberately and willfully hopping over the enter line of Figures 4.3.1 and 4.3.2 and combining the boxes in the two ellipses of either figure into one large global ellipse that encompasses the entire creative process. This type of achievement was noticeable as Einstein was struggling to generalize his special theory of relativity, as Bell Labs scientists were exploring multiple quantum-well devices based on their earlier single quantum-well devices, or as they struggled to optimize DWDM technology from WDM technology.

Convergence of Goals: In the realm of inventing, the convergence of goals is appropriate as the scientist seeks to achieve a multiplicity of goals in numerous directions. In the context of this book, an invention is explored or perfected to satisfy at least three goals: (a) the finding of basic underlying axioms of science, (b) the best way to the solution of one or more social needs, and (c) the attainment of elegance in achieving a result. Earlier philosophers searching for truth, virtue, and beauty would have had their goals converge in an axiom of wisdom rather than in an invention.

Convergence of Thoughts: Type B (Figure 4.3.1) approach to consciously force a convergence of goals (see above) during the process of inventing.

Corporate Research Environments: Four scenarios of research corporate environments are identified: creative managers unobtrusively managing creative scientists, inexperienced managers managing creative scientists, creative managers unobtrusively managing unskilled research scientists, and inexperienced managers managing unskilled research scientists. The first scenario, with an adaptive blending of the creative talents in both the managerial and scientific domains, is likely to be ideal for breakthroughs, inventions, and innovations. The migration of scenarios 2, 3, and 4 to the first scenario over a period is likely to painful and expensive—or it can even turn unstable.

Creation of Novelty: This process occurs in the practice of (Type B) structured creativity. It is essentially a process of trial and error and a process of elimination of what is already known. In the practice of unstructured creativity (Type A), the creation of novelty appears spontaneously and unexpectedly. The interleaving of the structured and unstructured creativities, the Type B activity monitors the Type A activity if novelty has resulted. Since these processes (A and B) are ill defined, the willful creation of novelty also remains unresolved.

Creativity and Communication: Communication is seen as a catalyst for creativity. The environment (via its diversified norms of communication) provides positive feedback to offer progressively evolved inventions, such as the successive generations of aircraft. In some perverted forms, communication also provides negative feedback to enhance misery and suffering, such as the deadly crops of bombs and mines. The conditions for stability of the creative activity appear to be more complex than the conditions for stability for economic, engineering, or commercial activity. Creativity is generally practiced on an individual basis, even though it may be invoked by "concept mining" or by the refined management tasks of managing talented scientist in classrooms and in corporations.

Creativity and the Information Age: With modern society leaning towards Internet tools and knowledge bases, the evolving information age is (in our opinion) likely to a quicken and intensify the feedback/feed-forward in the (learn → invent → profit → relearn) cycle. The microeconomics of knowledge transactions governs the oscillation in this cyclic process. In our opinion, the macroeconomics that evolves from the microeconomics of knowledge is likely to become quite different and more influential over the global economy than the macroeconomics based on the Adam Smith (or production based) and the Keynsian (from John Maynard Keynes, or money based) economic models of the previous generations.

Creativity and Intelligent Agents: Intelligent agents that monitor the boundaries of global knowledge are just starting to surface in the Internet arena. More recently, new search engines facilitate the functionality of intelligent agents in the knowledge domain. These agents are more likely to enhance and intensify the feedback/feed-forward processes (see Creativity and the Information Age) in modern society.

Creativity and the Internet: See Creativity and Information Age and also Creativity and Intelligent Agents.

Creativity and Logic: Logic plays a dominant role in Type B scientists as they mentally process every clue or observation that may lead to an invention. In Type A scientists, the logical functions are taken over by the subconscious mind in process of inventing. The emphasis varies in Type A/B and B/A students (see Figure 4.3.1) and C/D and D/C scientists (see Figure 4.3.2).

Darwinism in Inventions: The substitution of one invention over its predecessor because of the new invention's greater rewards, social benefit, economic advantage, greater reliability, etc. An example of this type of situation occurred when CMOS technology started to displace TTL technology or as fiber-optic technology is overtaking twisted-pair high-speed technology. Whereas true Darwinism is a natural phenomenon, creative Darwinism can be manipulated slightly by creating the right environment for inventions of a given type. In a sense, it is akin to genetic engineering in the evolution of future species.

Derived objects: Objects and processes in research that appear as by-products or intermediate objects in the investigation of the subject matter of a dissertation or a new marketable product (e.g., wafer technology as IC's were being developed or glass purification techniques as fiber-optic technology was being developed).

Dewey Decimal System: A numerical code for classifying subject matter or knowledge. Developed by Dewey in 1873, the code is still operative in some institutions for arranging books in libraries.

DWDM: Dense wave division multiplexing, a technique to constrain the flow of information-bearing optical pulses to be within a particular bandwidth of light. Thus, numerous channels of optical energy can coexist in the same fiber and carry independent channels of information, since the fiber can carry light extending over a reasonably large bandwidth. This technique has increased the data-carrying capacity of fibers by two or three orders of magnitude.

Edison and the electric chair: A negative invention of Edison when he was trying to electrocute death-row prisoners. Some of his experiments resulted in cruelty and severe burns. The practice was soon stopped.

Educators: In the context of this book, educators are elevated to the status of mentors who play a significant role in preparing students to contribute to society by making positive inventions, innovations, and devices that are novel and original.

Elegance/Beauty: The conscious aim of some inventors in making their inventions and innovations. Typically, an inventor's striving for elegance or beauty enhances the practicality of the concept, process, or device after the basic invention or innovation has been made. While revisiting the invention with a new perspective, the inventor reexamines the entire process to find out the optimality of the steps, the algorithms, and their combinations. The process may bring a new perspective, a more refined concept, or even a mathematical identity that would offer a more universal, beautiful, or elegant solution to a social or technological need within the society.

Evolution of Inventing: In the context of this book, the evolution of inventing refers to the change in the thought processes of the early scientists from utilizing their raw talent to be creative to adopting a more rigorous and systematic way of making inventions. Because inventing is a unique and an undocumented process, we do not have a measure by which the scientist conceived or thought during the inventive process. However, by studying the implications of the inventors' own words, the trend appears that the formality of science, inference, logic, and mathematics is more ingrained in modern inventions than ever before. Nevertheless, there are genuine exceptions.

Expert Systems (ES): A branch of AI wherein the computer system suggests possible solutions to a problem on the basis of the opinions of a number of experts who have provided opinions and ways of solving similar problems in the past. Barring a strict mathematical solution to any specific problem, ES provides opinion-based solutions with a level of confidence.

Explosion of Knowledge: The recent rapid expansion of knowledge and information due to the increased pace of research, information technology, computers, and high-speed networks. High-speed transmission technologies, including DSL, ISDN, SONET, ATM, and DWDM, have contributed to the explosion. High-capacity switching and transoceanic fiber backbone provide the information highways. Access technologies practiced by communications service providers, including the ISP, have also contributed to the explosive growth of knowledge in both positive and negative directions.

Explosion of Technology: An explosive growth in technology, noticeable within the last few decades. The positive feedback between the learn-node and relearn-node of an expanding knowledge (learn → invent → profit → relearn) cycle, leads to new technologies and is fueled by the advent of the knowledge society and information highways. The whole process is hastened especially because financial organizations favor and encourage quick profits at the profit-node.

Fiber Optics: The science of deploying optical fiber to carry photonic packets as the information-bearing signals to the various nodes within a network.

Flow graph of an Invention: Graph that illuminates the process that constructs an optimal representation of an invention in achieving its result. The process consists of dismantling the current invention or previous inventions into smaller and discrete sub-inventions or fragments of concepts that can then be constructed as new, better, or more optimal inventions.

Generalization of an Invention: The process of analyzing or relaxing the laws of physics (under special conditions) to find out whether the basic truth underlying any given invention (if it is unique and novel) is also applicable to other circumstances or disciplines. Systematizing, analogizing, categorizing, conceptualizing, rationalizing, and then generalizing the invention generally facilitate this highly directed Type B creativity.

Gnome engine: An early and rather specialized implementation of the IC engine technology for aircraft.

Group creativity: Property of a group whereby it seeks to solve a particularly hard problem or consciously invent a new solution.

Hard work: In the context of inventing, hard work is concerted Type B activity.

Hierarchy of Concepts: Sequential or other relational order inherent in the numerous steps, procedures, or precursors leading to an invention. The underlying basis is that every important step, procedure, or event has an essential role in the overall achievement of the goal of invention.

Improvisation: Improvisation is a notch above reengineering. Improvisation may come in numerous ways and in many combinations, and it fits between adaptation and reengineering, both of which are within the realm of routine engineering activities of most modern corporations.

Inference Engine: An artificially intelligent set of programs that detects the cause–effect relationship in events or observations and tries to predict the outcome with some known degree of confidence. Some applications include the MYCIN and Intern types of medical diagnosis programs.

Innovation: The contribution made by altering the composition of earlier inventions or by adapting to circumstances that has led to known inventions without making a major discovery or finding any universal truth in the basic sciences. Typically, innovations are numerous and exist in the technological fields. Largely, a social, economic, or procedural benefit is necessary before patent examiners will grant a patent for an innovation.

Innovention: A word (coined by us) to refer to the invention or innovation that is conceived by the student or scientist while dealing with a novelty during his or her research. Generally, an innovention is reclassified as an invention or an innovation after a team of experts validates the nature of the contribution. Early in the research career, it is easy to become convinced that any novelty is an invention. However, the maturity of the mentor (for the student) or of the scientist can correctly categorize the novelty within an innovention as breakthrough, invention, innovation, adaptation, improvisation, reengineering or simply a CAD-based optimization.

Inspiration: A flash of genius leading to an optimum or near-optimum solution to a problem of significant value. Unlike poets, philosophers, musicians, and other artists, scientists contribute to society in a mathematically precise environment. In Type A creativity, the subconscious mind processes the problem in a mental domain just as a computer system would function in a background mode. However, computer systems work according to a programmable concept (e.g., exhaustive search, steepest descent, random search, or combinations thereof.), but the mind has no directly discernible algorithm for conceiving an inspiration that leads to the right solution.

Intellectual honesty: A prerequisite to being a researcher with integrity looking for socially beneficial inventions. Intellectual honesty becomes especially important while the scientist seeks some basic truth or law of nature to become an integral part of the invention.

Intelligent Networks: Networks that perform intelligent routing functions based on data stored in a managed database. The function of simple intelligent networks does not overlap the functions of AI-based machines, even though segments of the more recent embodiments of intelligent networks can perform AI functions.

Internet: The ongoing socially significant and catalyzing application of the current network technology that has bought the information revolution to the public with the active participation of the Internet Service Providers, or ISP's.

Intuition Capture: Like inspiration, intuition may not be understood completely, but the capture of an intuition, especially if it deals with a major invention, needs to be frozen in time, quickly written down, or at least moved to a more accessible part of the human memory, to be investigated further. For this reason, many creative people train themselves to be able to benefit from intuition even though it is not totally understood.

Invention: The act of creating something from nothing. The goal is to benefit society with a device, a concept, an identity, a poem, or something else that has lasting value.

Inventions, Classic: The inventions of early scientists who made significant contributions to society. Sixteen major inventions (in electrical engineering, computer science, and communications) and two generic concepts (dealing with packet switching and computers) are discussed in the book.

Inventions, Derived or Secondary: The adaptations of major inventions in order to apply or to suit the invention to a similar or an associated problem. This is a Type B process of using inductive and inferential logic.

Inventions, Recent: The profusion of inventions in the recent era marks a distinction from the classic inventions because the newer patents tend to become specific to technology and process. Their scope is thus constrained. In contrast, the older, fewer patents were founded on a universal or basic law of nature, and they formed the core concepts for many other patents that followed.

Inventions, Socially Significant: Social significance is based on the use and value that an invention brings to a society. Novelty is the primary concern, and social significance is a secondary concern, for the patent office in issuing a patent. However, the social significance of an invention is based on what follows after the patent is issued. Numerous extraneous factors influence the prominence that the patent and the inventor attain. Historically, binary algebra was never patented, and more recently, DSL was never patented even though it was studied as early as 1977.

Jet Propulsion: A propulsion technique for aircraft and missiles based on partial and controlled ignition of the fuel as it passes through a series of turbines within the jet engine. An entirely new concept beyond the previous propeller aircraft designs, jet propulsion has totally revolutionized the industry and brought significant benefits to society.

Knowledge Transactions: Scenarios depicting the exchange or sale of critical or scarce knowledge that can swing the economic balance between individuals and nations. The short-term gain from such transactions can destabilize the economic base of advanced nations by transferring their knowledge bases or skill sets to another nation. Earlier examples of such transactions are the transfer of VLSI technology to Japan, Korea, etc. Current examples are the transfer of software technology to India, and future examples are the transfer of production and robotic technologies to China. The long-term effect of such knowledge transactions on the major economic cycles of nations is still to be investigated.

Language of ideas: An unarticulated language that uses rules of syntax and grammar (as they are used in the spoken language) to parse concepts, contributions, and techniques relating to inventions into substructures and flowcharts within inventions. The rules of such a language do not need to be standardized or written down, since inventing is a personal experience. The utility and the rewards of evolved tools in the analysis of ideas (in our opinion) can become an integral part of practicing Type B creativity.

Lexical Analysis of ideas: The process of identifying the objects (ideas, concepts, techniques, equations, laws of nature, etc.) embedded in papers and patents that disclose major inventions and contributions.

Magnetic Domain Devices: Magnetic domains are also known as "bubbles." Under properly adjusted bias fields, magnetic domains can exist as tiny regions of magnetic charge that can be propelled, navigated, and logically gated to perform logical functions (such as conjunction and disjunction). See bubble devices.

Management of Scientists: See Corporate Research Environments. Four extreme scenarios are identified, and conditions for the creative management of creative scientists are explored in Chapter 5.

Mathematical Basis of Creativity: A refined Type B creativity in which the equations of physics and mathematical identities and constants (embedded in patents) are scrutinized to see whether a corollary or a new variation of the equation, identity, or constant has not been overlooked or cannot be enhanced. The mathematical basis of creativity can also take the form of investigating such equations, identities, or constants in the light of recent research results dealing with similar constants or parameters. The corollary to this field of research is to write an invention in its mathematical form and keep altering constants, parameters, and boundary conditions to find out when the equation fails. The next corollary is to investigate whether there are circumstances or other environmental conditions when similar equations may be written down. We suspect that Maxwell was using this approach while attempting to generalized versions of his equations. The analogies between the EM and ES field distribution equations from his own earlier versions would have triggered the generalization, much as Einstein's later theories evolved from his special theory. We suggest that Maxwell's equations provide an excellent platform for the initiation of this Type B creativity exercise. Furthermore, we also see how Maxwell's equations are modified and enhanced by Einstein's theories.

Medical Imaging: The process of creating a 2D or a 3D image within a specimen or a body based on MRI or even X-ray techniques. Computer software is invariably involved to generate and validate the images.

Mentor: A faculty or research member who interacts with a (Ph.D.) student to guide and monitor the students' research, culminating in contributions, papers, a dissertation, or devices. Details of the men to his duties, responsibilities, roles, and relationships with the student are discussed in Chapter 2.

Models of Creativity: Creativity does not have a specific model. However, neither is it a random phenomenon. To the extent that there is some structure to creativity, it can be studied and enhanced. From observing our doctoral students, we present two extreme modes of activity (Type A and Type B, discussed in detail in Chapter 4) of those students. We have labeled these extreme modes as unstructured creativity and structured creativity. The more mature students typically blend these two extremes as an adaptive and flueid mental activity. However, the mentor can supplement either mode of creativity for the individual student who is predisposed to other mode.

Motivation of Inventors: The passion to contribute, to be a catalyst of change (for the better), and to express one's creativity are perhaps the prime motives of inventors. Reward, recognition, prestige, and financial gains (in our opinion) are secondary factors. In many instances, the classic contributions live beyond the scientist, and to this extent, the inventions and masterpieces of truth, elegance, and social benefit bear the distinctive imprint of the creative genius of the scientist. For some noteworthy scientists, such as Mendel, Fleming, and Tesla, this character of their genius is apparent from their contributions.

MRI: The capacity to deploy the modern computer and imaging technologies in conjunction with the measurement of the timing information in the echo waves that carry data relating to spin–lattice and spin–spin relaxation times.

Mutations of Inventions: These inventions are derived (in our opinion) by commuting and mutating the objects of established inventions and by deploying the Type B tools and techniques of creativity (discussed in Chapter 4). For gifted scientists, it is likely that Type A activity may (subconsciously) take over and lead to inventions more generic than the original invention. The realm and boundary of true creativity is thus pushed one step further.

Natural Intelligence: In the context of this book, the human thought process, as opposed to artificial intelligence or machine intelligence.

NIH Syndrome: The resistance to an idea imported from other institutions bearing a different insignia for solving a technological or social problem or having a different mindset.

Parsing of Ideas: In the context of this book, the process of identifying different ideas and processes linking the various objects embedded in an invention that make the invention practical and feasible.

Passion for Learning/Truth: In the context of this book, the passion for (utmost) truth, passion for (maximal) social benefit, or passion for (profound) elegance in the deployment of the natural laws of sciences in the realization of an invention.

Patent: A document that acknowledges the novelty in the submission of an application made to the patent office. A patent reinforces the idea that the ownership of the novelty is assigned to the inventor, and patent rights are granted to the applicant for 17 or 20 years in the United States.

Pattern Recognition: A branch of AI in which a set of computer programs analyzes objects and their attributes in order to find similarities and differences among them. Typically, these AI routines yield specific information about the similarities and differences.

Personal Conflicts: In the context of this book, conflicts that drain the emotional energy required to sustain the effort necessary for prolonged investigations leading to a significant contribution. The importance of resolving these conflicts faced by doctoral students is discussed in Chapter 2.

Personal Traits of Students: Traits that tend to make students lean towards Type A or Type B creativity. Either type of practice of creativity can be genetic or acquired. However, with enough introspection and some help from the mentor, the ideal blending of the two types may be developed and enhanced. At critical instances, a fragment of a Type B activity can offer several strides in a Type A creativity, and conversely, a small intuition from a Type A activity can save a lot of toil in a Type B activity. For this and other elusive traits of scientists, the overall practice of individual creativity retains an aura of mystery.

Personality Attributes: In the context of this book, the personality traits required for research are presented in Chapter 2.

R&D Environments: Within universities, the office of academic research oversees some of the administrative functions in sponsored research. Performing individual research is still the role of faculty and research staff, provided that they follow the reigning code of ethics and contribute to social welfare. Strict adherence is highly desirable to the practice of creativity, which includes the three basic and intertwined attempts to search for scientific truth, social value, and mathematical elegance. In the corporate environment, the commercial value of research and development starts to surface. With skilled management, a harmonious blend of multiple goals can be pursued.

Reengineering: A routine corporate activity aimed at adapting and modifying an existing design to suit a new or an enhanced product.

With the current software and database technology, reengineering appears to be one notch above CAD/CAM. Numerous examples are presented in Chapter 2.

Representation and Imagery: For some researchers, concepts and ideas come back as objects and images, however rudimentary they might be. Initially, such objects are used as placeholders in understanding an invention, including the researcher's own inventions. However, with the Type B activity that follows, the linkages between the objects and images can also facilitate a firm grasp of the invention, leading to a new invention or a much greater invention. If the linkages are derived from laws of physics, natural phenomena, scientific truth, or proven mathematical relationships, then the validity of the invention is firmly established.

Research and Management: In the context of this book, the interdependence of management and research is explored in Chapter 5. Stable and unstable forms of relationships are delineated.

Robotic Environments: Software-driven computerized facilities to govern the response of machines to predefined and predictable input conditions.

Role of Wisdom: Wisdom is perhaps the highest form of invention. The goals of wisdom may generally overlap and reach beyond the goals of positive inventions. We refrain from exploring the boundary between wisdom and creativity, but we do believe that it is possible to build "wisdom machines" with the intelligence to guide massively parallel processed AI-based computers.

Scholarship of Students: In the context of this book, the academic scores of students at the entry level. However, creativity is devoid of academic scores for many scientists (e.g., Mendel, Edison, Einstein). For this reason, scholarship scores are not the only factors considered in the student's quest for admission to a doctoral program.

Semantic Analysis of Ideas: The deliberate process of establishing linkages between ideas and other objects embedded in (particular or extraneous) inventions. This type of analysis is akin to the semantic analysis performed by compilers of higher-level languages. In either case, intellectual rigor is essential to performing the semantic analysis of ideas at any reasonable level. Even a partial effort, however (in our opinion), can lead to tremendous rewards when used in conjunction with the syntactic analysis of ideas.

Sensitivity: In the context of this book, the innovative response of students to a mentor, members of examining committees, or reviewers. In many cases, the mentor learns the cues from the students to invoke Type A responses during a group discussion or a concept-mining session.

Social Needs: In the context of this book, this component of Maslow's need hierarchy sometimes prompts students to enter a doctoral program. In an obscure way, corporations and even nations also undertake research projects (e.g., genetic engineering, optical computing, and even space exploration) to satisfy a corporate or a national fancy.

Structured Creativity: We refer to this mode of activity as Type B creativity, discussed in Chapter 4 of the book.

Symbolic Manipulation of Concepts: After a lexical scan of an invention, which searches for concepts, symbols, objects, and components that make the invention work, the symbolic manipulation of concepts, objects or components can be attempted to optimize the invention by itself or in conjunction with similar inventions. Typically, a complete analysis (including syntactic and semantic analyses) of an invention is necessary to attempt a symbolic manipulation of the invention.

Synergistic ideas: Ideas that, when conjoined, yield positive feedback for positive inventions that account for the super-inventions (e.g., Maser, transistor, optical fiber) of the past. It is not necessary that these ideas come from different members of a research team; instead, the ideas of the same inventor may be invoked by visiting the problem later with a different perspective. Some of earlier scientists practiced this approach by purposely abandoning research for some time.

Syntactic Analysis of Ideas: The parsing of an invention with an eye toward isolating the main ideas or themes upon which the invention is based. After some experience, a mental (or even a logical) database is developed by an individual or corporation to make up a dictionary of innovative ideas to recompose inventions as unique and distinct combinations and permutation of ideas in the dictionary. The coherence of such ideas leads (in our opinion) to the coherence of a language for stating and synthesizing inventions and patents on demand. This practice is common in the pharmaceutical industry to synthesize and formulate new vaccines and new drugs.

Systematize: When the rules for reconstituting ideas are developed, then the flow graphs of inventions are partitioned into smaller subsystems, and a type of systems-inventing becomes possible. Systems-inventing deploys a language of ideas to invent things. System-engineering deploys a set of equations to build things.

Tagore: The Nobel Laureate Indian poet philosopher from Bengal with significant contributions (e.g., Gitanjali, Where the mind is without fear, Fireflies) to literature.

Thought process, conscious: In the context of this book, this thought process is the basis of structured, or Type B, creativity.

Thought process, subconscious: In the context of this book, this thought process is the unstructured, or Type A, creativity.

Truth, Pursuit of: One of the three simultaneous activities involved in creating a positive invention. The other two goals are the pursuit of virtue and the pursuit of elegance.

Ultra large scale integration (ULSI): Follows VLSI (very large scale integration) in the fabrication of chips for computers and for specialized devices.

Unstructured Creativity: In the context of this book, the response of the subconscious mind to offer reasonably satisfactory solutions to a scientific problem that has remained unsolved by the usual problem-solving techniques including the methodologies of the purely structured creativity (Type B) discussed in Chapter 4. Poets, philosophers, painters, musicians, and artists are known to experience spontaneous and uncontrolled bursts of unstructured creativity.

Virtue, Pursuit of: The second of the three simultaneous goals of bringing social benefit by making a positive invention. The other two goals are the pursuit of truth and of elegance.

Wavelength division multiplexing (WDM): A technique of enhancing the information-bearing capacity of optical fiber by confining the spectrum of light for each channel to occupy a smaller segment of the entire spectral capacity of the fiber

Willful Control of Creative Thought: The practice of Type B creativity by being extremely meticulous and systematic about each step in a scientific investigation. The control extends to the blending of Type A creativity to yield an entire spectrum of thought to suit the scientific problem at hand. The blending of the two types of creativities is discussed in Chapter 4.

X rays: Accidentally discovered in 1895 by Wilhelm Conrad Roentgen, X rays are also known as Roentgen rays with smaller wavelengths than those of ultraviolet waves. As the wavelengths of light decrease, the light absorbs more energy. X rays have a very small wavelength and may sometimes behave as particles rather than waves. X-ray detectors collect actual photons of X-ray light.

CASES OF CLASSIC CREATIVITY

Ghiyath al-Din Abu'l-Fath Umar ibn Ibrahim Al-Nisaburi al-Khayyami (1048–1131)

OMAR KHAYYAM: THE POET WHO STITCHED THE TENTS OF SCIENCE

We end this book with a tribute to Persian poet, philosopher, and mathematician Omar Khayyam (1048–1131) by quoting one of his verses. The translation is by Edward FitzGerald (1809–1883):[1]

> Ah Love! could you and I with Him conspire
> To grasp this sorry scheme of things entire,
> Would not we shatter it to bits—and then
> Remould it nearer to the heart's desire!

[1]E. FitzGerald, "From His Paraphrase of the Rubáiyát of Omár Khayyám," in E. C. Stedman, *A Victorian Anthology*. Cambridge: Riverside Press, 1895. Online edition published April 2003, bartleby.com: http://www.bartleby.com/br/246.html.

The meaning of the name al-Khayyami is "tent maker," but Omar's thoughts strayed far beyond the tents he would stitch. In fact, his contributions to astronomy are as remarkable as his contributions to poetry and to mathematics. Suffice it to say that creativity has never been confined to the era of science as we know it today. Did he think that the sky was a tent and that it was within the realm of his duty as a tent-maker to reach up to the sky with his creative thoughts? Was he on the same wavelength eight hundred years before Einstein as he would choose to trace the curvature of light with the precision of his thought?

This book helps researchers make the transition from passive thinking to creative thinking and challenges them to overcome barriers to creative thought. The book helps individuals progress beyond mathematical or academic viewpoints regarding scientific problems by encouraging them to make contributions to society as a persistent and prevalent knowledge worker. In Chapter 4, we describe the process of inventing as either unstructured or structured creativity. These two extremes of the creative spectrum can be blended in different ways, depending on the personalities of the scientists researchers and on the goals that have been set. In both cases, we predict that the kinds and ways of demonstrating of creativity will undergo major changes in view of the new tools and techniques used in today's information society. The major inventions discussed in this text are based on patents in electrical engineering, computers, and communication.

About the Authors

Syed V. Ahamed received the Ph.D. and D. Sc. (E.E.) degrees from the University of Manchester and the MBA (Econ.) from New York University. He taught at the University of Colorado for two years before joining AT&T Bell Laboratories in 1966. He became a professor of computer science at the City University of New York in 1982 and a member of the doctoral faculty in 1985. He taught at New York Polytechnic at Brooklyn as a visiting professor of computer science from 1982 to 1986. Professor Ahamed has been a telecommunications consultant to Bell Communications Research, AT&T Bell Laboratories, Lucent Technologies, Timplex Corporation, Telecom Australia, ETH Zurich, and Celcom, Malaysia. In Malaysia, he worked as a management consultant to establish an IT platform and design a high-speed backbone network. He was the corecipient of the 1981 IEEE Communications Society Leonard G. Abraham Prize in the Field of Communications Systems, given annually to the best original paper published in the *IEEE Journal on Selected Areas in Communications.* He was also the corecipient of the 1964 Honorable Citation for the IEEE Aerospace Society paper. With Professor Michael J. Miller, he has written two books: *Digital Transmission Systems and Networks,* volumes I and II, in 1987 and 1988. In 1992 he became a Fellow of the IEEE "for his seminal contribution to simulation and design studies of high-speed digital subscriber lines." The results published in the *Bell System Technical Journal* in 1982 form the basis for the later versions of the DSL. He is listed in Marquis 1995's, *Who's Who in the World* and holds many such honors. He has also written two other books: *Intelligent Broadband Multimedia Networks* and *Design and Engineering of Intelligent Communications Systems,* with V. B. Lawrence. Currently, he works with an excellent team of Ph.D. students at the Graduate Center of the City University of New York. The doctoral students continue to contribute to knowledge-processing systems and "wisdom machines" proposed by Dr. Ahamed in 1999 and 2002. He holds over 20 U.S. and European patents, ranging from slip meters for induction motors to medical networks for hospitals. He has also published over 150 papers and numerous chapters in books.

Victor B. Lawrence received the Ph.D. degree from the University of London in the United

Kingdom. After teaching at Kumasi University of Science and Technology in Ghana, he joined AT&T Bell Laboratories in 1974. Currently, he is vice president, Advanced Communications Technology, Bell Laboratories, Lucent Technologies, where he leads the development of technologies that go into the most innovative, reliable, and cost-effective communications networks for the world's leading telecommunications service providers. He has taught at Columbia University, Princeton University, and Rutgers University as an adjunct professor, and he delivered the Chancellor's Distinguished Lecture Series at the University of California, Berkeley in 1986. Dr. Lawrence has served as chairman of the IEEE Awards Board (1994–1995), as editor-in-chief of the *IEEE Transactions on Communications* (1987–1991), and as a member of the IEEE Fellow Committee (1988–1991) and the IEEE Communications Society Board of Governors (1990–1992). He was also special rapporteur on coding (1982–1984) and on transmission impairments (1984) for CCITT (now ITU). He is a member of the National Academy of Engineering and a Fellow of both the Institute of Electrical and Electronics Engineers (IEEE) and AT&T Bell Laboratories. For his scientific achievements, he has received numerous awards, including the 2004 IEEE Award in International Communication "for extraordinary contributions and leadership in the development of international data communication systems and standards, and for fostering international cooperation" and a 1994 Emmy Award for the HDTV Grand Alliance Standard. He was also the corecipient of the 1984 J. Harry Karp Best Paper Award and the 1981 Gullemin–Cauer Prize Award. He holds over 20 U.S. and international patents, has published over 45 technical papers, and is the coauthor of four books: *Intelligent Broadband Multimedia Networks*, *Design and Engineering of Intelligent Communications Systems*, *Introduction to Digital Filters*, and *Tutorials on Modem Communications.*

Index